Biosaline Agriculture and Salt Tolerance of Plants

Biosaline Agriculture and Salt Tolerance of Plants

Editors

Hans Raj Gheyi
Claudivan Feitosa de Lacerda
Devinder Sandhu

Basel • Beijing • Wuhan • Barcelona • Belgrade • Novi Sad • Cluj • Manchester

Editors

Hans Raj Gheyi
Universidade Federal de
Campina Grande
Campina Grande, Brazil

Claudivan Feitosa de Lacerda
Federal University of Ceará
Fortaleza, Brazil

Devinder Sandhu
USDA-ARS
Riverside, CA, USA

Editorial Office
MDPI
St. Alban-Anlage 66
4052 Basel, Switzerland

This is a reprint of articles from the Special Issue published online in the open access journal *Agriculture* (ISSN 2077-0472) (available at: https://www.mdpi.com/journal/agriculture/special_issues/biosaline_agriculture_salt_tolerance_plant).

For citation purposes, cite each article independently as indicated on the article page online and as indicated below:

Lastname, A.A.; Lastname, B.B. Article Title. *Journal Name* **Year**, *Volume Number*, Page Range.

ISBN 978-3-0365-9130-8 (Hbk)
ISBN 978-3-0365-9131-5 (PDF)
doi.org/10.3390/books978-3-0365-9131-5

© 2023 by the authors. Articles in this book are Open Access and distributed under the Creative Commons Attribution (CC BY) license. The book as a whole is distributed by MDPI under the terms and conditions of the Creative Commons Attribution-NonCommercial-NoDerivs (CC BY-NC-ND) license.

Contents

About the Editors vii

Preface ix

Hans Raj Gheyi, Devinder Sandhu and Claudivan Feitosa de Lacerda
Fields of the Future: Pivotal Role of Biosaline Agriculture in Farming
Reprinted from: *Agriculture* 2023, *13*, 1774, doi:10.3390/agriculture13091774 1

Henderson Castelo Sousa, Geocleber Gomes de Sousa, Thales Vinícius de Araújo Viana,
Arthur Prudêncio de Araújo Pereira, Carla Ingryd Nojosa Lessa,
Maria Vanessa Pires de Souza, et al.
Bacillus aryabhattai Mitigates the Effects of Salt and Water Stress on the Agronomic Performance of Maize under an Agroecological System
Reprinted from: *Agriculture* 2023, *13*, 1150, doi:10.3390/agriculture13061150 7

Rodrigo Rafael da Silva, José Francismar de Medeiros, Gabriela Carvalho Maia de Queiroz,
Leonardo Vieira de Sousa, Maria Vanessa Pires de Souza,
Milena de Almeida Bastos do Nascimento, et al.
Ionic Response and Sorghum Production under Water and Saline Stress in a
Semi-Arid Environment
Reprinted from: *Agriculture* 2023, *13*, 1127, doi:10.3390/agriculture13061127 27

Carla Ingryd Nojosa Lessa, Claudivan Feitosa de Lacerda, Cláudio Cesar de Aguiar Cajazeiras,
Antonia Leila Rocha Neves, Fernando Bezerra Lopes, Alexsandro Oliveira da Silva, et al.
Potential of Brackish Groundwater for Different Biosaline Agriculture Systems in the Brazilian
Semi-Arid Region
Reprinted from: *Agriculture* 2023, *13*, 550, doi:10.3390/agriculture13030550 41

Valeska Karolini Nunes Oliveira, André Alisson Rodrigues da Silva,
Geovani Soares de Lima, Lauriane Almeida dos Anjos Soares, Hans Raj Gheyi,
Claudivan Feitosa de Lacerda, et al.
Foliar Application of Salicylic Acid Mitigates Saline Stress on Physiology, Production, and
Post-Harvest Quality of Hydroponic Japanese Cucumber
Reprinted from: *Agriculture* 2023, *13*, 395, doi:10.3390/agriculture13020395 63

Jorge F. S. Ferreira, Xuan Liu, Stella Ribeiro Prazeres Suddarth, Christina Nguyen
and Devinder Sandhu
NaCl Accumulation, Shoot Biomass, Antioxidant Capacity, and Gene Expression of
Passiflora edulis f. Flavicarpa Deg. in Response to Irrigation Waters of Moderate to High Salinity
Reprinted from: *Agriculture* 2022, *12*, 1856, doi:10.3390/agriculture12111856 87

Gabriel O. Martins, Stefane S. Santos, Edclecio R. Esteves, Raimundo R. de Melo Neto,
Raimundo R. Gomes Filho, Alberto S. de Melo, et al.
Salt Tolerance Indicators in 'Tahiti' Acid Lime Grafted on 13 Rootstocks
Reprinted from: *Agriculture* 2022, *12*, 1673, doi:10.3390/agriculture12101673 101

Hayam I. A. Elsawy, Khadiga Alharbi, Amany M. M. Mohamed, Akihiro Ueda,
Muneera AlKahtani, Latifa AlHusnain, et al.
Calcium Lignosulfonate Can Mitigate the Impact of Salt Stress on Growth, Physiological,
and Yield Characteristics of Two Barley Cultivars (*Hordeum vulgare* L.)
Reprinted from: *Agriculture* 2022, *12*, 1459, doi:10.3390/agriculture12091459 131

Monaliza Alves dos Santos, Maria Betânia Galvão Santos Freire, Fernando José Freire, Alexandre Tavares da Rocha, Pedro Gabriel de Lucena, Cinthya Mirella Pacheco Ladislau and Hidelblandi Farias de Melo
Reclamation of Saline Soil under Association between *Atriplex nummularia* L. and Glycophytes Plants
Reprinted from: *Agriculture* 2022, 12, 1124, doi:10.3390/agriculture12081124 151

Pablo Rugero Magalhães Dourado, Edivan Rodrigues de Souza, Monaliza Alves dos Santos, Cintia Maria Teixeira Lins, Danilo Rodrigues Monteiro, Martha Katharinne Silva Souza Paulino and Bruce Schaffer
Stomatal Regulation and Osmotic Adjustment in Sorghum in Response to Salinity
Reprinted from: *Agriculture* 2022, 12, 658, doi:10.3390/agriculture12050658 169

Eduardo Santos Cavalcante, Claudivan Feitosa Lacerda, Rosilene Oliveira Mesquita, Alberto Soares de Melo, Jorge Freire da Silva Ferreira, Adunias dos Santos Teixeira, et al.
Supplemental Irrigation with Brackish Water Improves Carbon Assimilation and Water Use Efficiency in Maize under Tropical Dryland Conditions
Reprinted from: *Agriculture* 2022, 12, 544, doi:10.3390/agriculture12040544 181

About the Editors

Hans Raj Gheyi

Dr. Hans Raj Gheyi graduated in Agriculture from the University of Udai Pur/India in 1963 and holds a Master's degree in Soil Science from the Punjab Agricultural University/India (1965) and a Doctorate in Agricultural Science from the Université Catholique de Louvain/Belgium (1974). He is Emeritus Professor of Federal University of Campina Grande, PB – Brazil and Research Fellow (Category 1A) of the National Council for Scientific and Technological Development - CNPq. During 1977–2010, he served as Titular Professor at the Federal University of Campina Grande (UFCG)/Brazil and as Visiting Professor (2010-2020) at the Federal University of Recôncavo da Bahia (UFRB/Brazil) and Vice Coordinator of National Institute of Science and Technology in Salinity (2010 to 2017), and since 1997 has served as Editor of the Brazilian Journal of Agricultural and Environmental Engineering (Agriambi) published by the Federal University of Campina Grande. During 1965–1970, he was a Junior Lecturer in Soil Science and Agricultural Chemistry at S.K.N. College of Agriculture of the University of Udai Pur/India, and from 1974 to 1977, he was an Assistant Professor in Soil Science at the Institut Nacional Agronomique at El- Harrach/Algiers. He has teaching and research experience in Soil Science and Agricultural Engineering, with an emphasis on Soil and Water Engineering, working mainly on the topics soil and water salinity, irrigation, soil fertility, soil–water–plant relationship, salt stress, plant tolerance to water deficit, diagnosis and reclamation of salt-affected soils, and the use of brackish water and wastewater. He is the author of several books, book chapters, and scientific articles, and has also guided dissertations and theses of students from various universities.

Claudivan Feitosa de Lacerda

Prof. Claudivan Feitosa de Lacerda graduated in Agronomy from the Federal University of Ceará (Brazil, 1991) and holds a Master's degree in Agronomy (Soils and Plant Nutrition) from the Federal University of Ceará (Brazil, 1995), PhD in Agricultural Sciences (Plant Physiology) from the Federal University of Viçosa (Brazil, 2000), and postgraduate internship by University of California, Riverside (UCR) and US Salinity Laboratory/ARS/USDA (USA, 2014). He is a researcher and Member of the Management Committee of the National Institute of Science and Technology in Salinity (2010 to 2017, Brazil). He is also a Full Professor in the Department of Agricultural Engineering at the Center for Agricultural Sciences at the Federal University of Ceará. His main research topics include biosaline agriculture, water and nutrient use efficiency, salt tolerance of plants, supplemental irrigation, the use of wastewater in irrigation, management of saline and sodic soils, and plant ecophysiology.

Devinder Sandhu

Dr. Devinder Sandhu is a Research Geneticist at the USDA-ARS US Salinity Laboratory located in Riverside, CA. He boasts a rich academic foundation, having earned both his B.S. and M.S. degrees in Plant Breeding from Punjab Agriculture University, India. Further, he obtained a Ph.D. in Agronomy from the University of Nebraska Lincoln. Before his current role, he served as a Professor of Biology at the University of Wisconsin-Stevens Point. With a research career marked by dedication, Dr. Sandhu has been pivotal in advancing our grasp on the complex genetics underlying salinity tolerance in crop plants. He has consistently leveraged advanced molecular techniques, aiming to pinpoint the genes crucial for salinity tolerance in crops. Such insights are imperative, especially given the challenges of improving crop yields on salinity-impacted terrains. Throughout his illustrious career, Dr. Sandhu's exceptional contributions to agricultural genetics have garnered him numerous awards and honors. He has an impressive publication record, with over 80 articles in peer-reviewed journals and 13 chapters in esteemed books. Additionally, he has guided 119 undergraduate students in their research endeavors, with 80 of them having the distinction of co-authoring scientific journal articles under his mentorship. Today, Dr. Sandhu is recognized as a foremost expert and a leading authority in plant research on salinity tolerance.

Preface

Climate change, water scarcity, and soil degradation are among the many challenges facing global agriculture today. It is imperative to explore, understand, and devise new agricultural practices that can ensure food security, environmental sustainability, and resilient farming communities. In particular, the age-old problem of soil and water salinization, exacerbated by anthropogenic actions and climatic shifts, demands integrated, multi-faceted management and research approaches.

The articles in this volume offer a deep dive into various facets of biosaline agriculture. Collectively, these studies, representing diverse scientific approaches, geographical contexts, and agricultural systems, underscore a unified theme—the pressing need for innovative strategies to tackle salinity. It is my hope and expectation that this Special Issue will attract and inspire agronomists, policymakers, farmers, and students to embrace the hard challenges and work towards a future or innovation and resilience, ensuring that our soils remain fertile, our waters wisely used, and our crops abundant.

I congratulate the authors and editors on their tremendous achievement and thank them for putting together this valuable collection.

Todd H. Skaggs
United States Salinity Laboratory, USDA-ARS
Riverside, California, USA

Editorial

Fields of the Future: Pivotal Role of Biosaline Agriculture in Farming

Hans Raj Gheyi [1,*], Devinder Sandhu [2] and Claudivan Feitosa de Lacerda [3]

1. Agricultural Engineering Department, Federal University of Campina Grande, Campina Grande 58428-830, Brazil
2. USDA-ARS, US Salinity Laboratory, Riverside, CA 92507, USA; devinder.sandhu@usda.gov
3. Agricultural Engineering Department, Federal University of Ceará, Fortaleza 60356-001, Brazil; cfeitosa@ufc.br
* Correspondence: hans.gheyi@ufcg.edu.br

Citation: Gheyi, H.R.; Sandhu, D.; de Lacerda, C.F. Fields of the Future: Pivotal Role of Biosaline Agriculture in Farming. *Agriculture* **2023**, *13*, 1774. https://doi.org/10.3390/agriculture13091774

Received: 24 August 2023
Revised: 5 September 2023
Accepted: 6 September 2023
Published: 7 September 2023

Copyright: © 2023 by the authors. Licensee MDPI, Basel, Switzerland. This article is an open access article distributed under the terms and conditions of the Creative Commons Attribution (CC BY) license (https://creativecommons.org/licenses/by/4.0/).

Worldwide, groundwater quality is in decline, growing progressively saltier. This is attributed to seawater incursions in coastal zones and other factors, such as concentration due to evaporation in farming and the introduction of salts due to human activities. Salt-affected soils occupy about 1.0 billion hectares in coastal and continental areas worldwide, and approximately one million hectares per year are added, mainly in arid and semi-arid regions of countries in Asia, Oceania, Europe, North America, and South America. In this context, biosaline agriculture presents a promising solution for utilizing brackish waters and salt-affected soils for productive systems in rural areas.

Biosaline agriculture is a broad term used to describe agriculture under a range of salinity levels in groundwater, soils, or both [1]. This approach aligns with the concept of saline agriculture, which emphasizes profitable and enhanced farming methods on saline lands using saline irrigation water. The goal is to maximize production through the holistic utilization of genetic resources—including plants, animals, fish, insects, and microorganisms—while sidestepping costly soil reclamation techniques [2]. Regardless of the term used (biosaline or saline), this sector can include other types of activities, since salinity is associated with other problems typical of arid and semi-arid regions, including water shortages, which can be intensified due to global climate change.

Advancements in research have made soil restoration in salinized areas not only more achievable but also a promising avenue for sustainable agriculture. Phytoremediation is an efficient technique for the rehabilitation of salt-affected areas, improving the physical, chemical, and biological aspects of soils [3,4]. In this Special Issue, entitled "Biosaline Agriculture and Salt Tolerance of Plants", a study evaluated the potential of the halophyte *Atriplex nummularia* for the reclamation of soils affected by salts, either alone or in association with glycophytes adapted to semi-arid environments, like *Mimosa caesalpiniifolia* Benth, *Leucaena leucocephala* (Lam.) de Wit, and *Azadirachta indica* [4]. The results indicated that *A. nummularia* alone was the most efficient treatment, with reductions of 80%, 63%, and 84%, respectively, in the electrical conductivity and the sodium adsorption ratio of saturation paste extract and the exchangeable sodium percentage of soil after 18 months. Therefore, the use of *A. nummularia* and species adapted to semi-arid regions promoted beneficial effects on the soil quality after the establishment of the plants. According to dos Santos et al. [4], the reclamation of degraded soils with species adapted to semi-arid regions would be suitable agronomic practice to improve soil quality and sustainability, contributing to the increased infiltration of water and carbon sequestration in soil.

The high consumption of water in irrigated agriculture and the scarcity of good-quality (low-salinity) water to meet the multiple growing demands of the population have increased pressure on the sector and have even made the expansion and or implementation of several agricultural enterprises unfeasible. Notably, many plants that thrive in water-limited environments still require significant freshwater resources. But these plants could be aptly cultivated using brackish water. This realization has amplified the interest in the use of brackish water and wastewater as well as in diversifying water sources in agricultural activities. Sources of brackish and saline water are very common in coastal regions and inland areas, especially in arid and semi-arid regions.

Despite the large number of studies in which brackish water has been used, little is known about the productive potential of these water sources in arid and semi-arid regions. While brackish water is abundant in many of these areas, its exact impact on soil health, crop yield, and overall ecosystem balance is not comprehensively understood. Historically, the primary concern has regarded the potential risks of salinization of soils, which can degrade the soil structure and reduce its agricultural viability. Along these lines, a study conducted by Lessa et al. [5], published in this Special Issue, demonstrated the potential of brackish groundwater in several biosaline agriculture systems. The results demonstrate that the simultaneous use of data from water sources (discharge rate and electrical conductivity) and biosaline systems (water demand and salt tolerance) generates more realistic information related to the potential of brackish water for agricultural purposes, and this type of evaluation should be recommended for semi-arid regions worldwide. The results indicate that the salt tolerance of crops is important, but it is not the only method of addressing salinity problems in the Brazilian semi-arid region. The joint analysis of the data shows that plant production systems with lower water requirements (forage palm, supplementary irrigation, seedling production, hydroponic cultivation, and multiple systems) have greater potential for biosaline agriculture than more salt-tolerant species (such as coconut). The study also indicated the need for diversification and the use of integrated systems as a way of guaranteeing the sustainability of biosaline agriculture in semi-arid regions, especially for small holdings. According to Lessa et al. [5], the data should serve as a basis for formulating public policies aimed at the economic and social sustainability of family farming in tropical drylands.

While the potential of brackish water remains under-explored and its use by farmers is limited [5], it is evident that the use of brackish water depends on management strategies which allow the use of these water sources with little or no impact on crops and soils. In view of the low water discharge rates of most wells with brackish water, there is a need to expand this water supply. Using supplemental irrigation with brackish water has been shown to offer economic benefits, including increased value and farmers' revenues. A paper published in this Special Issue also demonstrated the importance of supplemental irrigation for the sustainability of biosaline agriculture in semi-arid regions [6]. The results suggested that the water stress associated with dry spells is more deleterious to the carbon assimilation and water use efficiency of maize plants compared to the salt stress associated with the use of supplemental irrigation with brackish water. Dry spells compromised the photosynthetic capacity of maize even under the normal water scenario, but the effects became drastic, particularly under drought and severe drought scenarios due to stomatal and nonstomatal effects. The supplemental irrigation of maize with brackish water with electrical conductivity of 4.5 dS m^{-1} reduced water stress and did not result in excessive salt accumulation in sandy loam soil.

In a parallel context, sorghum, despite yield reductions under water stress, remained resilient to saline conditions, tolerating irrigation waters with salinity up to 6 dS m^{-1} [7]. The combined action of osmotic adjustment and stomatal regulation enabled sorghum to thrive in saline environments [8]. The studies presented in this Special Issue [7,8] showed that a concomitant decrease in transpiration rate with a decline in the photosynthesis rate as the soil salinity escalated ensured that the water use efficiency remained constant. This discovery emphasizes the untapped potential of saline waters for irrigating crops like sorghum, especially in water-scarce regions. Furthermore, in a comprehensive study on 'Tahiti' acid lime grafted onto 13 distinct rootstocks, it was observed that water salinity levels of 4.8 dS m^{-1} adversely affected plant performance, primarily through osmotic impacts on photosynthesis, transpiration, and stomatal conductance [9]. However, the core photosynthetic mechanism remained intact. Notably, certain genotypes demonstrated resilience against increased salinity. The authors recommend using water with electrical conductivity of up to 2.4 dS m^{-1} for optimal acid lime cultivation, emphasizing the use of salt-resistant rootstocks and adopting a 0.10 leaching fraction [9]. Hence, the use of brackish water represents an important strategy that can be employed in biosaline agriculture for semi-arid regions, which are increasingly impacted by the shortage of good-quality water. Considering the great spatial variability in rainfall in tropical semi-arid regions and the increase in drought years associated with global climate change scenarios, long-term studies are required to evaluate this strategy in other important crop systems as well as on different soil types [6].

One article featured in this Special Issue investigated the impact of the foliar application of salicylic acid (SA) in mitigating salinity stress in cucumbers cultivated in a hydroponic system [10]. Salicylic acid (SA) is a phytohormone that is crucial in mitigating both biotic and abiotic stresses in plants. It is a phenolic compound that not only regulates plant growth but also manages reactive oxygen species metabolism, contributing to a plant's antioxidant system. Its efficacy varies depending on its concentration, plant species, developmental stage, and application method. In cucumbers, the foliar application of salicylic acid in concentrations between 1.4 and 2.0 mM positively influenced the synthesis of photosynthetic pigments, leaf gas exchange, and the quantum efficiency of photosystem II, in addition to reducing the percentage of electrolyte leakage in the leaf blade, increasing the production, and improving the post-harvest quality (soluble solids, ascorbic acid content, and titratable acidity) of cucumber fruits [10].

A study published in this Special Issue showed that the addition of calcium lignosulfonate significantly enhanced salt tolerance in barley, bolstering its resilience to elevated salt stress levels [11]. Lignosulfonates, byproducts from the paper industry, are complex polymers formed by solubilizing lignin under alkaline conditions, resulting in various chelated forms like Fe-, Ca-, and K-chelated lignosulfonates. These compounds, especially calcium lignosulfonate (Ca-LIGN), have demonstrated positive effects on plant growth, fruit expression, nutrient efficiency, and overall soil health. Physiological parameters analysis revealed that the difference in growth caused by adding Ca-LIGN was primarily due to the higher activity of the antioxidant enzyme peroxidase in the leaves and roots of barley [11]. Furthermore, adding Ca-LIGN to barley plants under various salinity levels boosted the content of chlorophyll *a, b*, relative water content, and grain yield production as well as protein content, while the electrolyte leakage was decreased [11].

Inoculation with microorganisms is another strategy that can be used, which can mitigate the effects of salinity, improving the soil microbiota and the absorption of nutrients by plants. Inoculation with microorganisms can accelerate the release of non-available inorganic or organic phosphorus into the rhizosphere and enrich the soil biologically, promoting benefits even in plants under salt stress. In this Special Issue, Castelo Sousa et al. [12] showed that inoculation with *Bacillus aryabhattai*, a plant-growth-promoting rhizobacteria, mitigates the effect of abiotic stress (salt and water) in maize plants, making it an option in regions with a scarcity of low-salinity water. According to the authors, further studies are needed to understand how *B. aryabhattai* acts on morphophysiological and production characteristics under stress conditions to develop efficient strategies to mitigate the harmful effects of salt and water stress [12].

In this Special Issue, a study on a single cultivar of yellow passion fruit (*Passiflora edulis* f. cv flavicarpa) investigated the impact of salinity on leaf antioxidant potential and biomass accumulation in grown plants, correlating these findings to genetic responses [13]. The study showed tissue damage, instances of plant mortality, and a significant reduction in shoot biomass when irrigation water salinity (EC$_w$) reached 12 dS m^{-1} [13]. Furthermore, by comparing sequences with model plants, homologs of various proteins and cotransporters involved in Na$^+$ and Cl$^-$ uptake from the soil, their extrusion from roots, and their movement from root to shoot were identified in yellow passion fruit. The gene expression analyses of six genes encoding Na$^+$ transporters and six genes encoding Cl$^-$ transporters indicated that the efflux of Na$^+$ from roots to the soil, the loading of Cl$^-$ from root to xylem, and the sequestration of Cl⁻ into vacuoles of mesophyll cells are vital components of salinity tolerance in passion fruit. The authors concluded that comprehensive insights into the salinity responses of yellow passion fruit would necessitate the exploration of diverse cultivars and the examination of the effects of saline waters rich in either sodium or chloride salts to discern which has the most detrimental impact on the plant [13].

Our understanding of salt tolerance has progressed, revealing intricate genetic mechanisms underlying this trait. These mechanisms, although well understood, present challenges in manipulation. Encouragingly, the genetic underpinnings of salinity tolerance seem conserved across various plant species, a fact evident from the successful gene transfers between Arabidopsis and other species.

In conclusion, faced with growing global challenges, breakthroughs in efficient water management and salt tolerance will be crucial for upholding food security and protecting the environment. It is also important to emphasize that there are great differences in terms of saline resources (plant, soil, and water) in dryland regions around the world, considering quantitative and qualitative aspects. In this sense, applying global knowledge to local realities is a major challenge for biosaline agriculture.

Conflicts of Interest: The authors declare no conflict of interest.

References

1. Masters, D.G.; Benes, S.E.; Norman, C. Biosaline agriculture for forage and livestock production. *Agric. Ecosyst. Environ.* **2007**, *119*, 234–248. [CrossRef]
2. Negacz, K.; Melek, Z.; de Vos, A.; Vellinga, P. Saline soils worldwide: Identifying the most promising areas for saline agriculture. *J. Arid Environ.* **2022**, *203*, e104775. [CrossRef]
3. Gheyi, H.R.; Lacerda, C.F.; Freire, M.B.G.S.; Costa, R.N.T.; Souza, E.R.d.; Silva, A.O.d.; Fracetto, G.G.M.; Cavalcante, L.F. Management and reclamation of salt-affected soils: General assessment and experiences in the Brazilian semiarid region. *Rev. Ciênc. Agron.* **2022**, *53*, e20217917. [CrossRef]
4. dos Santos, M.A.; Freire, M.B.G.S.; Freire, F.J.; da Rocha, A.T.; de Lucena, P.G.; Ladislau, C.M.P.; de Melo, H.F. Reclamation of saline soil under association between *Atriplex nummularia* L. and glycophytes plants. *Agriculture* **2022**, *12*, 1124. [CrossRef]
5. Lessa, C.I.N.; de Lacerda, C.F.; Cajazeiras, C.C.d.A.; Neves, A.L.R.; Lopes, F.B.; Silva, A.O.d.; Sousa, H.C.; Gheyi, H.R.; Nogueira, R.d.S.; Lima, S.C.R.V.; et al. Potential of brackish groundwater for different biosaline agriculture systems in the Brazilian semi-arid region. *Agriculture* **2023**, *13*, 550. [CrossRef]
6. Cavalcante, E.S.; Lacerda, C.F.; Mesquita, R.O.; de Melo, A.S.; da Silva Ferreira, J.F.; dos Santos Teixeira, A.; Lima, S.C.R.V.; da Silva Sales, J.R.; de Souza Silva, J.; Gheyi, H.R. Supplemental irrigation with brackish water improves carbon assimilation and water use efficiency in maize under tropical dryland conditions. *Agriculture* **2022**, *12*, 544. [CrossRef]
7. da Silva, R.R.; de Medeiros, J.F.; de Queiroz, G.C.M.; de Sousa, L.V.; de Souza, M.V.P.; de Almeida Bastos do Nascimento, M.; da Silva Morais, F.M.; da Nóbrega, R.F.; Silva, L.M.e.; Ferreira, F.N.; et al. Ionic response and sorghum production under water and saline stress in a semi-arid environment. *Agriculture* **2023**, *13*, 1127. [CrossRef]
8. Dourado, P.R.M.; de Souza, E.R.; Santos, M.A.d.; Lins, C.M.T.; Monteiro, D.R.; Paulino, M.K.S.S.; Schaffer, B. Stomatal regulation and osmotic adjustment in sorghum in response to salinity. *Agriculture* **2022**, *12*, 658. [CrossRef]
9. Martins, G.O.; Santos, S.S.; Esteves, E.R.; de Melo Neto, R.R.; Gomes Filho, R.R.; de Melo, A.S.; Fernandes, P.D.; Gheyi, H.R.; Soares Filho, W.S.; Brito, M.E.B. Salt tolerance indicators in 'Tahiti' acid lime grafted on 13 rootstocks. *Agriculture* **2022**, *12*, 1673. [CrossRef]
10. Oliveira, V.K.N.; Silva, A.A.R.d.; Lima, G.S.d.; Soares, L.A.d.A.; Gheyi, H.R.; de Lacerda, C.F.; Vieira de Azevedo, C.A.; Nobre, R.G.; Garófalo Chaves, L.H.; Dantas Fernandes, P.; et al. Foliar application of salicylic acid mitigates saline stress on physiology, production, and post-harvest quality of hydroponic Japanese cucumber. *Agriculture* **2023**, *13*, 395. [CrossRef]
11. Elsawy, H.I.A.; Alharbi, K.; Mohamed, A.M.M.; Ueda, A.; AlKahtani, M.; AlHusnain, L.; Attia, K.A.; Abdelaal, K.; Shahein, A.M.E.A. Calcium lignosulfonate can mitigate the impact of salt stress on growth, physiological, and yield characteristics of two barley cultivars (*Hordeum vulgare* L.). *Agriculture* **2022**, *12*, 1459. [CrossRef]

2. Castelo Sousa, H.; Gomes de Sousa, G.; de Araújo Viana, T.V.; Prudêncio de Araújo Pereira, A.; Nojosa Lessa, C.I.; Pires de Souza, M.V.; da Silva Guilherme, J.M.; Ferreira Goes, G.; da Silveira Alves, F.G.; Primola Gomes, S.; et al. *Bacillus aryabhattai* mitigates the effects of salt and water stress on the agronomic performance of maize under an agroecological System. *Agriculture* **2023**, *13*, 1150. [CrossRef]
3. Ferreira, J.F.S.; Liu, X.; Suddarth, S.R.P.; Nguyen, C.; Sandhu, D. NaCl accumulation, shoot biomass, antioxidant capacity, and gene expression of *Passiflora edulis* f. flavicarpa Deg. in response to irrigation waters of moderate to high salinity. *Agriculture* **2022**, *12*, 1856. [CrossRef]

Disclaimer/Publisher's Note: The statements, opinions and data contained in all publications are solely those of the individual author(s) and contributor(s) and not of MDPI and/or the editor(s). MDPI and/or the editor(s) disclaim responsibility for any injury to people or property resulting from any ideas, methods, instructions or products referred to in the content.

Article

Bacillus aryabhattai Mitigates the Effects of Salt and Water Stress on the Agronomic Performance of Maize under an Agroecological System

Henderson Castelo Sousa [1,*], Geocleber Gomes de Sousa [2,*], Thales Vinícius de Araújo Viana [1], Arthur Prudêncio de Araújo Pereira [3], Carla Ingryd Nojosa Lessa [1], Maria Vanessa Pires de Souza [1], José Marcelo da Silva Guilherme [1], Geovana Ferreira Goes [1], Francisco Gleyson da Silveira Alves [4], Silas Primola Gomes [2] and Fred Denilson Barbosa da Silva [2]

1. Agricultural Engineering Department, Federal University of Ceará, Fortaleza 60455-760, Brazil; thales@ufc.br (T.V.d.A.V.); ingrydnojosa@alu.ufc.br (C.I.N.L.); vanessa.pires@alu.ufc.br (M.V.P.d.S.); josemarcelo01@alu.ufc.br (J.M.d.S.G.); geovanagoes@alu.ufc.br (G.F.G.)
2. Institute of Rural Development, University of International Integration of Afro-Brazilian Lusofonia, Redenção 62790-000, Brazil; silas.primola@unilab.edu.br (S.P.G.); freddenilson@unilab.edu.br (F.D.B.d.S.)
3. Soil Science Department, Federal University of Ceará, Fortaleza 60355-636, Brazil; arthur.prudencio@ufc.br
4. Department of Animal Science, Federal University of Ceará, Fortaleza 60356-000, Brazil; gleyson@ufc.br
* Correspondence: henderson@alu.ufc.br (H.C.S.); sousagg@unilab.edu.br (G.G.d.S.)

Abstract: The use of plant-growth-promoting rhizobacteria (PGPR) can be one option for mitigating the impact of abiotic constraints on different cropping systems in the tropical semi-arid region. Studies suggest that these bacteria have mechanisms to mitigate the effects of water stress and to promote more significant growth in plant species. These mechanisms involve phenotypic changes in growth, water conservation, plant cell protection, and damage restoration through the integration of phytohormone modulation, stress-induced enzyme apparatus, and metabolites. The aim of this study was to evaluate the growth, leaf gas exchange, and yield in maize (*Zea mays* L.—BRS Caatingueiro) inoculated with *Bacillus aryabhattai* and subjected to water and salt stress. The experiment followed a randomised block design, in a split-plot arrangement, with six repetitions. The plots comprised two levels of electrical conductivity of the irrigation water (0.3 dS m^{-1} and 3.0 dS m^{-1}); the subplots consisted of three irrigation depths (50%, 75%, and 100% of the crop evapotranspiration (ETc)); while the sub-subplots included the presence or absence of *B. aryabhattai* inoculant. A water deficit of 50% of the ETc resulted in the principal negative effects on growth, reducing the leaf area and stem diameter. The use of *B. aryabhattai* mitigated salt stress and promoted better leaf gas exchange by increasing the CO_2 assimilation rate, stomatal conductance, and internal CO_2 concentration. However, irrigation with brackish water (3.0 dS m^{-1}) reduced the instantaneous water-use efficiency of the maize. Our results showed that inoculation wiht PGPR mitigates the effect of abiotic stress (salt and water) in maize plants, making it an option in regions with a scarcity of low-salinity water.

Keywords: *Zea mays*; abiotic stress; microorganisms; salinity; water deficit

1. Introduction

Maize (*Zea mays* L.), with its origin in Central America, is of great economic importance and is cultivated worldwide. In Brazil it is one of the main cereals produced (21,581.9 million hectares), with an emphasis on food for human and animal consumption as well as for bioenergy production [1–4]. The crop has gradually expanded into arid and semi-arid regions, where it helps to solve problems related to food security in places that have limited water resources [5,6]. It is worth noting that maize is considered moderately sensitive to salinity, with a threshold of 1.1 and 1.7 dS m^{-1} for water and soil electrical conductivity, respectively [7].

The semi-arid region of Brazil is considered one of the largest semi-arid regions, with approximately 27 million inhabitants [8], where irrigation is an important tool for ensuring food security [9]. The characteristics of the region are high temperatures, high evapotranspiration, and a low rainfall rate [10,11]. Water shortages and high salt concentrations in the groundwater are problems that limit agricultural production in this region [9,12].

An excess of salts in the soil solution reduces water absorption by plants and alters metabolic and morphological structures, causing a reduction in seed germination, growth, and productivity in agricultural crops [13–15]. Water and salt stress reduce the soil water potential, making the soil solution unavailable, or not readily available, for nutrient uptake by plants. These stresses have a negative effect on physiological processes, causing partial closure of the stomata, limiting the internal CO_2 concentration, reducing the rates of photosynthesis and transpiration, and consequently the water-use efficiency and agricultural crop yields worldwide [16–18]. Evaluating the interaction between salt and water stress in the courgette, [19] found a reduction in photosynthesis and transpiration. Similarly, [20] found a reduction in the productivity of peanuts under salt and water stress.

It should be noted that various strategies have been used in the scientific environment to mitigate salt and water stress. One alternative to mitigate the effects of such stress and ensure production in agroecological systems is the use of microbial inoculants formulated with plant-growth-promoting bacteria (PGPB) [21–23]. These microorganisms can offer protection to plants against water deficits by maintaining moisture levels and providing better root development and nutrient supply. Researchers are seeking to identify microorganisms, together with their action mechanisms, that are able to mitigate abiotic stress [24,25]. Various promising studies have found that inoculating maize with beneficial microorganisms results in greater productivity [26].

In this scenario, the use of plant-growth-promoting rhizobacteria (PGPR), especially from the *Bacillus* genus, stands out in plant development. Some of the known mechanisms by which PGPRs can improve plant development include beneficial effects on promoting plant emergence and growth [27], antagonistic activity against phytopathogenic fungi [28], improvement of soil structure (by bacterial exopolysaccharides), provision of N to plants through biological nitrogen fixation, solubilization and mineralization of nutrients, particularly phosphate, and improvement of resistance to non-biological stresses [29]. The strain of B. aryabhattai CMAA 1363 was able to provide drought tolerance in maize plants [24].

Given this promising scenario, the present study tested the hypothesis that the use of plant-growth-promoting bacteria mitigates the effect of abiotic stress (salt and water) on the agronomic performance of maize. The aim of this study, therefore, was to evaluate the growth, leaf gas exchange, and production parameters of maize inoculated with *Bacillus aryabhattai* under water and salt stress.

2. Material and Methods

2.1. Location and Characterisation of the Experimental Area

The experiment was conducted from 25 August to 17 November 2022 (dry season) under field conditions at the Piroás Experimental Farm (PEF) (04°14′53″ S; 38°45′10″ W, at a mean altitude of 240 m), belonging to the Universidade da Integração Internacional da Lusofonia Afro-Brasileira (UNILAB), in Redenção, in the state of Ceará.

The climate in the region is of type BSh′ (tropical semi-arid climate), characterized by very hot temperatures, a rainy season during the summer and autumn (February to May), strong insolation, and high evaporation rates [30]. The amount of rainfall and the maximum and minimum air temperature were recorded daily throughout the experiment, as well as the average relative humidity (Figure 1), monitored by means of a data logger (HOBO® U12-012 Temp/RH/Light/Ext).

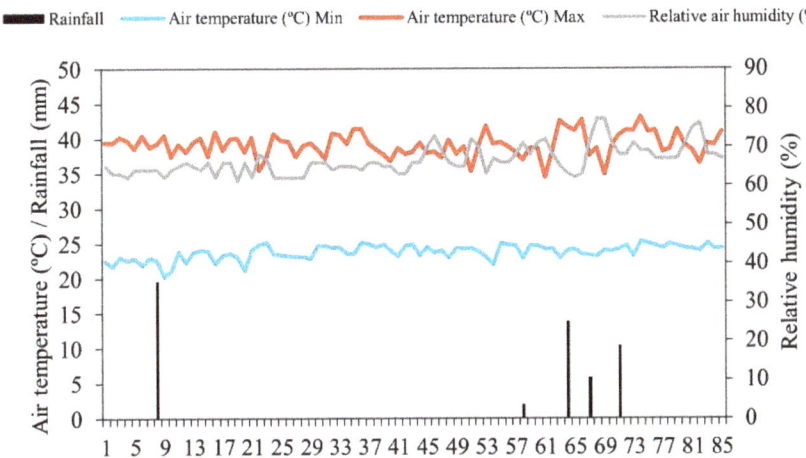

Figure 1. Mean values for maximum (Max) and minimum (Min) temperature and relative humidity obtained during the experimental cycle.

The soil in the experimental area is classified as an ultisol. Samples were collected from the surface layer (0–20 cm) and sent to the laboratory to determine the physical and chemical attributes (Table 1), as per the methodology described by [31].

Table 1. Chemical and physical characteristics of the soil sample before applying the treatments (0–20 cm).

pH	OM	N	C	P	Ca	Mg	Na	Al	H + Al	K	ECse	ESP	C/N	V
H$_2$O	g kg^{-1}			mg kg^{-1}			cmol$_c$ dm^{-3}				dS m^{-1}	%		%
5.6	11.59	0.71	6.72	20	3.20	2.60	0.07	0.35	2.15	0.17	0.76	1	9	74
SD (g cm^{-3})			CS		FS			Silt		Clay			Textural Classification	
Bulk	Particle					g kg^{-1}								
1.31	2.61		507		283			133		77			Loamy Sand	

OM—Organic matter; ESP—Percentage of exchangeable sodium; ECse—Electrical conductivity of the soil saturation extract; V—Base saturation; SD—Soil density; CS—Coarse sand; FS—Fine sand.

2.2. Experimental Design and Treatments

The experimental design was randomised blocks in a split-plot arrangement, with six repetitions. The plots comprised two levels of electrical conductivity of the irrigation water (ECw): water supply (0.3 dS m^{-1}) and a brackish solution (3.0 dS m^{-1}). The sub-plots consisted of three irrigation depths (ID1 = 50%, ID2 = 75%, and ID3 = 100% of the crop evapotranspiration [ETc]). The sub-subplots included the presence or absence of *B. aryabhattai* inoculant (Figure 2).

2.3. Irrigation Management

A drip irrigation system was used at a spacing of 0.3 m, corresponding to one emitter per plant. Emitters of 4, 6, and 8 L h^{-1} were used to standardise the irrigation time, affording water regimes of 50%, 75%, and 100% of the *ETc*, respectively. Uniformity tests were carried out, returning a distribution coefficient of 92%.

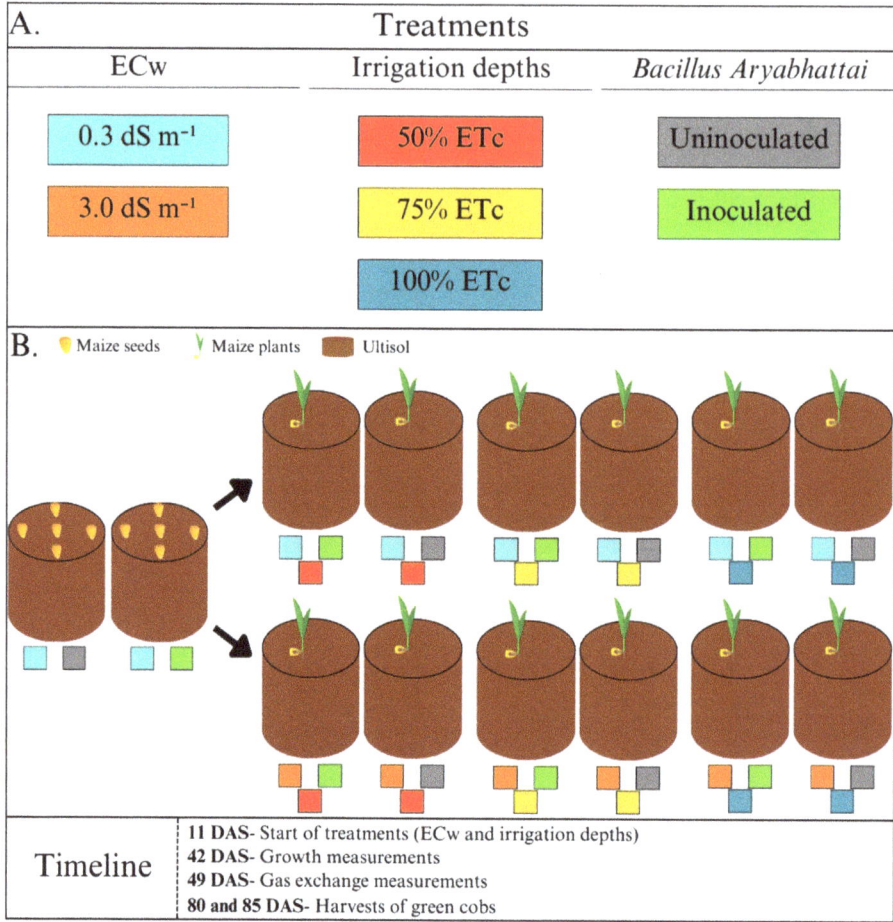

Figure 2. Diagram of the experimental design showing (**A**) the composition and interaction of the study factors—electrical conductivity of the water, irrigation depths and inoculation—and (**B**) a timeline of the procedures carried out during the experiment.

Irrigation management was estimated daily from the reference evapotranspiration using data from a Class A evaporimeter pan. The crop evapotranspiration, in mm day^{-1}, was calculated from the evaporation measured in the Class A pan, as per Equation (1).

$$ETc = ECA \times Kp \times Kc \qquad (1)$$

where:

ETc—Crop evapotranspiration, in mm day^{-1};
ECA—Evaporation measured in the class A pan, in mm/day^{-1};
Kp—Class A pan coefficient, dimensionless;
Kc—Crop coefficient, dimensionless.

The following crop coefficients (Kc) were adopted: 0.86 (up to 40 days after sowing—DAS); 1.23 (from 41 to 53 DAS); 0.97 (from 54 to 73 DAS), and 0.52 (from 74 DAS to the end of the cycle) [28]. A leaching fraction of 15% was added to the applied irrigation depth [32]. The irrigation time was obtained using Equation (2):

$$It = \frac{ETc \times Sd}{Af \times q} \times 60 \qquad (2)$$

where:

It—Irrigation time (min);
ETc—Crop evapotranspiration for the period (mm);
Sd—Spacing between emitters;
Af—Application efficiency (0.92);
q—Flow rate (L h^{-1}).

Table 2 shows the total irrigation depth applied during the experiment throughout the crop cycle based on each treatment.

Table 2. Total irrigation depth applied in each treatment.

ECw (dS m^{-1})	ETc (%)	Total Depth Applied (mm)	
		Uninoculated	Inoculated
0.3	50	260.4	260.4
	75	390.6	390.6
	100	520.8	520.8
3.0	50	260.4	260.4
	75	390.6	390.6
	100	520.8	520.8

Fresh water (0.3 dS m^{-1}) from the dam belonging to FEP was used to irrigate the plants of the control treatment. This same water source was stored in 500 L tanks and used in preparing the 3.0 dS m^{-1} saline solution by dissolving sodium chloride (NaCl), calcium chloride (CaCl$_2$2H$_2$O), and magnesium chloride (MgCl$_2$6H$_2$O), maintaining the proportions predominantly found in the principal water sources of the northeast of Brazil of 7:2:1 [33] and based on the relationship between the ECw and its molar concentration (mmol$_c$ L^{-1} = CE × 10). The electrical conductivity of the water was periodically monitored using a bench conductivity meter (AZ® 806,505 pH/Cond./TDS/Salt). The water was sent for its chemical characteristics to be determined following the methodology of [34] and was classified using the methodology described by [35]. The results are shown in Table 3.

Table 3. Chemical characterisation and classification of the irrigation water used in the experiment.

ECw	Ca^{2+}	Mg^{2+}	K$^+$	Na$^+$	Cl$^-$	HCO$_3^-$	pH	CE	SAR	Classification [1]
dS m^{-1}	mmol$_c$ L^{-1}				mmol L^{-1}		in H$_2$O	dS m^{-1}	(mmol$_c$ L^{-1})$^{0.5}$	
0.3	0.6	1.4	0.2	0.4	2.5	0.1	6.9	0.3	0.4	C$_2$S$_1$
3.0	6.33	7.64	2.0	15.6	25	1.0	7.79	3.0	5.9	C$_4$S$_2$

[1]—[35]; SAR—Sodium adsorption ratio.

The experiment was irrigated daily with water of 0.3 dS m^{-1} up to 10 days after sowing (DAS) with a water depth of 100% of the ETc. The treatments, including the water regimes and ECw, were started at 11 DAS.

2.4. Agroecological Maize Production System (Plant Material, Inoculation, and Fertilisation)

Seeds of the maize (Zea mays L.) 'BRS Caatingueiro' variety were used, sown manually with five seeds per hole at a spacing of 0.8 × 0.2 m between the rows and plants. This cultivar is used by producers in the region and has a super-early cycle. At 10 DAS, with the plant stand already established, thinning was carried out to leave one plant per hole.

The inoculation was carried out using the commercial product Auras® (Embrapa and NOOA Agricultural Science and Technology, Patos de Minas–Minas Gerais, Brazil) formulated with *Bacillus aryabhattai* CMAA 1363, licensed by the Brazilian Agricultural Research Corporation (Embrapa, Jaguariúna–São Paulo, Brazil), obtained from the rhizosphere of *Cereus jamaracu*, a cactus present in the Caatinga biome of the Brazilian semi-arid region [36]. The seeds were immersed in the bacterial solution immediately before planting, applying 4 mL kg^{-1} of maize seeds. The rhizobacterium belongs to the inoculant class, with a concentration of 1×10^8 UFC/mL.

Fertiliser management was based on the chemical analysis of the soil (Table 1) and used organic fertiliser (cattle manure and cattle biofertiliser) applied as a base and topdressing as recommended by [37] for irrigated maize in the state of Ceará, equal to 90 kg ha^{-1} N, 40 kg ha^{-1} P_2O_5, and 30 kg ha^{-1} K_2O.

The chemical characteristics of the cattle manure and cattle biofertiliser were determined as per the methodology of [31] and are shown in Table 4.

Table 4. Chemical characterisation of the organic fertilisers used in the experiment.

Organic Source	N	P	K$^+$	Ca^{2+}	Mg^{2+}
			g L^{-1}		
Cattle manure	0.96	0.47	0.59	1.10	0.25
Cattle biofertiliser	0.82	1.4	1.0	2.5	0.75

2.5. Variables under Analysis

2.5.1. Growth

At 42 DAS, the following variables were evaluated: plant height (PH, cm), using a tape, measuring from the soil to the apex of the plant; number of leaves (NL), by directly counting the fully expanded leaves; stem diameter (SD, mm), measured two centimetres from the ground using a pachymeter; leaf area (LA, cm^2), using an area integrator (Area meter, LI-3100, Li-Cor, Inc., Lincoln, NE, USA).

2.5.2. Leaf Gas Exchange

At 49 DAS, gas exchange measurements were taken using the third fully expanded leaf from the apex of the plant. The net photosynthetic rate (A, μmol CO_2 m^{-2} s^{-1}), stomatal conductance (g_s, mol m^{-2} s^{-1}), rate of transpiration (E, mmol m^{-2} s^{-1}), and internal CO_2 concentration (C_i, μmol mol^{-1}) were measured using an infrared gas analyser (Li-6400XT, LICOR, Lincoln, NE, USA) under the following conditions: ambient air temperature, CO_2 of 400 ppm, photosynthetically active radiation of 1800 μmol m^{-2} s^{-1}, between 09:00 and 11:00. The instantaneous water-use efficiency (WUEi) was estimated from the photosynthesis and transpiration data. The relative chlorophyll index (RCI, SPAD) was measured on the same leaves using a portable meter (SPAD—502 Plus, Minolta, Tokyo, Japan).

2.5.3. Yield

To determine the production parameters, two harvests of green ears were carried out (80 and 85 DAS), when the following were evaluated: ear length (EL, cm), measuring longitudinally using a ruler; ear diameter (ED, mm), measuring transversely using a digital pachymeter; ear yield with straw (EYWS, kg ha^{-1}) and ear yield without straw (EYWoS, kg ha^{-1}), estimated from the mean weight of the ear and the stipulated plant stand per hectare (62,500 plants ha^{-1}).

2.6. Data Analysis

The data obtained were subjected to the Kolmogorov–Smirnov test of normality at a level of 0.05 probability. After verifying the normality, analyses of variance were applied using the F-test ($p < 0.05$). In cases of statistical significance, the mean values were compared with Tukey's test ($p < 0.05$) using the Assistat 7.7 Beta software [38].

3. Results and Discussion

3.1. Growth

The analyses of variance revealed that the leaf area and stalk diameter were significantly influenced by the water regime alone and by the interaction between the electrical conductivity of the water and inoculation. The leaf area was significantly affected by the electrical conductivity of the water and by the interaction between the water regime and inoculation. The triple interaction of the factors ECw × ID × INOC had a significant influence on plant height. The number of leaves was not significantly influenced by any of the factors (Table 5).

Table 5. Summary of the analysis of variance for plant height (PH), number of leaves (NL), stem diameter (SD), and leaf area (LA) in maize plants under different levels of electrical conductivity of the irrigation water (ECw), irrigation depth (ID), and inoculation (INOC) 42 days after sowing.

Source of Variation	DF	Mean Square			
		PH	NL	SD	LA
Blocks	5	29.67 ns	1.95 ns	1.53 ns	207.38 ns
ECw	1	0.06 ns	2.60 ns	87.96 **	5613.37 *
Residual (ECw)	5	27.97	0.61	2.74	412.05
Irrigation depths (ID)	2	21.30 ns	0.40 ns	56.08 **	10,546.35 **
Residual (ID)	20	15.37	0.43	2.49	974.55
Inoculation (INOC)	1	9.56 ns	0.33 ns	35.25 **	10,360.73 **
Residual (INOC)	30	16.57	0.54	3.05	1161.23
ECw × ID	2	160.75 **	0.25 ns	1.87 ns	2864.11 ns
ECw × INOC	1	149.91 **	0.004 ns	0.0007 *	2064.59 ns
ID × INOC	2	0.24 *	0.16 ns	0.27 ns	5769.978 *
ECw × ID × INOC	2	70.49 *	1.42 ns	4.56 ns	654.36 ns
CV (%)—Ecw		5.46	9.49	12.68	5.47
CV (%)—ID		4.05	7.97	12.08	8.41
CV (%)—INOC		4.20	8.95	13.37	9.18

DF: Degrees of freedom; CV: Coefficient of variation; ns, *, and **: not significant, significant at $p \leq 0.05$, and significant at $p \leq 0.01$, respectively.

The height of the maize plants under low electrical conductivity and full irrigation (100% of the ETc) using water of lower salinity (0.3 dS m^{-1}) was greater regardless of inoculation; however, under irrigation at 75% of the ETc, the inoculated plants differed statistically from the uninoculated plants, showing higher values (97.44 cm). Similarly, under irrigation with water of higher salinity (3.0 dS m^{-1}), there was a significant difference from the water regime only, of 75%, with the inoculated plants obtaining the highest mean value (100.73 cm) (Figure 3).

The maize plants showed greater height when *B. aryabhattai* was used under a moderate deficit (75% of the ETc), regardless of the quality of the water used, indicating the beneficial effect of this stress condition. Rhizosphere bacteria show beneficial effects in various crops, possessing several mechanisms that help mitigate water stress, especially in relation to strengthening phytohormone activity (abscisic acid, gibberellins, cytokinins, and auxins) [24,39].

The mitigating effect of water stress in maize by bacteria of the genus *Bacillus* was also reported by [40] under the conditions of a reduced water supply (30% of field capacity), where inoculated plants were taller by around 27.29% compared to uninoculated plants. Reference [41] found that the optimal irrigation regime (100%) had a positive influence on the height of maize plants compared to lower percentages (50% and 75%).

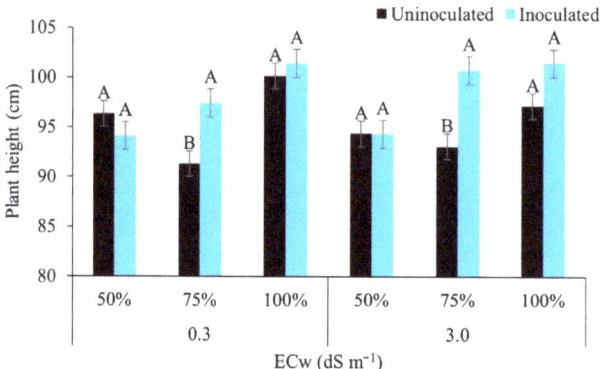

Figure 3. Height of maize plants under different levels of electrical conductivity of the irrigation water, different irrigation depths, with and without inoculation, 42 days after sowing. Uppercase letters compare mean values between plants with and without inoculants for the same electrical conductivity and irrigation depth using Tukey's test ($p \leq 0.05$). Error bars represent the standard error of the mean (n = 6).

From Figure 4A, it can be seen that between the water regimes, ID1 and ID2, the stem diameter did not differ statistically at the lower values (11.55 and 12.77 mm, respectively), whereas ID3, at 100%, resulted in larger diameters (14.85 mm). Optimal water conditions contribute to turgor pressure, allowing plant cells to develop internal hydrostatic pressure in the cell walls that is essential for cell expansion; on the other hand, a water deficit mainly inhibits leaf expansion and stem growth due to a reduction in pressure [42].

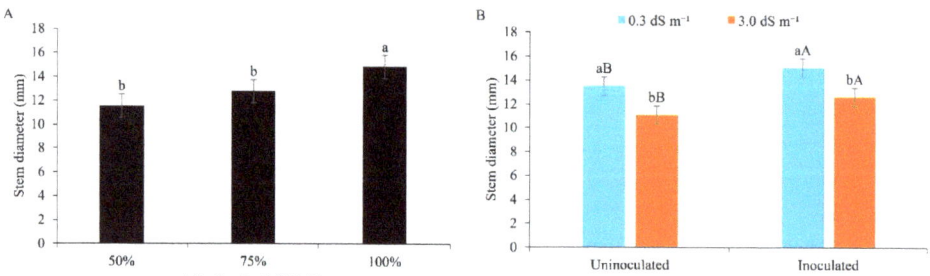

Figure 4. Stem diameter of maize plants under different water regimes (**A**) and different levels of electrical conductivity of the irrigation water, with and without inoculation (**B**), 42 days after sowing. (**A**): Lowercase letters compare mean values with Tukey's test ($p \leq 0.05$). (**B**): Lowercase letters compare mean values between ECw levels within each type of inoculation; uppercase letters compare mean values for the type of inoculation within each ECw with Tukey's test ($p \leq 0.05$). Error bars represent the standard error of the mean (n = 6).

This result is similar to that of [41], who used different irrigation rates estimated by a class A pan (50%, 75%, 100%, and 125% of the ETc), where the greatest stalk diameter for green maize (11.72 mm) was obtained using the highest rate. Evaluating different irrigation depths in a subsurface drip system, [43] found that reductions starting at 80% of the required depth caused a reduction in the stalk diameter of maize.

The stem diameter was statistically greater when applying water of lower salinity (0.3 dS m^{-1}) to inoculated plants, with a mean value of 15.04 mm. When using brackish water, the stem diameter was smaller regardless of inoculation, but showed higher values in inoculated plants in a direct comparison (12.61 mm) (Figure 4B).

The harmful effects of salinity on water and nutrient uptake resulted in a reduction in the stem diameter; however, these effects were mitigated when using *B. aryabhattai*. The presence of rhizobacteria may have mitigated the osmotic effects imposed by salt stress via biochemical changes in the plant or rhizosphere, increasing the physiology of the exposed plants and facilitating water uptake [27,44]. Inoculation with PGPBs during the early stages of maize crop under drought conditions significantly improved the stem diameter [45].

Studying different levels of electrical conductivity for the water (0.2, 1.3, 2.6, 3.9, and 5.2 dS m^{-1}), [46] found a linear reduction in the stalk diameter of maize with the increasing salinity of the irrigation water. Similar results were found by [47], who reported that the use of brackish water up to 30 DAS reduced the stem diameter in the cowpea.

It can be seen that the leaf area differed statistically between the levels of electrical conductivity of the irrigation water, with a higher mean value for irrigation water of lower conductivity (0.3 dS m^{-1} = 380.97 cm^2) (Figure 5A). The reduction in leaf elongation is a mechanism of survival and water conservation, where under stress conditions, the plants close their stomata and reduce transpiration. In addition, osmotic effects directly interfere with the water uptake of plants [17,48].

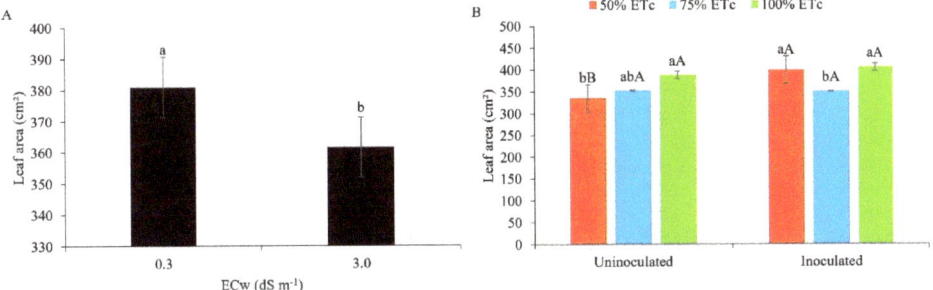

Figure 5. Leaf area of maize plants under different levels of electrical conductivity of the irrigation water (**A**) and different irrigation depths, with and without inoculant (**B**), 42 days after sowing. (**A**): Lowercase letters compare mean values using Tukey's test ($p \leq 0.05$). (**B**): Lowercase letters compare mean values between irrigation depths within each type of inoculation; uppercase letters compare mean values for the type of inoculation within each irrigation regime using Tukey's test ($p \leq 0.05$). Error bars represent the standard error of the mean ($n = 6$).

Similar results with maize under salt stress were obtained by [49], where the leaf area underwent a significant reduction of 19.9% in relation to the lowest level of salt (0.5 dS m^{-1}). Working with maize, [17] saw a reduction in the leaf area of 15.3% 45 DAS under salt stress (3 dS m^{-1}).

As shown in Figure 5B, the leaf area was statistically smaller when associating the water regime of 50% with no inoculant (334.47 cm^2); however, in inoculated plants, the ID of 50% (398.19 cm^2) and 100% (405.19 cm^2) of the Etc were statistically superior to the ID of 75% of the Etc (349.93 cm^2).

This result reflects the behaviour of plants subjected to water stress, i.e., they tend to reduce their leaf area as a mechanism for reducing water loss by transpiration, since the water absorption capacity of plants is directly affected by the water content of the soil [5,42]. However, the use of inoculants may have increased colonisation in the soil adhering to the roots, increasing the moisture and improving the ratio of root-adhering soil to root tissue, promoting greater resistance to water stress and consequently, greater leaf area development [25]. Ref. [50], evaluating maize seeds treated with exopolysaccharide-producing bacteria, found an increase in the soil moisture content and a greater leaf area.

3.2. Leaf Gas Exchange

As shown in the summary of the analysis of variance of the physiological variables (Table 6), the net photosynthetic rate and the internal CO_2 concentration were significantly influenced by the interaction between the electrical conductivity of the water and inoculation. Transpiration, on the other hand, was influenced by the Ecw × ID interaction, while the chlorophyll index was independently influenced by the same factors. The Ecw × ID and ID × INOC interactions influenced the leaf temperature. On the other hand, the electrical conductivity of the water was the single significant factor for water-use efficiency. The triple interaction of the factors Ecw × ID × INOC had a significant influence on stomatal conductance.

Table 6. Summary of the analyses of variance for photosynthesis (A), stomatal conductance (gs), internal CO_2 concentration (Ci), transpiration (E), relative chlorophyll index (RCI), leaf temperature (LT), and instantaneous water-use efficiency (WUEi) in maize plants under different levels of electrical conductivity of the irrigation water (ECw), different irrigation depths (ID), and inoculation (INOC), 49 days after sowing.

Source of Variation	DF	Mean Square						
		A	gs	Ci	E	RCI	LT	WUEi
Blocks	5	9.03 ns	0.67 ns	798.40 ns	0.43 **	10.05 ns	3.67 ns	0.006 *
ECw	1	361.19 **	0.15 ns	4504.68 *	36.83 **	191.12 **	74.72 **	11.07 **
Residual (ECw)	5	7.25	0.16	303.63	0.00 **	2.94	0.61	0.28
Irrigation depths (ID)	2	2.14 ns	0.26 ns	315.25 ns	0.25 ns	92.35 *	0.77 *	0.22 ns
Residual (ID)	20	8.90	0.14	101.83	0.11	23.43	0.15	0.15
Inoculation (INOC)	1	2.13 ns	3.60 **	336.02 ns	0.11 ns	39.45 ns	0.04 *	0.01 ns
Residual (INOC)	30	3.13	0.18	110.47	0.16	9.78	0.02	0.21
ECw × ID	2	2.86 ns	1.46 **	9.75 ns	0.58 *	66.17 ns	0.67 *	0.01 ns
ECw × INOC	1	6.97 *	2.48 **	595.02 *	0.47 ns	0.09 ns	0.04 ns	0.008 ns
ID × INOC	2	1.88 ns	0.04 ns	234.33 ns	0.27 ns	4.97 ns	0.10 *	0.006 ns
ECw × ID × INOC	2	0.78 ns	2.57 **	110.47 ns	0.18 ns	0.87 ns	0.02 ns	0.27 ns
CV (%)—ECw		11.20	11.85	6.34	0.77	5.54	2.59	9.60
CV (%)—ID		12.41	10.04	3.67	7.88	15.62	1.32	7.03
CV (%)—INOC		7.37	14.05	3.82	9.11	10.09	0.50	8.31

DF: Degrees of freedom; CV: Coefficient of variation; ns, *, and **: not significant, significant at $p \leq 0.05$, and significant at $p \leq 0.01$, respectively.

From Figure 6, it can be seen that the photosynthetic rate was higher when the maize crop was subjected to irrigation with brackish water (3.0 dS m^{-1}), with and without inoculation. This behaviour is possibly linked to the presence of magnesium chloride in the irrigation water of higher salinity, since chloride is a crucial micronutrient in capturing light, helping the function of the enzyme that catalyses the photolysis of water in photosystem II, while magnesium is the central macronutrient of chlorophyll, a molecule located in the chloroplasts, which are responsible for capturing sunlight during photosynthesis [51,52].

When observing the effect of salt stress without the use of inoculants, [26], evaluating the photosynthetic rate of maize in pots under irrigation with brackish water, found a reduction in this variable in the presence of salt stress 45 days after sowing. Reference [53], investigating the effects of water salinity on photosynthesis in peanut plants inoculated with *Bradyrhizobium* sp., found similar results to the present study regarding the mitigating effect of the inoculant in plants grown under salt stress.

Figure 7 shows that in water of lower salinity, the inoculated plants achieved a greater stomatal conductance under the irrigation regimes of 50% and 75%, with no statistical difference for the regime of 100% (full irrigation). At the highest level of salinity, the opposite occurred, where plants with the inoculant achieved a greater stomatal conductance under the irrigation regime of 100% only, the other regimes showing no statistical difference. Reference [54] emphasises that this result may be linked to the participation of resistance-promoting bacteria, i.e., those that have the ability to branch the roots and release exudates that increase the relative water content of the rhizosphere, thereby better coping with stress conditions.

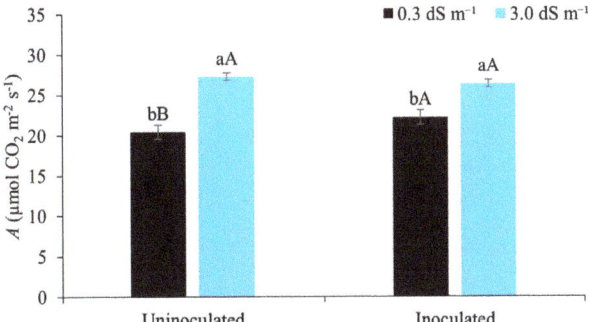

Figure 6. Net photosynthetic rate (*A*) in maize plants under different levels of electrical conductivity of the irrigation water, with and without inoculation, 49 days after sowing. Lowercase letters compare mean values between ECw levels within each type of inoculation; uppercase letters compare means values for the type of inoculation within each ECw with Tukey's test ($p \leq 0.05$). Error bars represent the standard error of the mean ($n = 6$).

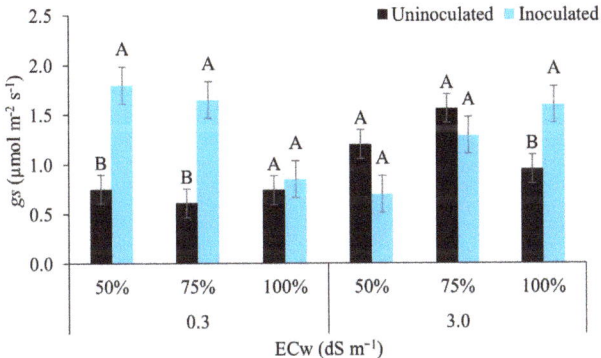

Figure 7. Stomatal conductance (*gs*) of maize plants under different levels of electrical conductivity of the irrigation water and different water regimes, with and without inoculation, 49 days after sowing. Uppercase letters compare mean values between plants with and without inoculant within the same electrical conductivity and irrigation depth with Tukey's test ($p \leq 0.05$). Error bars represent the standard error of the mean ($n = 6$).

Studying the courgette irrigated with brackish water under water stress, found that the isolated effect of irrigation water with increasing levels of salts was lower stomatal conductance [19]. However, when using strains of growth-promoting bacteria in maize, [55] reported similar results to the present study. According to those authors, inoculated plants were better able to adjust to stress, showing greater conductance compared to uninoculated plants. Reinforcing the above, ref. [56] described how resistance-promoting bacteria promote a significant increase in osmoprotectants under salt stress, improving the water potential and hydraulic conductivity that positively affect stomatal opening.

It can be seen from Figure 8 that the internal CO_2 concentration was higher when the maize was irrigated with water of lower salinity (0.3 dS m^{-1}), demonstrating the negative effects of salt stress, which interferes in the osmotic, toxic, and nutritional processes and affects the net CO_2 assimilation [18].

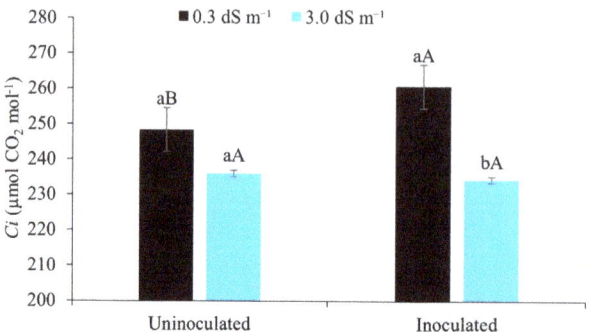

Figure 8. Internal CO_2 concentration (Ci) of maize plants under different levels of electrical conductivity of the irrigation water, with and without inoculation, 49 days after sowing. Lowercase letters compare mean values between ECw levels within each type of inoculation; uppercase letters compare mean values for the type of inoculation within each ECw using Tukey's test ($p \leq 0.05$). Error bars represent the standard error of the mean ($n = 6$).

Similar trends were observed by [57] studying salt stress in okra, where an increase in the electrical conductivity of the irrigation water promoted a reduction in the internal CO_2 concentration. The same authors confirm that salt stress induces partial stomatal closure as an attempt by the plant to minimise water loss, which in return reduces the entry of CO_2 from the atmosphere into the leaf mesophyll and, since no exchange takes place, reduces its concentration in the substomatal cavity.

According to Figure 9, there was no significant difference in plant transpiration between the water regimes when irrigated with water of lower salinity. However, when compared to higher levels of salinity, the 75% and 100% regimes promoted greater transpiration. Salt and water stress induce osmotic adjustment, which is considered an important mechanism for the maintenance of water uptake and cell turgor under stress conditions [58].

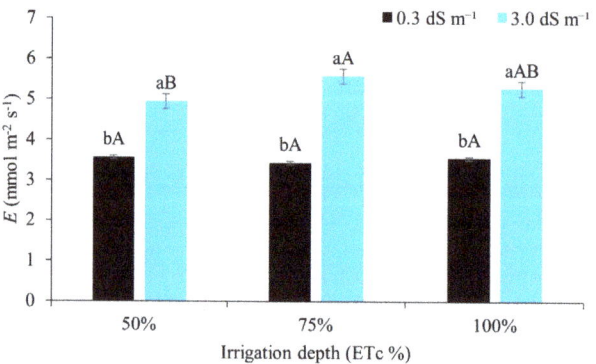

Figure 9. Transpiration (E) in maize plants under different water regimes with and without inoculant, 49 days after sowing. Lowercase letters compare mean values between ECw levels within each water regime; uppercase letters compare mean values between water regimes at the same ECw with Tukey's test ($p \leq 0.05$). Error bars represent the standard error of the mean ($n = 6$).

Different results, with a reduction in plant transpiration when the electrical conductivity of the irrigation water was increased, were found by [59], cultivating irrigated maize under a water regime of 100% of the ETc. Reference [15] also showed a reduction in transpiration in maize irrigated with brackish water.

The chlorophyll index of the maize was higher under an electrical conductivity of 3.0 dS m^{-1} and was statistically different from the lower salinity (0.3 dS m^{-1}) (Figure 10A).

This response may be related to the conditions of low CO_2 availability due to stomatal closure and physiological imbalances linked to the high salt content, reaffirming that the chlorophyll content is influenced by both biotic and abiotic factors [42].

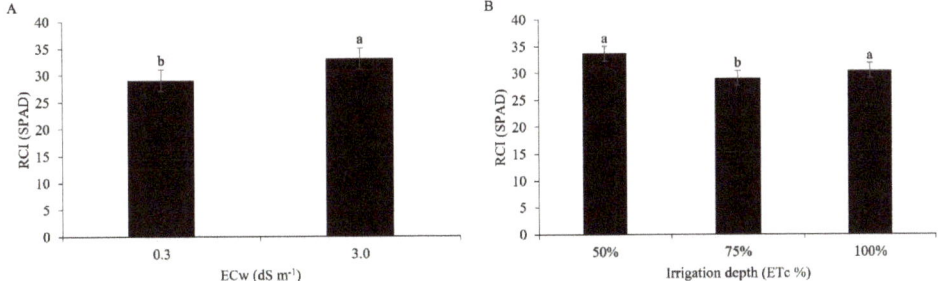

Figure 10. Relative chlorophyll index (RCI) of maize plants under irrigation with water of different levels of electrical conductivity (**A**) and different water regimes (**B**), 49 days after sowing. Lowercase letters compare mean values using Tukey's test ($p \leq 0.05$). Error bars represent the standard error of the mean ($n = 6$).

Under the influence of the applied water regimes, the chlorophyll index reached the highest value when 50% of the ETc was used, in relation to the other regimes, showing that under the conditions of the present study the water deficit did not negatively affect this variable (Figure 10B). The opposite result was found in [60], where a reduction in the chlorophyll index followed a reduction in the irrigation depth for five irrigation depths at different sampling times.

The internal leaf temperature (Figure 11A) significantly increased when using water of higher salinity (3.0 dS m^{-1}), with an increase of 2.82, 2.64, and 2.04 °C for the irrigation depths of 50%, 75%, and 100%, respectively, compared to water of lower salinity (0.3 dS m^{-1}). Following the trend for transpiration under salt stress, the leaf temperature gradually increased. It should be noted that plants under salt stress show great difficulty in absorbing water from the soil; as such, there is an increase in internal temperature, since water helps in the thermal regulation of plants, even under conditions of high transpiration [42–59]. It is worth noting that transpiration via movement of the stomata helps in reducing the leaf temperature (cooling), which is crucial during the day when the leaf absorbs large amounts of energy from the sun [42].

Irrigating peanut plants with brackish water (1, 2, 3, 4, and 5 dS m^{-1}), [53] reported a linear increase in the internal leaf temperature. Reference [61], studying maize, also found that salinity afforded an increase in the leaf temperature, reaching 36.7 °C.

It can be seen from Figure 11B that only inoculated plants under the ID of 100% of the ETc differed statistically from the other treatments, with the highest values (30.7 °C). The symbiosis between plants and microorganisms tends to afford better osmotic adjustment, improving transpiration and reducing leaf temperature [62]. Under the conditions of the present study, the heat-dissipation mechanism of the inoculated plants under full irrigation (100%) was possibly not compromised, since the recorded temperatures are within the range for plants with a C4 metabolism, such as maize [42].

Evaluating the peanut under inoculation with *Bradyrhizobium* sp., [53] obtained different results to those of the present study, where inoculated plants showed a lower leaf temperature.

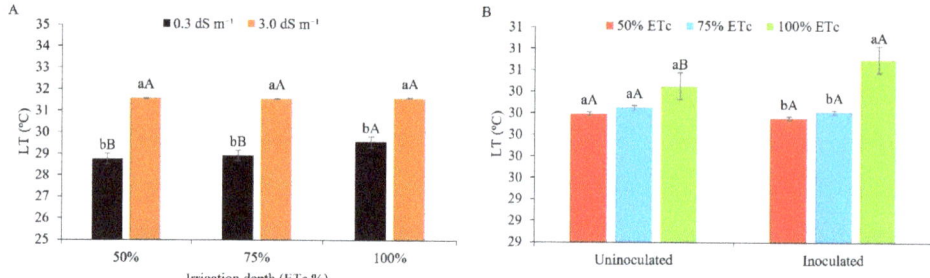

Figure 11. Leaf temperature (LT) in maize plants under different levels of electrical conductivity of the irrigation water, different irrigation depths (**A**), and different water regimes, with and without inoculant (**B**), 49 days after sowing. (**A**): Lowercase letters compare mean values between ECw levels within each irrigation depth; uppercase letters compare mean values between irrigation depths at the same ECw with Tukey's test ($p \leq 0.05$). (**B**): Lowercase letters compare mean values between irrigation depths within each type of inoculation; uppercase letters compare mean values between the types of inoculation within each irrigation depth with Tukey's test ($p \leq 0.05$). Error bars represent the standard error of the mean ($n = 6$).

The instantaneous water-use efficiency in maize plants under irrigation at the higher level of salinity (3.0 dS m^{-1}) was lower by around 17.6% in relation to irrigation at 0.3 dS m^{-1} (Figure 12). Salt stress induced by irrigation results in limited water uptake due to osmotic and physiological effects, in addition to biochemical changes, which result in a reduced water-use efficiency [15].

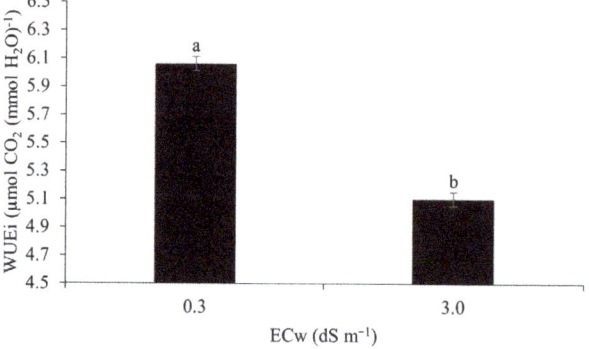

Figure 12. Instantaneous water-use efficiency (WUEi) in maize plants under different levels of electrical conductivity of the irrigation water 49 days after sowing. Lowercase letters compare mean values with Tukey's test ($p \leq 0.05$). Error bars represent the standard error of the mean ($n = 6$).

A reduction in the instantaneous water-use efficiency of maize at 49 DAS was also reported by [63] when irrigating the crop with brackish water of 4.5 dS m^{-1} under field conditions in the northeast of Brazil.

3.3. Yield

The summary of the analyses of variance of the yield parameters (Table 7) shows the significant influence of the ECw × INOC and ID × INOC interactions on the ear length, while the diameter was not affected by any of the factors. On the other hand, for ears with straw and ears without straw, the yield was significantly influenced by the interaction of the factors under study (ECw × ID × INOC).

Table 7. Summary of the analyses of variance for ear length (EL), ear diameter (ED), ear yield with straw (EYWS), and ear yield without straw (EYWoS) in maize plants under different levels of electrical conductivity of the irrigation water (ECw), different irrigation depths (ID), and inoculation (INOC).

Source of Variation	DF	Mean Square			
		EL	ED	EYWS	EYWoS
Blocks	5	3.09 ns	17.63 ns	6,772,473.13 ns	3,162,178.48 ns
ECw	1	5.15 ns	9.90 ns	40,862,788.82 **	14,079,648.00 **
Residual (ECw)	5	1.69	4.58	1,711,897.76	820,739.41
Irrigation depths (ID)	2	1.35 ns	5.27 ns	3,121,961.40 ns	1,275,678.96 *
Residual (ID)	20	1.49	3.63	1,464,975.55	351,695.38
Inoculation (INOC)	1	0.14 ns	0.38 ns	3,533,704.78 **	450,274.80 *
Residual (INOC)	30	0.86	1.96	363,835.20	309,105.44
ECw × ID	2	0.34 ns	1.57 ns	5,385,095.95 *	1,002,466.33 ns
ECw × INOC	1	3.78 *	2.51 ns	1,257,793.16 ns	69,497.37 ns
ID × INOC	2	6.11 **	1.65 ns	956,622.93 ns	225,383.24 ns
ECw × ID × INOC	2	1.27 ns	5.31 ns	1,678,902.68 *	1,118,325.91 *
CV (%)—ECw		10.92	6.24	29.91	29.74
CV (%)—ID		10.24	5.56	27.67	19.47
CV (%)—INOC		7.80	4.08	13.79	18.25

DF: Degrees of freedom; CV: Coefficient of variation; ns, *, and **: not significant, significant at $p \leq 0.05$, and significant at $p \leq 0.01$, respectively.

The ear length was not statistically affected when using water of lower or higher salinity in the presence of *Bacillus aryabhattai*; however, salt stress promoted greater ear length in inoculated maize plants and was statistically superior to water of lower salinity in the absence of the inoculant (Figure 13A). These data reveal the possible mitigating effect of *Bacillus* for maize in saline environments. The applied PGP bacteria may have generated a mechanism of plant protection against salt stress through the production of auxins and increased nitrogen fixation [21,64]. In addition, they assist the plant in combating physiological drought under salt stress by increasing the water content in the cell [65], reflecting in greater performance for ear length.

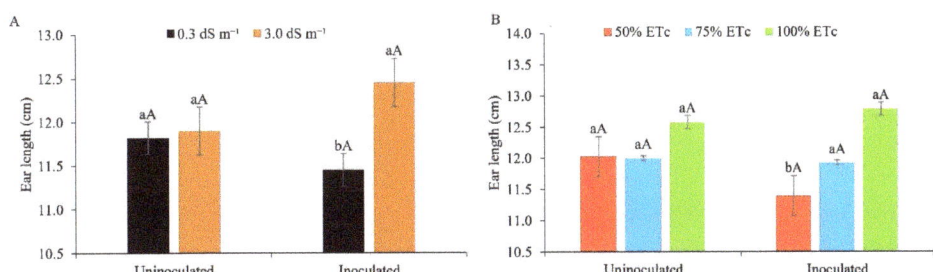

Figure 13. Ear length in maize plants under different levels of electrical conductivity of the irrigation water, with and without inoculation (**A**), and different irrigation depths, with and without inoculation (**B**). (**A**): Lowercase letters compare mean values between ECw levels within each type of inoculation; uppercase letters compare mean values for the type of inoculation within each ECw, with Tukey's test ($p \leq 0.05$). (**B**): Lowercase letters compare mean values between irrigation depths within each type of inoculation; uppercase letters compare mean values between the types of inoculation within each water regime with Tukey's test ($p \leq 0.05$). Error bars represent the standard error of the mean ($n = 6$).

Studies using maize seeds inoculated with 'Graminante®', a commercial biotechnological product based on *Azospirillum* spp., and irrigated with low-salinity water returned similar results to the present study for ear length [66]. A reduction in the ear length of maize plants irrigated with brackish water with an electrical conductivity of 3.0 dS m^{-1} under field conditions was also found by [67]. In this study, the effect of poultry biofertiliser—a mixture of live microorganisms (bacteria, yeasts, algae, and filamentous fungi), which, when available to plants, colonise the rhizosphere and/or the interior of the plant, and promote growth by increasing the supply of primary nutrients—was investigated [68].

Figure 13B shows that the ears of the inoculated plants were superior once irrigated with 75% and 100% of the ETc, differing statistically from the deficit irrigation of 50%, with a superiority of 4.6% and 12.3%, respectively. In the absence of the PGPB, however, there was no difference among the water regimes under study.

The use of *B. Aryabhattai* produces compatible osmolytes, small organic molecules such as betaine that assist during environmental stress [31,69,70], and biofilm formation [71], which acts by forming a hydrated microenvironment around the root, retaining water, and making it available for longer [72], thereby mitigating water stress in maize grown under field conditions in the semi-arid region of the northeast of Brazil. Studies conducted with halotolerant rhizobacteria (HT-PGPR) report that they can also help saline soils to recover their natural balance, promoting benefits for plants grown under saline conditions [73]. The opposite effect to that seen in the present study was reported by [74] when irrigating uninoculated maize with brackish water at 100% of the ETc. The same authors found no significant effect for the ear length.

Irrigation with water of lower salinity at 100% of the ETc in inoculated maize plants (Figure 14A) afforded the highest ear yield with straw (4058 kg ha^{-1}), higher than the treatment with 50% and 75% of the ETc. The inoculated plants irrigated with brackish water at 75% of the ETc were statistically superior to those from the other regimes. This effect may be related to the protection imposed on the soil by *B. aryabhattai*, which plays an important role in the rhizosphere, improving the soil structure by increasing the volume of macropores, increasing water availability, and binding cations such as Na$^+$ that help mitigate salt stress [65,71,75]. Growing uninoculated maize under field conditions, irrigated with low-salinity water at 100% of the ETc, recorded superior results to those of the present study, achieving a productivity of 10 t ha^{-1} [76].

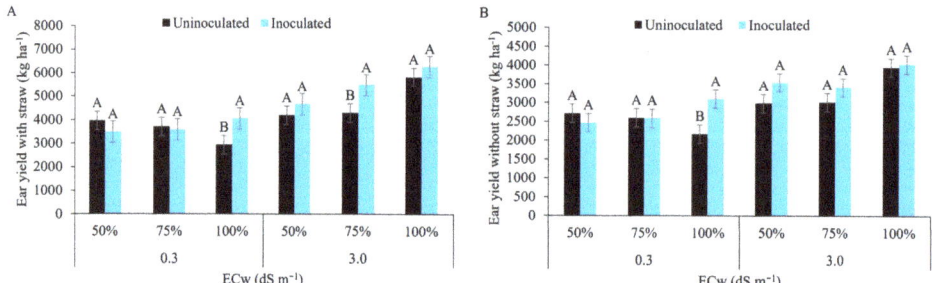

Figure 14. Ear yield with straw (**A**) and ear yield without straw (**B**) in maize plants under different levels of electrical conductivity of the irrigation water, different irrigation depths, with and without inoculation. Uppercase letters compare mean values between plants with and without inoculant within the same electrical conductivity and irrigation depth with Tukey's test ($p \leq 0.05$). Error bars represent the standard error of the mean ($n = 6$).

For the ear yield without straw, the water of lower salinity together with the irrigation depth of 100% of the ETc (3098 kg ha^{-1}) was statistically superior to the other treatments, while salt stress showed no statistical difference between the factors under study (Figure 14B). The results obtained in plants under salt and water stresses showed that these stresses, alone or combined, can reduce the productive performance of maize crop. Supporting the findings of this study, [77] describes that the combination of salt and water stress during the reproductive stage can negatively affect the productivity of maize crops. Similar trends to the data found in the present study were reported by [78] when evaluating the use of *Bacillus subtilis* in seeds of the 'Pioneer 3431' simple hybrid maize cultivar irrigated at 100% of the ETc with low-salinity water. Reference [79], growing uninoculated maize with low-salinity water, found a higher yield than in the present study (6150 kg ha^{-1}).

4. Conclusions

A water deficit of 50% of the ETc resulted in the principal negative effects on growth, reducing the leaf area and stem diameter. The use of *B. aryabhattai* mitigated salt stress and promoted a better performance in leaf gas exchange by increasing the CO_2 assimilation rate, stomatal conductance, and internal CO_2 concentration. However, irrigation with brackish water (3.0 dS m^{-1}) reduced the instantaneous water-use efficiency of the maize.

Overall, inoculation partially reduced the effects of abiotic stress by means of morphophysiological characteristics, such as increased leaf area and plant height, as well as with no salt stress. These observations reinforce the hypothesis that inoculation mitigates the effect of abiotic stress (salt and water) in maize plants, making it an option in regions with a scarcity of low-salinity water. However, further studies are needed to understand how *B. aryabhattai* acts on morphophysiological and production characteristics under stress conditions in order to develop efficient strategies to mitigate the harmful effects of salt and water stress in the semi-arid region of the northeast of Brazil.

Author Contributions: Conceptualization, H.C.S., G.G.d.S., T.V.d.A.V., A.P.d.A.P. and F.D.B.d.S.; methodology, H.C.S., G.G.d.S., T.V.d.A.V., A.P.d.A.P., M.V.P.d.S., F.G.d.S.A. and S.P.G.; investigation, H.C.S., C.I.N.L., M.V.P.d.S., G.G.d.S., G.F.G. and J.M.d.S.G.; writing—original draft preparation, H.C.S., G.G.d.S., C.I.N.L., T.V.d.A.V. and S.P.G.; writing—review and editing, H.C.S., A.P.d.A.P., G.G.d.S., F.G.d.S.A. and F.D.B.d.S.; project administration, G.G.d.S. and T.V.d.A.V. All authors have read and agreed to the published version of the manuscript.

Funding: This research received no external funding.

Institutional Review Board Statement: Not applicable.

Data Availability Statement: Not applicable.

Acknowledgments: Acknowledgments are due to the Conselho Nacional de Desenvolvimento Científico e Tecnológico (CNPq), Improvement of Higher Level Personnel Agency (CAPES), for the financial support provided for this research and award of a fellowship to the first author. We would like to thank Itamar Soares de Melo (EMBRAPA Meio Ambiente) for providing the *Bacillus aryabhattai* strains.

Conflicts of Interest: The authors declare there to be no conflict of interest.

References

1. Fornasieri Filho, D. *Manual da Cultura do Milho*; FUNEP: Jaboticabal, Brazil, 2007; 576p.
2. Dantas Junior, E.E.; Chaves, L.H.G.; Fernandes, J.D. Lâminas de irrigação localizada e adubação potássica na produção de milho verde, em condições semiáridas. *Rev. Espac.* **2016**, *37*, 1–9.
3. Lopes, J.R.F.; Dantas, M.P.; Ferreira, F.E.P. Identificação da influência da pluviometria no rendimento do milho no semiárido brasileiro. *Rev. Bras. Agric. Irrig.* **2019**, *13*, 3610–3618. [CrossRef]
4. CONAB–Companhia Nacional de Abastecimento. Ministério da Agricultura, Pecuária e Abastecimento. Safra Brasileira de Grãos: Boletim de Grãos 2019/2020. Available online: https://portaldeinformacoes.conab.gov.br/safra-estimativa-de-evolucao-graos.html (accessed on 5 March 2023).
5. Song, L.; Jin, J.; He, J. Effects of Severe Water Stress on Maize Growth Processes in the Field. *Sustainability* **2019**, *11*, 5086. [CrossRef]
6. Sah, R.P.; Chakraborty, M.; Prasad, K.; Pandit, M.; Tudu, V.K.; Chakravarty, M.K.; Narayan, S.C.; Rana, M.; Moharana, D. Impact of water deficit stress in maize: Phenology and yield components. *Sci. Rep.* **2020**, *10*, 2944. [CrossRef]
7. Ayers, R.S.; Westcot, D.W. *Water Quality for Agriculture*; Food and Agriculture Organization of the United Nations (FAO): Rome, Italy, 1985; p. 174.
8. SUDENE–Superintendência do Desenvolvimento do Nordeste. Nova Delimitação do Semiárido. 2017. Available online: http://antigo.sudene.gov.br/images/arquivos/semiarido/arquivos/Rela%C3%A7%C3%A3o_de_Munic%C3%Adpios_Semi%C3%A1rido.pdf (accessed on 9 January 2023).

9. Frizzone, J.A.; Lima, S.C.R.V.; Lacerda, C.F.; Mateos, L. Socio-Economic Indexes for Water Use in Irrigation in a Representative Basin of the Tropical Semiarid Region. *Water* **2021**, *13*, 2643. [CrossRef]
10. Marengo, J.A.; Bernasconi, M. Regional differences in aridity/drought conditions over Northeast Brazil: Present state and future projections. *Clim. Change* **2015**, *129*, 103–115. [CrossRef]
11. Cavalcante Júnior, R.G.; Freitas, M.A.V.; Silva, N.F.; Azevedo Filho, F.R. Sustainable groundwater exploitation aiming at the reduction of water vulnerability in the Brazilian semi-arid region. *Energies* **2019**, *12*, 904. [CrossRef]
12. Lessa, C.I.N.; de Lacerda, C.F.; Cajazeiras, C.C.D.A.; Neves, A.L.R.; Lopes, F.B.; Silva, A.O.D.; Sousa, H.C.; Gheyi, H.R.; Nogueira, R.D.S.; Lima, S.C.R.V.; et al. Potential of Brackish Groundwater for Different Biosaline Agriculture Systems in the Brazilian Semi-Arid Region. *Agriculture* **2023**, *13*, 550. [CrossRef]
13. Rajabi Dehnavi, A.; Zahedi, M.; Ludwiczak, A.; Cardenas Perez, S.; Piernik, A. Effect of Salinity on Seed Germination and Seedling Development of Sorghum (*Sorghum bicolor* (L.) Moench) Genotypes. *Agronomy* **2020**, *10*, 859. [CrossRef]
14. Isayenkov, S.V.; Maathuis, F.J.M. Plant salinity stress: Many unanswered questions remain. *Front. Plant Sci.* **2019**, *10*, 80. [CrossRef]
15. Islam, A.T.M.T.; Koedsuk, T.; Ullah, H.; Tisarum, R.; Jenweerawat, S.; Cha-um, S.; Datta, A. Salt tolerance of hybrid baby corn genotypes in relation to growth, yield, physiological, and biochemical characters. *S. Afr. J. Bot.* **2022**, *147*, 808–819. [CrossRef]
16. Lacerda, C.F.; Ferreira, J.F.S.; Suarez, D.L.; Freitas, E.D.; Liu, X.; Ribeiro, A.A. Evidence of nitrogen and potassium losses in soil columns cultivated with maize under salt stress. *Rev. Bras. Eng. Agrícola Ambient.* **2018**, *22*, 553–557. [CrossRef]
17. Sousa, H.C.; Sousa, G.G.; Lessa, C.I.N.; Lima, A.F.S.; Ribeiro, R.M.R.; Costa, F.H.R. Growth and gas exchange of corn under salt stress and nitrogen doses. *Rev. Bras. Eng. Agrícola Ambient.* **2021**, *25*, 174–181. [CrossRef]
18. Stadnik, B.; Tobiasz-Salach, R.; Mazurek, M. Effect of silicon on oat salinity tolerance: Analysis of the egipegenetic and physiological response of plants. *Agriculture* **2023**, *13*, 81. [CrossRef]
19. Sousa, H.C.; Sousa, G.G.D.; Cambissa, P.B.C.; Lessa, C.I.N.; Goes, G.F.; Silva, F.D.B.D.; Abreu, F.D.S.; Viana, T.V.D.A. Gas exchange and growth of zucchini crop subjected to salt and water stress. *Rev. Bras. Eng. Agrícola Ambient.* **2022**, *26*, 815–822. [CrossRef]
20. Goes, G.F.; Sousa, G.G.; Santos, S.O.; Silva Júnior, F.B.; Ceita, E.D.A.R.; Leite, K.N. Produtividade da cultura do amendoim sob diferentes supressões da irrigação com água salina. *Irriga* **2021**, *26*, 210–220. [CrossRef]
21. Shilev, S. Plant-Growth-Promoting Bacteria Mitigating Soil Salinity Stress in Plants. *Appl. Sci.* **2020**, *10*, 7326. [CrossRef]
22. Armanhi, J.S.L.; Souza, R.S.C.; Biazotti, B.B.; Yassitepe, J.E.D.C.T.; Arruda, P. Modulating drought stress response of maize by a synthetic bacterial community. *Front. Microbiol.* **2021**, *12*, 747541. [CrossRef]
23. Poudel, M.; Mendes, R.; Costa, L.A.S.; Bueno, C.G.; Meng, Y.; Folimonova, S.Y.; Garrett, K.A.; Martins, S.J. The role of plant-associated bacteria, fungi, and viruses in drought stress mitigation. *Front. Microbiol.* **2021**, *12*, 743512. [CrossRef]
24. Kavamura, V.N.; Santos, S.N.; Silva, J.L.; Parma, M.M.; Ávila, L.A.; Visconti, A.; Zucchi, T.D.; TaketanI, R.G.; Andreote, F.D.; Melo, I.S.D. Screening of Brazilian cacti rhizobacteria for plant growth promotion under drought. *Microbiol. Res.* **2013**, *168*, 183–191. [CrossRef]
25. Niu, X.; Song, L.; Xiao, Y.; Ge, W. Drought-tolerant plant growth-promoting rhizobacteria associated with foxtail millet in a semi-arid agroecosystem and their potential in alleviating drought stress. *Front. Microbiol.* **2018**, *8*, 2580. [CrossRef] [PubMed]
26. Sousa, S.M.; De Oliveira, C.A.; Andrade, D.L.; Carvalho, C.G.D.; Ribeiro, V.P.; Pastina, M.M.; Marriel, I.E.; Lana, U.G.D.P.; Gomes, E.A.P. Tropical Bacillus Strains Inoculation Enhances Maize Root Surface Area, Dry Weight, Nutrient Uptake and Grain Yield. *J. Plant Growth Regul.* **2021**, *40*, 867–877. [CrossRef]
27. May, A.; Santos, M.D.S.; Silva, E.H.F.M.D.; Viana, R.D.S.; Vieira Junior, N.A.; Ramos, N.P.; Melo, I.S.D. Effect of Bacillus aryabhattai on the initial establishment of pre-sprouted seedlings of sugarcane varieties. *Res. Soc. Dev.* **2021**, *10*, 1–9. [CrossRef]
28. Bittencourt, P.P.; Alves, A.F.; Ferreira, M.B.; da Silva Irineu, L.E.S.; Pinto, V.B.; Olivares, F.L. Mechanisms and Applications of Bacterial Inoculants in Plant Drought Stress Tolerance. *Microorganisms* **2023**, *11*, 502. [CrossRef]
29. Etesami, H.; Beattie, G.A. Mining halophytes for plant growth-promoting halotolerant bacteria to enhance the salinity tolerance of non-halophytic crops. *Front. Microbiol.* **2018**, *9*, 148. [CrossRef]
30. Alvares, C.A.; Stape, J.L.; Sentelhas, P.C.; Gonçalves, J.L.M.; Sparovek, G. Köppen's climate classification map for Brazil. *Meteorol. Z.* **2013**, *22*, 711–728. [CrossRef]
31. Teixeira, P.C.; Donagemma, G.K.; Fontana, A.; Teixeira, W. *Manual de Métodos de Análise de Solo*, 3rd ed.; Embrapa: Brasília, Brazil, 2017; p. 573.
32. de Souza, L.S.B.; de Moura, M.S.B.; Sediyama, G.C.; da Silva, T.G.F. Requerimento hídrico e coeficiente de cultura do milho e feijão-caupi em sistemas exclusivo e consorciado. *Rev. Caatinga* **2015**, *28*, 151–160. [CrossRef]
33. Medeiros, J.F. Qualidade da água de Irrigação Utilizada nas Propriedades Assistidas pelo "GAT" nos Estados do RN, PB, CE e avaliação da Salinidade dos Solos. Master's Thesis, Universidade Federal da Paraíba, Campina Grande, Brazil, 1992. Available online: http://dspace.sti.ufcg.edu.br:8080/jspui/handle/riufcg/2896 (accessed on 8 January 2023).
34. Silva, F.C. *Manual de Análises Químicas de Solos, Plantas e Fertilizantes*; Embrapa Comunicação para Transferência de Tecnologia: Brasília, Brazil, 1999; p. 370.
35. Richards, L.A. *Diagnosis and Improvement of Saline and Alkali Soils*; US Department of Agriculture: Washington, DC, USA, 1954; p. 160.
36. Kavamura, V.N.; Santos, S.N.; Taketani, R.G.; Vasconcellos, R.L.; Melo, I.S. Draft genome sequence of plant growth-promoting drought-tolerant *Bacillus* sp. strain CMAA 1363 isolated from the Brazilian Caatinga biome. *Genome Announc.* **2017**, *5*, e01534-16. [CrossRef]

37. Fernandes, V.L.B. *Recomendações de Adubação e Calagem para o Estado do Ceará*; UFC: Fortaleza, Brazil, 1993; p. 248.
38. Silva, F.D.A.S.; Azevedo, C.A.V.D. The Assistat Software Version 7.7 and its use in the analysis of experimental data. *Afr. J. Agric. Res.* **2016**, *11*, 3733–3740.
39. Bulgarelli, D.; Schlaeppi, K.; Spaepen, S.; Van Themaat, E.V.L.; Schulze-Lefert, P. Structure and functions of the bacterial microbiota of plants. *Annu. Rev. Plant Biol.* **2013**, *64*, 807–838. [CrossRef]
40. Kavamura, V.N. Bactérias Associadas às Cactáceas da Caatinga: Promoção de Crescimento de Plantas sob Estresse Hídrico. Ph.D. Thesis, University of São Paulo, Piracicaba, Brazil, 2012. Available online: http://www.teses.usp.br/teses/disponiveis/11/11138/tde-25102012-095956/ (accessed on 10 February 2023).
41. Oliveira, E.J.; Melo, H.C.D.E.; Trindade, K.L.; Guedes, T.D.E.M.; Sousa, C.M. Morphophysiology and yield of green corn cultivated under different water depths and nitrogen doses in the cerrado conditions of Goiás, Brazil. *Res. Soc. Dev.* **2020**, *9*, e6179108857. [CrossRef]
42. Taiz, L.; Zeiger, E.; Moller, I.M.; Murphy, A. *Fisiologia e Desenvolvimento Vegetal*, 6th ed.; Artmed: Porto Alegre, Brazil, 2017; 858p.
43. Irmak, S.; Mohammed, A.T.; Kukal, M.S. Maize response to coupled irrigation and nitrogen fertilisation under center pivot, subsurface drip and surface (furrow) irrigation: Growth, development and productivity. *Agric. Water Manag.* **2020**, *263*, 107457. [CrossRef]
44. Qiu, Y.; Fan, Y.; Chen, Y.; Hao, X.; Li, S.; Kang, S. Response of dry matter and water use efficiency of alfalfa to water and salinity stress in arid and semiarid regions of Northwest China. *Agric. Water Manag.* **2021**, *254*, 106934. [CrossRef]
45. Lin, Y.; Wattes, D.B.; Kloepper, J.W.; Feng, Y.; Torbert, H.A. Influence of plant growth-promoting rhizobacteria on corn growth under drought stress. *Commun. Soil Sci. Plant Anal.* **2019**, *51*, 250–264. [CrossRef]
46. Ricardi, M.; Rosa, H.A. Desenvolvimento inicial do milho submetido a estresse salino. *Rev. Cultiv. Saber* **2018**, *1*, 174–184.
47. Sá, F.; Ferreira Neto, M.; Lima, Y.; Paiva, E.; Prata, R.; Lacerda, C.; Brito, M. Growth, gas exchange and photochemical efficiency of the cowpea bean under salt stress and phosphorus fertilisation. *Comum. Sci.* **2018**, *9*, 668–679. [CrossRef]
48. de Souza, M.V.; de Sousa, G.G.; da Silva Sales, J.R.; da Costa Freire, M.H.; da Silva, G.L.; de Araújo Viana, T.V. Água salina e biofertilisantes de esterco bovino e caprino na salinidade do solo, crescimento e fisiologia da fava. *Rev. Bras. Cienc. Agrar.* **2019**, *14*, 340–349.
49. Oliveira, F.D.; Medeiros, J.F.; Cunha, R.C.; Souza, M.W.; Lima, L.A. Uso de bioestimulante como agente amenizador do estresse salino na cultura do milho pipoca. *Cienc. Agron.* **2016**, *47*, 307–315.
50. Naseem, H.; Bano, A. Role of plant growth-promoting rhizobacteria and their exopolysaccharide in drought tolerance of maize. *J. Plant Interact.* **2014**, *9*, 689–701. [CrossRef]
51. Prado, R.M. *Nutrição de Plantas*, 2nd ed.; UNESP: São Paulo, Brazil, 2020; 414p.
52. Jaghdani, S.J.; Janhns, P.; Trankner, M. The impact of magnesium deficiency on photosynthesis and photoprotection in Spinacia oleracea. *Plant Stress* **2021**, *2*, 1–11. [CrossRef]
53. Lima, A.F.; Santos, M.F.; Oliveira, M.L.; Sousa, G.G.; Mendes Filho, P.F.; Luz, L.N. Physiological responses of inoculated and uninoculated peanuts under saline stress. *Rev. Ambiente Água* **2021**, *16*, e2643. [CrossRef]
54. Cao, M.; Narayanan, M.; Shi, X.; Chen, X.; Li, Z.; Ma, Y. Optimistic contributions of plant growth-promoting bacteria for sustainable agriculture and climate stress alleviation. *Environ. Res.* **2023**, *217*, 114924. [CrossRef] [PubMed]
55. Vishnupradeep, R.; Bruno, L.B.; Taj, Z.; Karthik, C.; Challabathula, D.; Kumar, A.; Freitas, H.; Rajkumar, M. Plant growth promoting bacteria improve growth and phytostabilization potential of *Zea mays* under chromium and drought stress by altering photosynthetic and antioxidant responses. *Environ. Technol. Innov.* **2022**, *25*, 102154. [CrossRef]
56. Mishra, P.; Mishra, J.; Arora, N.K. Plant growth promoting bacteria for combating salinity stress in plants–Recent developments and prospects: A review. *Microbiol. Res.* **2021**, *252*, 126861. [CrossRef] [PubMed]
57. Sales, J.R.; Magalhães, C.L.; Freitas, A.G.; Goes, G.F.; Sousa, H.C.; Sousa, G.G. Índices fisiológicos de quiabeiro irrigado com água salina sob adubação organomineral. *Rev. Bras. Eng. Agrícola Ambient.* **2021**, *25*, 466–471. [CrossRef]
58. Liao, Q.; Gu, S.; Kang, S.; Du, T.; Tong, L.; Wood, J.D.E.; Ding, R. Mild water and salt stress improve water use efficiency by decreasing stomatal conductance via osmotic adjustment in field maize. *Sci. Total Environ.* **2022**, *805*, 150364. [CrossRef]
59. Rodrigues, V.D.S.; Sousa, G.G.D.; Soares, S.D.C.; Leite, K.N.; Ceita, E.D.; Sousa, J.T.M.D. Gas exchanges and mineral content of corn crops irrigated with saline water. *Rev. Ceres* **2021**, *68*, 453–459. [CrossRef]
60. Nascimento, F.N.; Bastos, E.A.; Cardoso, M.J.; Júnior, A.S.A.; Ribeiro, V.Q. Parâmetros fisiológicos e produtividade de espigas verdes de milho sob diferentes lâminas de irrigação. *Rev. Bras. Milho Sorgo* **2015**, *14*, 167–181. [CrossRef]
61. Hessini, K.; Issaoui, K.; Ferchichi, S.; Saif, T.; Abdelly, C.; Siddique, K.H.M.; Cruz, C. Interactive effects of salinity and nitrogen forms on plant growth, photosynthesis and osmotic adjustment in maize. *Plant Physiol. Biochem.* **2019**, *139*, 171–178. [CrossRef]
62. Abdel Latef, A.A.H.; Abu Alhmad, M.F.; Kordrostami, M.; Abo-Baker, A.B.A.E.; Zakir, A. Inoculation with Azospirillum lipoferum or Azotobacter chroococcum Reinforces Maize Growth by Improving Physiological Activities Under Saline Conditions. *J. Plant Growth Regul.* **2020**, *39*, 1293–1306. [CrossRef]
63. Cavalcante, E.S.; Lacerda, C.F.; Mesquita, R.O.; de Melo, A.S.; da Silva Ferreira, J.F.; dos Santos Teixeira, A.; Lima, S.C.R.V.; da Silva Sales, J.R.; de Souza Silva, J.; Gheyi, H.R. Supplemental Irrigation with Brackish Water Improves Carbon Assimilation and Water Use Efficiency in Maize under Tropical Dryland Conditions. *Agriculture* **2022**, *12*, 544. [CrossRef]
64. Costa-Gutierrez, S.B.; Raimondo, E.E.; Lami, M.J.; Vincent, P.A.; Espinosa-Urgel, M.; Cristóbal, R.E. Inoculation of Pseudomonas mutant strains can improve growth of soybean and corn plants in soils under salt stress. *Rhizosphere* **2020**, *16*, 100255. [CrossRef]

65. Shukla, P.S.; Agarwal, P.K.; Jha, B. Improved salinity tolerance of *Arachis hypogaea* (L.) by the interaction of halotolerant plant-growth-promoting rhizobacteria. *J. Plant Growth Regul.* **2012**, *31*, 195–206. [CrossRef]
66. Cavallet, L.E.; Pessoa, A.C.S.; Helmich, J.J.; Helmich, P.R.; Ost, C.F. Produtividade do milho em resposta à aplicação de nitrogênio e inoculação das sementes com *Azospirillum* spp. *Rev. Bras. Eng. Agric. Ambient.* **2000**, *4*, 129–132. [CrossRef]
67. Freire, M.H.; Viana, T.V.; Sousa, G.G.; Azevedo, B.M.; Sousa, H.C.; Goes, G.F.; Lessa, C.I.; da Silva, F.D. Organic fertilisation and salt stress on the agronomic performance of maize crop. *Rev. Bras. Eng. Agrícola Ambient.* **2022**, *26*, 848–854. [CrossRef]
68. Marrocos, S.T.P.; Junior, J.N.; Granjeiro, L.C.; Ambrosio, M.M.D.Q.; Da Cunha, A.P.A. Composição química e microbiológica de biofertilisantes em diferentes tempos de decomposição. *Rev. Caatinga* **2012**, *25*, 34–43.
69. Kumar, M.; Mishra, S.; Dixit, V.; Kumar, M.; Agarwal, L.; Chauhan, P.S.; Nautiyal, C.S. Synergistic effect of *Pseudomonas putida* and *Bacillus amyloliquefaciens* ameliorates drought stress in chickpea (*Cicer arietinum* L.). *Plant Signal. Behav.* **2016**, *11*, e1071004. [CrossRef]
70. EMBRAPA-Empresa Brasileira de Pesquisa Agropecuária. Boletim de Pesquisa e Desenvolvimento. Isolamento e Potencial Uso de Bactérias do Gênero Bacillus na Promoção de Crescimento de Plantas em Condições de Déficit Hídrico. 2019. Available online: https://www.infoteca.cnptia.embrapa.br/infoteca/bitstream/doc/1114738/1/bol192.pdf (accessed on 1 March 2023).
71. Qurashi, A.W.; Sabri, A.N. Bacterial exopolysaccharide and biofilm formation stimulate chickpea growth and soil aggregation under salt stress. *Braz. J. Microbiol.* **2012**, *43*, 1183–1191. [CrossRef]
72. Araújo, V.L.V.P.; Lira Júnior, M.A.; Souza Júnior, V.S.; Araújo Filho, J.C.; Fracetto, F.J.C.; Andreote, F.D.; Pereira, A.P.A.; Mendes Júnior, J.P.; Barros, F.M.R.; Fracetto, G.G.M. Bacteria fromm tropical semiarid temporary ponds promote maize growth under water stress. *Microbiol. Res.* **2020**, *24*, 126564. [CrossRef]
73. Kumar, V.; Raghuvanshi, N.; Pandey, A.K.; Kumar, A.; Thoday-Kennedy, E.; Kant, S. Role of halotolerant plant growth-promoting rhizobacteria in mitigating salinity stress: Recent advances and possibilities. *Agriculture* **2023**, *13*, 168. [CrossRef]
74. Rodrigues, V.S.; Bezerra, F.M.L.; Sousa, G.G.; Fiusa, J.N.; Viana, T.V.A. Yield of maize crop irrigated with saline waters. *Rev. Bras. Eng. Agric. Ambient.* **2020**, *24*, 101–105. [CrossRef]
75. Arora, N.K.; Tewari, S.; Singh, R. Multifaceted plant-associated microbes and their mechanisms diminish the concept of direct and indirect PGPRs. In *Plant-Microbe Symbiosis: Fundamentals and Advances*; Arora, N.K., Ed.; Springer: New Delhi, India, 2013; pp. 411–449.
76. Cavalcante Filho, H.A.; Helcias, R.P.; Pereira, R.G.; Cavalcante, M.; Medeiros, P.V.Q.; Andrade, A.R.S. Desempenho de cultivares de milho para produção de milho verde, minimilho e forragem. *Rev. Ciências Agroveterinárias* **2022**, *21*, 449–455. [CrossRef]
77. Zhang, Y.; Li, X.; Simunek, J.; Shi, H.; Chen, N.; Hu, Q. Optimizing drip irrigation with alternate use of fresh and brackish waters by analyzing salt stress: The experimental and simulation approaches. *Soil Tillage Res.* **2022**, *219*, 105355. [CrossRef]
78. Mazzuchelli, R.C.L.; Sossai, B.F.; Araujo, F.F. Inoculação de *Bacillus subtilis* e *Azospirillum brasilense* na cultura do milho. *Colloq. Agrar.* **2014**, *10*, 40–47. [CrossRef]
79. Silva, G.C.; Schmitz, R.; Silva, L.C.; Carpanini, G.G.; Magalhães, R.C. Desempenho de cultivares para pro-dução de milho verde na agricultura familiar do sul de Roraima. *Rev. Bras. Milho Sorgo* **2015**, *14*, 273–282. [CrossRef]

Disclaimer/Publisher's Note: The statements, opinions and data contained in all publications are solely those of the individual author(s) and contributor(s) and not of MDPI and/or the editor(s). MDPI and/or the editor(s) disclaim responsibility for any injury to people or property resulting from any ideas, methods, instructions or products referred to in the content.

Article

Ionic Response and Sorghum Production under Water and Saline Stress in a Semi-Arid Environment

Rodrigo Rafael da Silva [1], José Francismar de Medeiros [1,*], Gabriela Carvalho Maia de Queiroz [1], Leonardo Vieira de Sousa [1], Maria Vanessa Pires de Souza [2], Milena de Almeida Bastos do Nascimento [3], Francimar Maik da Silva Morais [1], Renan Ferreira da Nóbrega [4], Lucas Melo e Silva [5], Fagner Nogueira Ferreira [1], Maria Isabela Batista Clemente [1], Carla Jamile Xavier Cordeiro [1], Jéssica Christie de Castro Granjeiro [1], Dárcio Cesar Constante [1] and Francisco Vanies da Silva Sá [1]

1. Agricultural Sciences Center, Federal Rural University of Semi-Arid, Mossoro 59625-900, RN, Brazil; rodrigosilva_rafael@hotmail.com (R.R.d.S.)
2. Agricultural Engineering Department, Federal University of Ceará, Fortaleza 60455-760, CE, Brazil
3. Field Technician of the National Rural Learning Service, Saint Louis 65010-270, MA, Brazil
4. Master in Soil and Water Management—PPGMSA, UFERSA, Mossoro 59625-900, RN, Brazil
5. Department of Animal Sciences, Federal Rural University of Semi-Arid, Mossoro 59625-900, RN, Brazil
* Correspondence: jfmedeir@ufersa.edu.br

Abstract: The increase in water demand in regions with limited good-quality water resources makes it necessary to study the effect of low-quality water on plant metabolism. Therefore, the objective of this study was to evaluate the effect of water and salt stress on the levels of mineral elements and accumulation of toxic elements Na^+ and Cl^- in the leaves and their consequences on the production variables of the sorghum cultivar IPA SF-15. The design adopted was randomized blocks in a factorial scheme (4 × 4), with four salt concentrations (1.5; 3.0; 4.5, and 6.0 dS m^{-1}) and four irrigation depths (51.3; 70.6; 90.0, and 118.4% of crop evapotranspiration ETc) in three repetitions. To obtain nutrient, sodium, and chlorine contents in the leaf, we collected the diagnosis leaf from six plants per plot. For production data, we performed two harvests at 76 and 95 days after planting (silage point and for sucrose extraction). We evaluated the dry mass, fresh mass yield, and total dry mass for the two cutting periods and applied the F-test at the 5% significance level. There was an effect of water stress but not saline, making it possible to use saline water for sorghum irrigation. As for the toxicity of ions, the plant showed tolerance behavior to Na^+ and Cl^- ions. The grain filling phase was more sensitive than the final phase of the crop cycle.

Keywords: *Sorghum bicolor* (L.) Moench; nutrient accumulation; water deficit

1. Introduction

Sorghum (Sorghum bicolor) is an important food crop in many countries, mainly in Africa and Asia, and is used worldwide in the biofuel industry and livestock [1]. In the Brazilian semi-arid region, it has stood out in terms of forage production with its cultivation in the rainy season and during the dry season, under irrigation, for the high productivity achieved per volume of water spent on crop irrigation [2–4].

Sorghum can adapt to different edaphoclimatic conditions due to its rusticity and ability to adapt to water scarcity [5]. According to [6], in a study with sorghum subjected to irrigation depths, it was observed that a 20% reduction in the standard irrigation depth did not influence the productivity of the crop, and it was possible to expand the production area using the same volume of water. Similar studies [7,8] obtained similar results, demonstrating the tolerance of sorghum to water stress conditions.

The considerably high resistance to drought and salinity becomes essential compared to other crops in terms of yield. [9]; it is a critical factor affecting the entire production mechanism [10]. However, sorghum has the characteristic of adjusting the osmotic balance in the cell, which induces a salinity tolerance strategy. In this sense, sorghum's strategies to express its productive performance under water and saline restrictions have been studied [6,7,9,10], showing safe results for using these resources in sorghum cultivation.

The high use of water for human and agricultural consumption demands low-quality water for irrigation [11]. In this way, it is necessary to explore the knowledge of good practices for using these waters to maintain the productive potential of the culture and the quality of the soil and production. The difficulty in using these waters in agricultural production is because of high concentrations of soluble salts since they are in contact with soluble materials from soil and rocks [12]. Salinity affects plants because of its osmotic effect, reducing water availability and providing high concentrations of Na^+ and Cl^- in the soil solution, which, when absorbed at high levels by crops, can trigger a toxic effect on plants [10,13], besides that, an ionic imbalance in the generated soil can affect the overall nutrition of the plant.

According to [14], sorghum tolerates a water and soil salinity of 4.5 and 6.8 dS m^{-1}, respectively. For values above these limits, a reduction of around 16% is expected for each unit increase in soil salinity [15]. Sorghum stands out compared with other grasses, such as maize, which is moderately tolerant to water salinity up to 4.5 dS m^{-1} [14,16], and barley, which tolerates salinity up to 6.0 dS m^{-1} [17]. Studies also show that in a saline environment, sorghum can exclude Na^+ from the xylem to the roots and compartmentalize it in cell vacuoles as an osmolyte to adjust cell osmosis [18–22]. Under saline conditions, the selective uptake of Ca^{2+} and K^+ over Na^+ is an additional mechanism for salt stress tolerance [21]. Excess Na^+ can lead to accumulation in the leaf and affect the translocation of Ca^{2+}, K^+, and Mg^{2+} ions [22], impairing photosynthetic activity and plant development [23].

Different management strategies to ensure the use of saline water in irrigation are necessary. Using saline water associated with an irrigation depth that provides water availability for the plant seeks to minimize the effects of saline stress on plant metabolism [24]. Therefore, using plants tolerant to water and saline deficit is a way to deal with the problem of water salinity in the semi-arid region [25]. Given this, the objective was to evaluate the effect of water and salt stress on the levels of mineral elements and accumulation of toxic elements Na^+ and Cl^- in the leaves and their consequences on the production variables of the sorghum cultivar IPA SF-15.

2. Materials and Methods

The study was conducted in the experimental area at the Cumaru site (5°33′30″ S, 37°11′56″ W, altitude of 110 m) in the municipality of Upanema-RN. According to the Köppen classification, the region's climate is BSh—hot semi-arid with autumn rains and average monthly air temperature consistently above 18 °C. Because of the low latitudes, the region has two well-defined seasons: wet (January to May) and dry (June to December). The average annual precipitation is 650 mm, characterized by high space–time variability. The soil in the area was classified as Cambisol [26], with chemical and physical characteristics in the 0.00–0.20 cm layer before planting, as shown in Table 1.

The crop investigated as sorghum (*Sorghum bicolor* (L.) Moench), cultivar IPA SF-15, with an aptitude for forage production. The experiment was conducted in the dry season, from September to December 2019. The preparation of the area consisted of plowing followed by harrowing, opening the planting furrows, and carrying out the basal fertilization with 180 kg ha^{-1} of MAP (10–50–00). Soil fertilization was done according to soil analysis recommendations and crop nutritional requirements. In fertirrigation, 60 kg ha^{-1} of N was applied using urea. To meet the demand for potassium, 30 kg ha^{-1} of K_2O was applied, using KCl as fertilizer. Fertilizers were applied at 21, 28, and 35 days after planting.

Table 1. Physical and chemical soil attributes in the study area before the experiment.

Layer		Soil physics					
		Sand	Silt	Clay		Soil Density	
cm			$g\,g^{-1}$			$g\,cm^{-3}$	
0–20		0.780	0.060	0.160		1.620	
				Soil chemistry			
	EC	pH	Ca^{2+}	Mg^{2+}	Na^+	K^+	P
	$dS\,m^{-1}$			$cmol_c\,dm^{-3}$			$mg\,dm^{-3}$
0–20	0.07	8.10	7.70	0.60	0.10	0.51	8.60

EC—Electrical conductivity.

The experimental design was in randomized blocks with three replications in a 4 × 4 factorial scheme, with four concentrations of salts expressed in electrical conductivity of irrigation water (1.5, 3.0, 4.5, and 6.0 dS m^{-1}) and four irrigation depths (51.3, 70.6, 90.0, and 118.4% of ETc). The experimental units comprised two double rows of seven meters, totaling 168 plants. The outer rows of each plot were considered borders.

The water with the lowest salinity level (1.5 dS m^{-1}) came from a tube well, and the water with the highest salinity (6.0 dS m^{-1}) was defined based on the salinity tolerance of the sorghum crop for yield than 50% of its productive potential [27]. The two other levels were 3.0 and 4.5 dS m^{-1}, corresponding to intermediate and equidistant points of the two extreme values. To obtain the three highest salinity levels, stock solutions were prepared at a concentration of 200 g L^{-1} of NaCl (3.42 mol L^{-1}), CaCl$_2$.2H$_2$O (1.36 mol L^{-1}), and MgSO$_4$.7H$_2$O (0.81 mol L^{-1}) and added quantities of the stock solution so that the final proportion was 6.3:2.7:1 of Na, Ca, and Mg, which represents the average composition of the waters in the region which exploits the Jandaíra Limestone Aquifer [28] (Table 2). Salinity levels were monitored daily using a portable conductivity meter.

Table 2. Chemical composition of natural water and after adding salt solutions used in the experiment.

EC	Na^+	Ca^{2+}	Mg^{2+}	K^+	Cl^-	SO_4^{2-}	HCO_3^-
$dS\,m^{-1}$				(mmol L^{-1})			
1.5	5.00	4.00	1.00	0.12	8.10	0.15	7.00
3.0	19.00	4.00	1.50	0.12	22.10	0.75	6.90
4.5	28.50	6.00	2.25	0.12	35.60	1.40	6.90
6.0	38.00	8.00	3.00	0.12	49.10	2.15	6.80

EC—Electrical conductivity, Na^+—Sodium, Ca^{2+}—Calcium, Mg^{2+}—Magnesium, K^+—Potassium, Cl^-—Chlorine, SO_4^{2-}—Sulfates, HCO_3^-—Bicarbonates.

The irrigation depths were estimated by the percentage of crop evapotranspiration (ETc) (Figure 1A), adjusting for field conditions and operation of the irrigation system. ETc was calculated daily from the calculation of daily reference evapotranspiration, using the Penman-Monteith method [29], and the daily crop coefficient (Kc) (Figure 1B) by the dual Kc method. Because it is a localized irrigation system (drip), we adopted an irrigation efficiency of 95% to calculate the standard irrigation depth. ETo was estimated from the data collected at a meteorological station near the experiment. To obtain the different depths, drip hoses spaced between lines of 1.65 m were used, with different spacing between emitters (20, 30, 40 cm) and flow rates (1.69, 1.65, 3.46, and 3.90 L h^{-1}), to provide flows per linear meter proportional to the required depths.

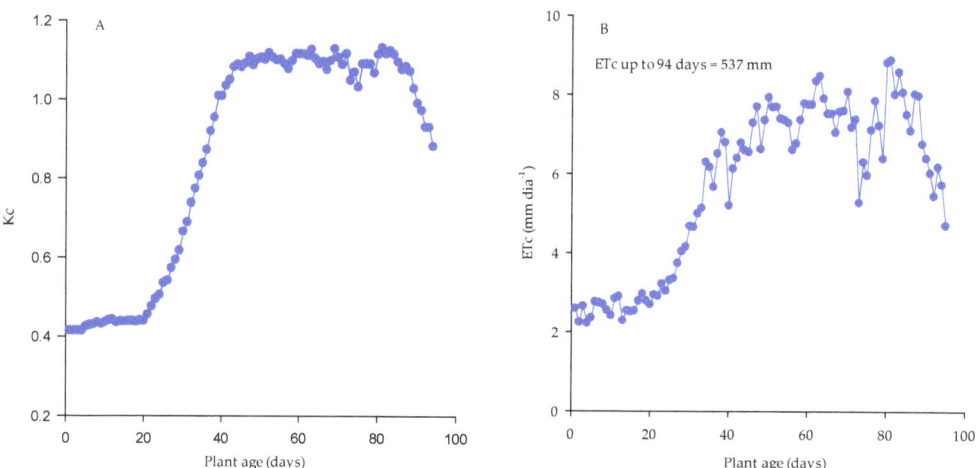

Figure 1. Crop coefficient (**A**) and crop evapotranspiration (**B**) were calculated and adopted. The total irrigation depth for 100% ETc was 6% more.

Sowing was carried out directly, placing five seeds per hole. Thinning was performed ten days after sowing, leaving three plants per hole spaced in double rows 1.40 × 0.25 × 0.30 m. In addition, two manual weedings and an application of Chlorantraniliprole and Imidacloprid were carried out through fertirrigation to control fall armyworms (*Spodoptera frugiperda*) and aphid (*Aphis gossypii*).

To obtain nutrient, sodium, and chlorine contents in the leaf, we collected the diagnosis leaf, the fourth leaf, from six different plants (Figure 2). First, the leaves were dehydrated in a forced circulation oven at a temperature of 65 °C until they reached constant mass; then, they were processed in a Willey SL-31 type mill to determine the P, K^+, Ca^{2+}, Mg^{2+} contents, Cl^-, and Na^+. The nutrient/ion contents were extracted using the dry digestion method [30]. The concentration of sodium and potassium was determined by the technique of flame photometry and phosphorus by the colorimetric method of molybdate-vanadate in a spectrophotometer. Ca^{2+} and Mg^{2+} were determined by atomic absorption spectrophotometry [30]. Chlorine concentration was determined by the MOHR method, extracted by calcium nitrate solution ($Ca(NO_3)_2 \cdot 4H_2O$), in the form of chloride ion, titrated with a standardized solution of silver nitrate ($AgNO_3$), in the presence of potassium dichromate (K_2CrO_4) as an indicator.

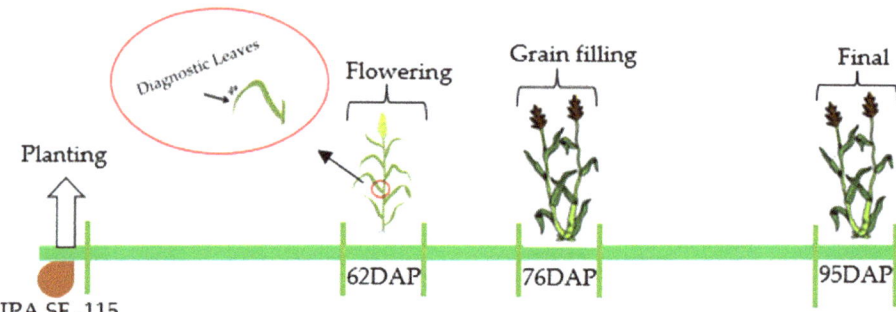

Figure 2. Timeline of evaluations during the experiment.

The first harvest occurred 76 days after planting (flowering), and the second was 95 days after planting (silage point and sucrose extraction). Production was quantified by counting the number of plants and weighing the plant material (leaves, stems, and inflorescences) collected in a 3 m central row. Weightings were carried out in the field using a portable digital hook scale with a capacity of 50 kg and a resolution of 10 g. Based on these data, the fresh mass yield was considered based on the average stand of plants in the area at the time of each harvest, and the values were expressed in Mg ha^{-1}.

The percentage of dry mass was obtained from six plants in the valuable area of the experimental plot. First, the plants were separated into leaves, stems, and inflorescences and weighed to obtain the total fresh mass. Then, samples of the respective fresh materials were placed in a forced circulation oven at a temperature of 65 °C until they reached the constant mass to obtain the dry mass. Each organ's dry mass percentage was estimated with these data, and the weighted average was calculated for the entire plant. The total dry mass, expressed in Mg ha^{-1}, resulted from the product between fresh mass yield and the dry mass content in the plant.

Data were analyzed for normal distribution using the Shapiro–Wilk test at a 5% significance level and performed the Pearson correlation matrix. Then, the results were submitted to analysis of variance by the F-test at the 5% significance level; in case of significant effects, regression analysis was carried out using the statistical software RStudio version 4.2.2.1 through the ExpDes.pt package.

3. Results

3.1. Content of Mineral Elements in Sorghum Diagnosis Leaf

Water deficit linearly interfered with the K$^+$ and P contents, in which the increase in the irrigation depth provided a reduction in the K$^+$ content ($p \leq 0.01$) (Figure 3A) concomitant with an increase in the P content in the leaves ($p \leq 0.01$) (Figure 3B). Salinity linearly reduced the Cl$^-$ content ($p \leq 0.01$) in the diagnosis leaf (Table 3), as can be seen in Figure 4, but did not affect the other ions studied. It is worth emphasizing that no interactive effect of salt × irrigation was observed in this study, which may indicate the adoption of brackish water as long as it is in good quantity or reduce the irrigation depth as long as the water used is of good quality.

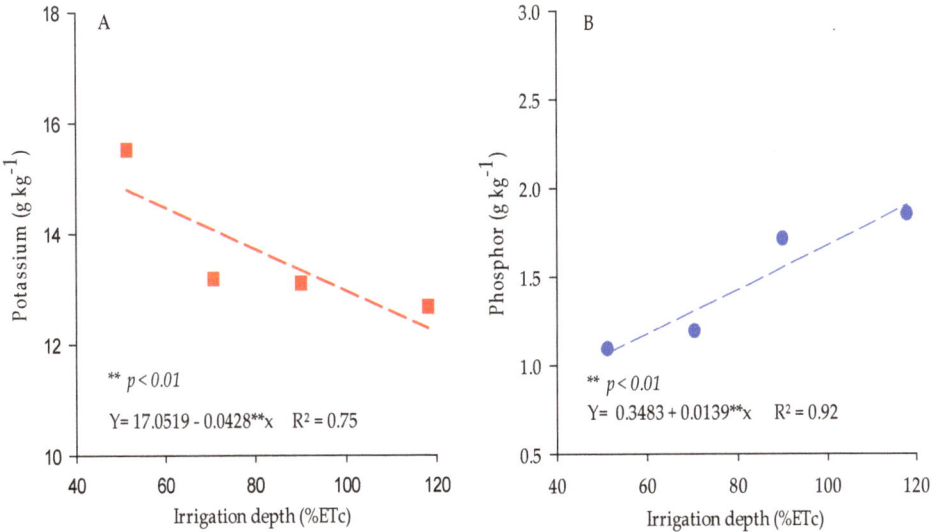

Figure 3. Potassium (**A**) and phosphorus (**B**) content in the diagnosis leaf for cultivar IPA SF-15 as a function of crop evapotranspiration levels.

Table 3. Summary of analysis of variance for nutrient content traits in sorghum (*Sorghum bicolor* (L.) Moench) IPA SF-15 diagnosis leaf as a function of water and salt stress.

SV	DF	Statistical Significance by F-Test					
		Na	K	Ca	Mg	P	Cl
Block	2	0.110	0.719	0.028	0.324	0.320	0.087
Salt	3	0.943	0.227	0.512	0.167	0.874	0.001
Comp. L	1	0.731	0.334	0.169	0.042	0.982	0.000
Comp. Q	1	0.628	0.093	0.816	0.543	0.405	0.732
Irrig.	3	0.421	0.002	0.152	0.326	0.000	0.388
Comp. L	1	0.288	0.001	0.768	0.152	0.000	0.153
Comp. Q	1	0.689	0.045	0.494	0.299	0.847	0.977
Salt × Irrig.	9	0.262	0.765	0.054	0.103	0.243	0.675
Residue	30	-	-	-	-	-	-
Total	47	-	-	-	-	-	-
CV (%)		11.30	12.82	16.06	20.97	24.92	11.00
Average		0.51	13.62	4.31	2.13	1.46	90.04

SV—Source of variation, DF—Degree of freedom, CV—Doefficient of variation, Salt—Effect of salinity; Irrig.—Effect of water stress; Comp. L and Q—Linear and quadratic component, significance ($0.01 \geq p \leq 0.05$) by F-test.

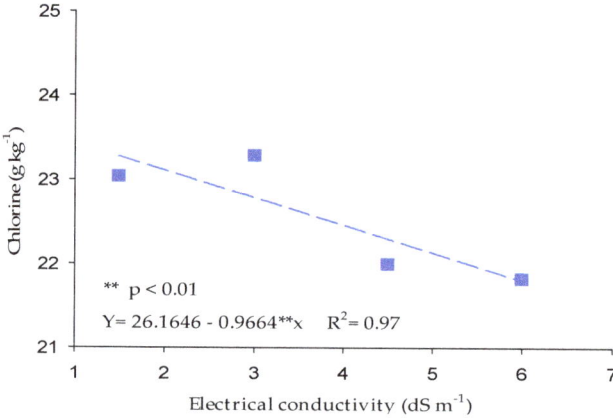

Figure 4. The chlorine content in the diagnosis leaf for the IPA SF-15 cultivar as a function of the electrical conductivity of the irrigation water.

3.1.1. Potassium and Phosphorus

Under an irrigation depth of 51.3% ETc, the observed potassium content was 14.86 g kg^{-1}, dropping sharply to 14.03 and 13.20 g kg^{-1} for the irrigation depths of 70.6 and 90.0% of ETc, respectively (Figure 3A). According to the linear model generated (K$^+$ = 17.0519 − 0.0428**x; R^2 = 0.75), for each unit increase in %ETc, the K$^+$ content decreased by approximately 0.04 g g kg^{-1} in dry mass (1.24%). The lowest K$^+$ content (12 g kg^{-1}) was obtained underwater control conditions of 118% ETc, approximately 19% lower than the maximum value observed.

Increasing the irrigation depth provided higher levels of P in the leaves (Figure 3B), with estimated values ranging from 1.06 to 1.99 g kg^{-1} of P in dry mass for a depth of 51.3 to 118% of ETc respectively, that is, an increase of 87% about lower water availability. From the regression analysis, a significant effect of the irrigation depth was observed with a linear rise in 0.014 g kg^{-1} of P in the dry mass (0.53%) in the leaf of the sorghum plant for each unit variation in the water depth irrigation.

3.1.2. Chlorine

Salt stress reduced the chlorine content in the leaf (Figure 4), which ranged from 24.73 to 20.37 g kg^{-1} for electrical conductivity from 1.5 to 6.0 dS m^{-1}, respectively, expressing a decrease of 18% about higher electrical conductivity. The regression model that best fit was linear decreasing (Cl$^-$ = 26.1646 − 0.9664**x; R^2 = 0.97), showing an estimated decrease of approximately 0.96 g kg^{-1} of Cl$^-$ in dry mass (−0.21%).

3.2. Production Variables

Regarding production analyses, analysis of variance (Table 4) identifies that salinity interfered only with the total fresh mass yield at 76 DAP ($p \leq 0.05$). Despite being significant, the tested linear and quadratic regression models did not fit the data in Figure 5A with very shallow R^2 values. This means some model was significant but would probably have no plausible physiological explanation for this behavior. As there was no adjustment of the models tested for the variables, the Tukey test was applied ($p \leq 0.05$) for the salinity factor for the variables' yield of fresh matter and total dry mass, where it was verified that there was no significant difference, as shown in Figure 5. As for the effect of the depth, it was observed that it interfered with the yield and total dry mass at 76 and 95 DAP ($p \leq 0.01$) since the interaction was not significant for any of the variables in any of the studied periods.

Table 4. Summary of analysis of variance: dry mass percentage (DMP), plant yield (Yield), total dry mass (TDM) at 76 and 95 days after planting (DAP) for sorghum (*Sorghum bicolor* L. Moench) IPA SF-15 as a function of water and salt stress.

		Statistical Significance by F-Test					
SV	DF	76 DAP			95 DAP		
		DMP	Yield	TDM	DMP	Yield	TDM
Block	2	0.000	0.000	0.463	0.985	0.108	0.362
Salt	3	0.841	0.042	0.192	0.525	0.928	0.656
Comp. L		0.902	0.132	0.263	0.699	0.968	0.701
Comp. Q		0.519	0.739	0.511	0.342	0.514	0.340
Lam	3	0.345	0.000	0.000	0.799	0.000	0.009
Comp. L		0.106	0.000	0.000	0.546	0.000	0.006
Comp. Q		0.547	0.212	0.403	0.422	0.342	0.342
Salt × Lam	9	0.263	0.088	0.629	0.175	0.766	0.454
Residue	30	-	-	-	-	-	-
Total	47	-	-	-	-	-	-
CV (%)		10.68	15.14	22.29	13.91	15.31	21.85
Average		30.00	65.88	19.58	38.00	83.14	31.74

SV—source of variation, DF—degree of freedom, CV—coefficient of variation, Salt—Effect of salinity; Irrig—Effect of water stress; Comp. L and Q—linear and quadratic component, significance (0.01 $\geq p \leq$ 0.05) by F-test.

It was observed that the fresh matter yields at 76 and 95 DAP were not affected by the saline levels applied with values ranging from 60.04, 71.02 and 81.72 to 84.55 Mg ha^{-1}, respectively Figure 5A. The total dry mass at 76 and 95 DAP also showed no significant difference for the salinity factor; the values ranged from 17.48, 21.49 and 29.97 to 33.57 Mg ha^{-1}, respectively, for saline levels (Figure 5B).

The yield of fresh matter and total dry mass for the two sorghum cutting periods increased linearly with the increase in the irrigation depth, regardless of the salinity level of the irrigation water (Figure 6A). It ranged from 52.92 to 83.00 Mg ha^{-1} for a water regime of 51.3% of ETc (244 mm) and 118% of Etc (547 mm), representing a 57% increase in yield compared to lower water availability. For each unit variation in %Etc, there is an estimated increase of approximately 0.46 Mg kg^{-1} in fresh mass yield (0.64%).

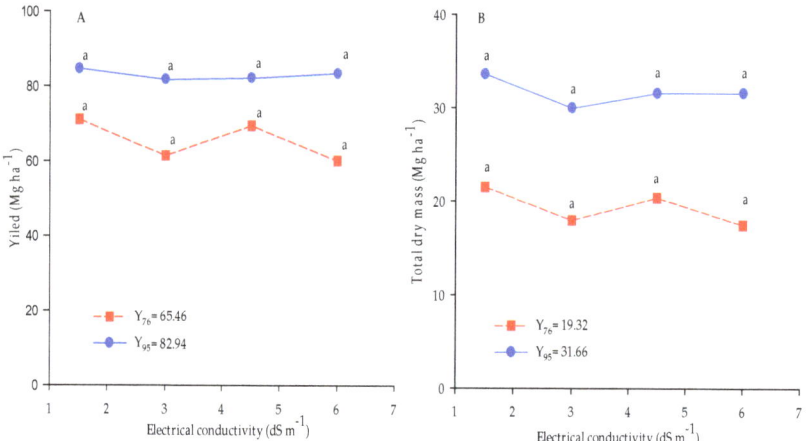

Figure 5. The yield of fresh matter (**A**) and total dry mass (**B**) for the sorghum cultivar (*Sorghum bicolor* (L.) Moench) IPA SF-15, as a function of the electrical conductivity of the irrigation water for cutting performed at 76 days (red) and 95 days (blue) after planting.

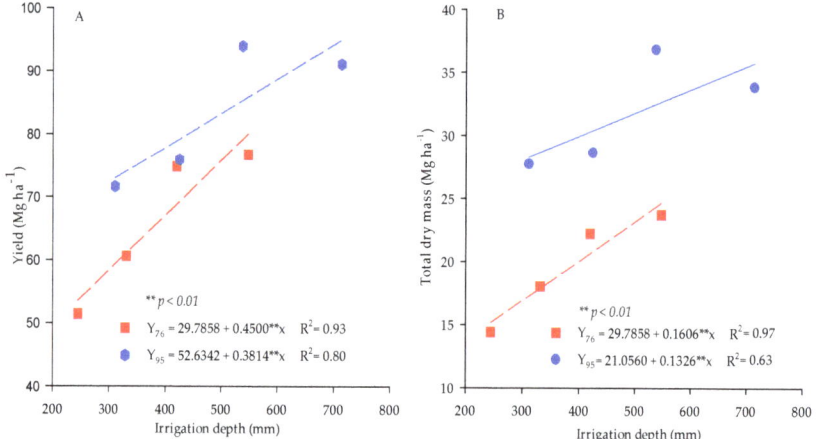

Figure 6. The yield of fresh matter (**A**) and total dry mass (**B**) for the sorghum cultivar (*Sorghum bicolor* (L.) Moench) IPA SF-15 as a function of the total irrigation depths for cutting performed at 76 days (red) and 95 days (blue) after planting.

Water availability significantly influenced dry mass production (Figure 6B), with a variation from 14.96 to 25.68 Mg ha^{-1} for the lowest and highest water availability condition. This variation promoted an increase of 72% in the production of total dry mass. Applying 90% of the ETc (419 mm), the dry mass production was 21.18 Mg ha^{-1}, only 7% lower than the production generated by the irrigation depth of 464 mm (100% of the ETc), which was 22.78 Mg ha^{-1}.

In the cutting period at 95 DAP, yield behavior was linearly increasing (Figure 6A). The variation ranged from 72.2 to 97.64 Mg ha^{-1}, representing a 35% increase in production from the lowest to the highest applied depth, representing an applied water volume of 311 to 713 mm, respectively. When comparing the two cutting periods, the increase in yield for the second cut was lower than the first, as was the volume of water applied. In regions with limited water volume, it was cutting 76 days after planting is necessary.

3.3. Pearson Correlation Matrix

The mineral elements K^+ and Mg^{2+} were strongly correlated with Na^+ and Ca^{2+} in the plant leaf (r = 0.53 and r = 0.80) (Table 5). As for the production variables, total dry mass and yield at 76 DAP correlated positively with the phosphorus content in the leaf (r = 0.52 and r = 0.50), while for the K^+ content, the correlations were negative for the percentage of mass, yield at 76 DAP (r = −0.36 and r = −0.42) and yield and total dry mass at 95 DAP (r = 0.34 and 0.41). In addition, other correlations were verified; however, the correlative effect was shallow both positively and negatively between the elements and the production variables for the two analyzed periods.

Table 5. Pearson correlation matrix for variables measured at 76 e 95 DAP.

	Na	K	Ca	Mg	P	Cl	DMP76	Yield76	TDM76	DMP95	Yield95	TDM95
Na	1											
K	0.53	1										
Ca	0.38	0.29	1									
Mg	0.50	0.33	0.80	1								
P	0.34	−0.04	0.34	0.35	1							
Cl	0.14	0.25	0.06	0.17	−0.06	1						
DMP 76	−0.34	−0.36	0.06	−0.17	0.04	−0.22	1					
Yield 76	0.09	−0.28	0.11	0.08	0.52	0.01	−0.04	1				
TDM 76	−0.06	−0.42	0.14	0.01	0.50	−0.08	0.43	0.88	1			
DMP 95	−0.03	−0.06	0.15	0.09	−0.09	0.11	0.04	0.13	0.14	1		
Yield 95	−0.06	−0.34	0.02	0.04	0.22	−0.15	−0.01	0.49	0.44	0.63	1	
TDM 95	−0.04	−0.41	−0.08	−0.01	0.35	−0.29	−0.05	0.53	0.45	0.07	0.81	1

Na^+—Sodium, K^+—Potassium, Ca^{2+}—Calcium, Mg^{2+}—Magnesium, P—Phosphorus, Cl^-—Chlorine, dry mass percentage (DMP), Yield plant yield (Yield), total dry mass (TDM) at 76 and 95 days after planting (DAP).

4. Discussion

In conditions of extreme water deficit, 51% of the water needs the K^+ content was maximum; this can be justified because the K^+ is used to control the opening and closing of the stomata. May notice that until water availability at 90% of the ETc, the K^+ content was within the limit of 13 g kg^{-1} in the dry mass recommended by [31] as a critical level of K^+ for the species. Despite the K^+ content being at the limit, no nutrient deficiency symptoms were observed in the plant.

According to water deficit, the increase in K^+ content in the leaf may be related to concentration effects, as the plant limits water loss through better stomatal regulation [32]. Furthermore, the response to water deficit of reducing shoot growth promotes root development, increasing the plant's resistance to water deficit [33–35]; therefore, this nutrient is more concentrated in the dry mass. Although researchers such as [36] have observed a reduction in K^+ in the aerial part of sorghum subjected to water stress, we understand that the different results found in this work can be explained by the fact that the K^+ evaluation was carried out directly from the diagnosis leaf.

Underwater scarcity, a form of plant defense, is the stomata's closing to avoid water loss to the atmosphere. In this sense, the increase in K^+ concentration is due to the greater demand by the plant to maintain photosynthesis and protect chloroplasts [37]. In addition, in leaves, K^+ acts in osmotic regulation, mainly under conditions of water deficiency, where the presence of the ion guarantees the turgor of the guard cells, reducing the osmotic potential [38] and, therefore, promoting the flow of water from the roots for the leaves. This osmotic adjustment function is widely known due to the high solubility of K^+ at the potential in the role of fundamental osmoregulation [39] for the osmotic control of the plant subjected to water stress. All these adjustments that the plant starts to execute are measures to reduce the stress in which it is submitted.

The reduction in P content according to water stress may be associated with its low mobility, given that, as it is absorbed by diffuse flow, it depends on soil moisture to be absorbed by the plant. Therefore, the increase in water availability contributes to the absorption of P by the plant. In underwater scarcity conditions, the root system tends to expand to increase P interception increasing plant uptake [40].

In a situation of salt stress, the plant reduces chlorine displacement to the leaf blade as a defense [41], despite being an essential nutrient for the plant. This is because, even though it is essential, an excess of the element can cause unpredictable damage to the plant [42]. In this way, potentially toxic ions are retained in the stem and root, showing the efficiency of sorghum in preventing displacement to the leaf depth [41–43] and contributing to greater tolerance to saline stress. Therefore, reducing Cl⁻ contents in the leaves with increased EC of water is a mechanism to avoid accumulating toxic levels for the plant.

Chlorine levels in the leaf are in the range of critical toxicity concentration in leaf tissues for tolerant plants, which varies from 15 to 50 g kg^{-1} [44]. However, it did not affect the production data. High concentrations of Cl⁻ induce nutritional disturbances [42], such as interference in the absorption of ions (Ca^{2+}, Mg^{2+}, Mn^{2+}, Zn^{2+}, and B), in addition to toxicity in foliar tissues [45]. In addition, the high concentration of Cl⁻ in the leaf is directly related to the increase in leaf succulence [24,46], an efficient protection mechanism to increase the partial dilution of salts by increasing the degree of juiciness. Another point to consider is that the Cl⁻ content absorbed by sorghum was higher than that of Na⁺, which may be due to the high mobility of this anion and its free transport in the plant.

The water stress condition caused by the irrigation deficit during early flowering results in stomatal closure, reducing transpiration, affecting plant development, and reducing yield [47]. Ref. [48] A reduction in production for wheat cultivars was observed when subjected to water deficit during the grain filling period. This may be related to lower nutrient availability in the solution caused by reduced soil moisture. Limited water availability justifies its use. According to [7], applying irrigation depth above 200 mm may be insignificant to increase yield. Reinforcing that even if it reduces production, it is justifiable to use the smallest blade possible to obtain an adequate yield of sorghum production.

The reduction in production when subjected to water deficit can be explained due to the closure of stomata and impairment of plant photosynthesis [49,50]; or by causing direct damage to the photosynthetic metabolism of the plant [51,52]. However, applying a depth of 50 and 71% of the crop's need, despite the reduction in production, there is a gain in water savings, mainly for regions where water availability is limited. Ref. [2] found a dry mass production of 27.02 Mg ha^{-1} using 75% of the reference evapotranspiration, values close to those found in this work, making possible better use of water in periods where water availability is limited.

Pearson's correlations showed few relationships between the variables; however, the mineral elements that presented a correlation were potassium with sodium, an effect that may be related to the high levels of potassium found in the soil and leaf. Potassium is responsible for the stomatal regulation of the plant in the process of opening and closing the plant's stomata. In addition, magnesium and calcium showed a strong correlation with each other. Calcium is fundamental in the plant's response to abiotic stress, acting as a messenger signaling the stress signal [22,53]. Despite the mineral elements provided in the irrigation water in more significant quantities, such as calcium, magnesium, and toxic sodium and chlorine ions, they did not affect sorghum production in the two cutting periods. However, phosphorus positively affected plant production at 76 DAP. As for potassium, its effect on all production variables was negative; this can be explained by the high levels of available potassium harming the plant.

5. Conclusions

The study showed saline stress did not affect sorghum production characteristics, making using saline water in irrigation possible. As for the toxicity of ions, the plant manifested a tolerance behavior to Na^+ and Cl^- ions, reducing the accumulation in the leaf and possibly directing it to the stem and roots, reinforcing that the crop is tolerant to salinity.

Water stress reduced the production of IPA SF-15; however, when seeking to reduce water costs and its limitation of water availability, the supply of 71% of the ETc of the crop can be adopted without significant losses for the sorghum crop. The threshold depth acceptable for the crop can be defined by the estimated depth of 90% of the maximum yield obtained in the experiment for the two planting times. The effect is more at grain filling than at the end of the crop cycle.

For semi-arid conditions where there are water limitations, the result of sorghum production for a water supply of up to 51% of the water requirement of the crop is an alternative to maintain the cultivation of sorghum IPA SF-15, mainly for small producers who live in these conditions of water scarcity and problems with the salinity of the available water sources.

Author Contributions: Conceptualization, R.R.d.S., G.C.M.d.Q. and J.F.d.M.; methodology, R.R.d.S., G.C.M.d.Q., J.F.d.M., L.V.d.S., M.V.P.d.S., M.d.A.B.d.N., F.M.d.S.M., R.F.d.N., L.M.e.S., F.N.F., M.I.B.C., C.J.X.C., J.C.d.C.G., D.C.C. and F.V.d.S.S.; validation, J.F.d.M., F.V.d.S.S.; formal analysis, R.R.d.S., G.C.M.d.Q. and J.F.d.M.; investigation, R.R.d.S., G.C.M.d.Q., J.F.d.M., L.V.d.S., M.V.P.d.S., M.d.A.B.d.N., F.M.d.S.M., R.F.d.N., L.M.e.S., F.N.F., M.I.B.C., C.J.X.C., J.C.d.C.G., D.C.C. and F.V.d.S.S.; resources, J.F.d.M.; data curation, R.R.d.S. and J.F.d.M.; writing—original draft preparation, R.R.d.S. and M.V.P.d.S.; writing—review and editing, R.R.d.S., J.F.d.M., G.C.M.d.Q.; supervision, J.F.d.M.; project administration, J.F.d.M.; and funding acquisition, J.F.d.M. All authors have read and agreed to the published version of the manuscript.

Funding: This research was funded by the National Council for Scientific and Technological Development, CNPq. Notice 01/2016 process (432.570/2016-0) and Fundação de Amparo à Pesquisa do Rio Grande do Norte—FAPERN, Process No.: 10910019.000263/2021-43.

Institutional Review Board Statement: Not applicable.

Data Availability Statement: The data used to support the findings of this study are included in the article.

Acknowledgments: The authors would like to thank the National Council for Scientific and Technological Development (CNPq), Coordination for the Improvement of Higher Education Personnel (CAPES), and the Federal Rural University of the Semi-Arid (UFERSA) for the financial support provided to this research.

Conflicts of Interest: The authors declare no conflict of interest.

References

1. Impa, S.M.; Perumal, R.; Bean, S.R.; John Sunoj, V.S.; Jagadish, S.V.K. Water Deficit and Heat Stress Induced Alterations in Grain Physico-Chemical Characteristics and Micronutrient Composition in Field Grown Grain Sorghum. *J. Cereal Sci.* **2019**, *86*, 124–131. [CrossRef]
2. Kirchner, J.H.; Robaina, A.D.; Peiter, M.X.; Torres, R.R.; Mezzomo, W.; Ben, L.H.B.; Pimenta, B.D.; Pereira, A.C. Funções de Produção e Eficiência No Uso Da Água Em Sorgo Forrageiro Irrigado. *Rev. Bras. Ciências Agrárias—Braz. J. Agric. Sci.* **2019**, *14*, e5646. [CrossRef]
3. Barros, B.G.D.F.; de Freitas, A.D.S.; Tabosa, J.N.; de Lyra, M.D.C.C.P.; Mergulhão, A.C.D.E.S.; da Silva, A.F.; Oliveira, W.D.S.; Fernandes-Júnior, P.I.; Sampaio, E.V.D.S.B. Biological Nitrogen Fixation in Field-Grown Sorghum under Different Edaphoclimatic Conditions Is Confirmed by N Isotopic Signatures. *Nutr. Cycl. Agroecosyst.* **2020**, *117*, 93–101. [CrossRef]
4. Tabosa, A.R.M.; Lima, L.E.; França, J.G.E.; Carvalho, E.X. Cadernos Do Semiárido Riquezas & Oportunidades /Conselho Regional de Engenharia e Agronomia de Pernambuco. In *Histórico e Importância do Sorgo*; CREA-PE: Recife, Brazil, 2020; pp. 1–84.
5. França, I.S.; de Sales Silva, J.C.; de Lima, P.Q. A Importância Do Sorgo Na Pecuária Bovina Leiteira No Brasil. *Nutr. Rev. Eletrônica* **2017**, *14*, 4964–4969.

6. Costa, A.R.F.C.; De Medeiros, J.F. Água Salina Como Alternativa Para Irrigação de Sorgo Para Geração de Energia No Nordeste Brasileiro. *Water Resour. Irrig. Manag.* **2017**, *2017*, 169–177.
7. Araya, A.; Kisekka, I.; Gowda, P.H.; Prasad, P.V.V. Grain Sorghum Production Functions under Different Irrigation Capacities. *Agric. Water Manag.* **2018**, *203*, 261–271. [CrossRef]
8. Klocke, N.L.; Currie, R.S.; Tomsicek, D.J.; Koehn, J.W. Sorghum Yield Response to Deficit Irrigation. *Trans. ASABE* **2012**, *55*, 947–955. [CrossRef]
9. Kaplan, M.; Kara, K.; Unlukara, A.; Kale, H.; Buyukkilic Beyzi, S.; Varol, I.S.; Kizilsimsek, M.; Kamalak, A. Water Deficit and Nitrogen Affects Yield and Feed Value of Sorghum Sudangrass Silage. *Agric. Water Manag.* **2019**, *218*, 30–36. [CrossRef]
10. Mansour, M.M.F.; Emam, M.M.; Salama, K.H.A.; Morsy, A.A. Sorghum under Saline Conditions: Responses, Tolerance Mechanisms, and Management Strategies. *Planta* **2021**, *254*, 24. [CrossRef]
11. Silva, J.L.D.A.; de Medeiros, J.F.; Alves, S.S.V.; Oliveira, F.D.A.D.; Junior, M.J.D.S.; Nascimento, I.B.D. Uso de Águas Salinas Como Alternativa Na Irrigação e Produção de Forragem No Semiárido Nordestino. *Rev. Bras. Eng. Agrícola Ambient.* **2014**, *18*, 66–72. [CrossRef]
12. Silva, J.J.F.; Migliorini, R.B. Caracterização das águas subterrâneas do aquífero furnas na região sul do estado de mato grosso. *Geociências* **2014**, *2*, 261–277.
13. Schossler, T.R.; Machado, D.M.; Zuffo, A.M.; Andrade, F.R.; Piauilino, A.C. Salinidade: Efeitos na fisiologia e na nutrição mineral de plantas. *Enciclopédia Biosf.* **2012**, *8*, 1563–1578.
14. Ayers, R.S.; Westcot, D. Salinity Problems. In *Water Quality for Agriculture*; Irrigation and Drainage, Paper 29; FAO: Roma, Italy, 1985.
15. Kenneth, K.T.; Neeltje, C.K. *Agricultural Drainage Water Management in Arid and Semi-Arid Areas*; Irrigation and Drainage, Paper 61; FAO: Roma, Italy, 2022.
16. Farooq, M.; Hussain, M.; Wakeel, A.; Siddique, K.H.M. Salt Stress in Maize: Effects, Resistance Mechanisms, and Management. A Review. *Agron. Sustain. Dev.* **2015**, *35*, 461–481. [CrossRef]
17. El Sabagh, A.; Hossain, A.; Islam, M.S.; Barutcular, C.; Hussain, S.; Hasanuzzaman, M.; Akram, T.; Mubeen, M.; Nasim, W.; Fahad, S.; et al. Drought and Salinity Stresses in Barley: Consequences and Mitigation Strategies. *Aust. J. Crop. Sci.* **2019**, *13*, 810–820. [CrossRef]
18. Tari, I.; Laskay, G.; Takács, Z.; Poór, P. Response of Sorghum to Abiotic Stresses: A Review. *J. Agron. Crop. Sci.* **2013**, *199*, 264–274. [CrossRef]
19. Almodares, A.; Hadi, M.R.; Kholdebarin, B.; Samedani, B.; Kharazian, Z.A. The Response of Sweet Sorghum Cultivars to Salt Stress and Accumulation of Na+, Cl and K+ Ions in Relation to Salinity. *J. Environ. Biol.* **2014**, *35*, 733–739.
20. Yan, K.; Xu, H.; Cao, W.; Chen, X. Salt Priming Improved Salt Tolerance in Sweet Sorghum by Enhancing Osmotic Resistance and Reducing Root Na+ Uptake. *Acta Physiol. Plant.* **2015**, *37*, 203. [CrossRef]
21. Shakeri, E.; Emam, Y.; Pessarakli, M.; Tabatabaei, S.A. Biochemical Traits Associated with Growing Sorghum Genotypes with Saline Water in the Field. *J. Plant Nutr.* **2020**, *43*, 1136–1153. [CrossRef]
22. Calone, R.; Sanoubar, R.; Lambertini, C.; Speranza, M.; Vittori Antisari, L.; Vianello, G.; Barbanti, L. Salt Tolerance and Na Allocation in Sorghum Bicolor under Variable Soil and Water Salinity. *Plants* **2020**, *9*, 561. [CrossRef]
23. Joardar, J.; Razir, S.; Islam, M.; Kobir, M. Salinity Impacts on Experimental Fodder Sorghum Production. *SAARC J. Agric.* **2018**, *16*, 145–155. [CrossRef]
24. De Lacerda, C.F.; Michael, H.; De Morais, M.; Prisco, J.T.; Gomes, E. Interação Entre Salinidade e Fósforo Em Plantas de Sorgo Forrageiro. *Rev. Ciência Agronômica* **2006**, *37*, 258–263.
25. Soares Filho, W.S.; Gheyi, H.R.; Brito, M.E.B.; Nobre, R.G.; Fernandes, P.D.; Miranda, R.D.S. Melhoramento Genético e Seleção de Cultivares Tolerantes à Salinidade. In *Manejo da Salinidade na Agricultura: Estudos Básicos e Aplicados*; Expressão Gráfica e Editora: Fortaleza, Brazil, 2016.
26. Santos, H.G.; Jacomine, P.K.T.; Anjos, L.H.C.; Oliveira, V.A.; Lumbreras, J.F.; Coelho, M.R. Sistema Brasileiro de Classificação de Solos. *Embrapa Solos* **2018**, *353*, 77.
27. Ayers, R.S.; Westcot, D. Water Quality for Agriculture. A Qualidade Da Água Na Agricultura. In *Estudos FAO: Irrigação e Drenagem 29—Revisado*; Gheyi, H.R.; de Medeiros, J.F.; Damasceno, e.F.A.V., Translators; UFPB: Campina Grande, Brazil, 1999; p. 153.
28. de A. Silva Júnior, L.G.; Gheyi, H.R.; de Medeiros, J.F. Composição Química De Águas Do Cristalino Do Nordeste Brasileiro. *Rev. Bras. Eng. Agrícola Ambient.* **1999**, *3*, 11–17. [CrossRef]
29. Allen, R.G.; Pereira, L.S.; Raes, D.; Smith, M. *FAO 2006. Evapotranspiración del Cultivo—Guías Para la Determinación de Los Requerimientos de Agua de Los Cultivos*; FAO: Roma, Italy, 2006; p. 298.
30. Chapman, H.; Pratt, P. *Methods of Analysis for Soils, Plants and Waters*; University of California: Riverside, CA, USA, 1961.
31. Martinez, H.E.P.; Carvalho, J.G.; Souza, R.B.; Diagnose, F.; Ribeiro, A.C.; Guimarães, P.T.G.; Alvarez, V.V.H. *Recomendações Para o Uso de Corretivos e Fertilizantes em Minas Gerais*; UFV: Viçosa, Brazil, 1999; p. 168.
32. Martineau, E.; Domec, J.C.; Bosc, A.; Denoroy, P.; Fandino, V.A.; Lavres, J.; Jordan-Meille, L. The Effects of Potassium Nutrition on Water Use in Field-Grown Maize (*Zea mays* L.). *Environ. Exp. Bot.* **2017**, *134*, 62–71. [CrossRef]
33. Hsiao, T.C. Water Stress, Growth, and Osmotic Adjustment. *Phil. Trans. R Soc. Lond.* **1976**, *273*, 479–500.
34. Turner, N.C. Adaptation to Water Deficits: A Changing Perspective. *Funct. Plant Biol.* **1986**, *13*, 175–190. [CrossRef]

35. Chaves, M.M.; Pereira, J.S.; Maroco, J.; Rodrigues, M.L.; Ricardo, C.P.P.; Osório, M.L.; Carvalho, I.; Faria, T.; Pinheiro, C. How Plants Cope with Water Stress in the Field. Photosynthesis and Growth. *Ann. Bot.* **2002**, *89*, 907–916. [CrossRef]
36. Leão, D.A.S.; Freire, A.L.O.; De Miranda, J.R.P. Estado Nutricional De Sorgo Cultivado Sob Estresse Hídrico E Adubação Fosfatada. *Pesqui. Agropecuária Trop.* **2011**, *41*, 74–79. [CrossRef]
37. Catuchi, T.A.; Guidorizzi, F.V.C.; Guidorizi, K.A.; Barbosa, A.D.M.; Souza, G.M. Respostas Fisiológicas de Cultivares de Soja à Adubação Potássica Sob Diferentes Regimes Hídricos. *Pesqui. Agropecu. Bras.* **2012**, *47*, 519–527. [CrossRef]
38. Mendes, H.S.J.; Paula, N.F.D.; Scarpinatti, E.A.; de Paula, R.C. Respostas Fisiológicas de Genótipos de Eucalyptus Grandis × E. Urophylla à Disponibilidade Hídrica e Adubação Potássica. *Cern. Lavras* **2013**, *19*, 603–611. [CrossRef]
39. Silva, A.R.A.; Bezerra, F.M.L.; de Lacerda, C.F.; de S. Miranda, R.; Marques, E.C. Ion Accumulation in Young Plants of the "green Dwarf" Coconut under Water and Salt Stress. *Rev. Cienc. Agron.* **2018**, *49*, 249–258. [CrossRef]
40. Kuwahara, F.A.; Souza, G.M.; Guidorizzi, K.A.; Costa, C.; Meirelles, P.R.D.L. Phosphorus as a Mitigator of the Effects of Water Stress on the Growth and Photosynthetic Capacity of Tropical C4 Grasses. *Acta Sci.—Agron.* **2016**, *38*, 363–370. [CrossRef]
41. Acosta-Motos, J.R.; Álvarez, S.; Barba-Espín, G.; Hernández, J.A.; Sánchez-Blanco, M.J. Salts and Nutrients Present in Regenerated Waters Induce Changes in Water Relations, Antioxidative Metabolism, Ion Accumulation and Restricted Ion Uptake in *Myrtus communis* L. Plants. *Plant Physiol. Biochem.* **2014**, *85*, 41–50. [CrossRef] [PubMed]
42. Parihar, P.; Singh, S.; Singh, R.; Singh, V.P.; Prasad, S.M. Effect of Salinity Stress on Plants and Its Tolerance Strategies: A Review. *Environ. Sci. Pollut. Res.* **2015**, *22*, 4056–4075. [CrossRef] [PubMed]
43. de Queiroz, G.C.M.; de Medeiros, J.F.; da Silva, R.R.; da Silva Morais, F.M.; de Sousa, L.V.; de Souza, M.V.P.; da Nóbrega Santos, E.; Ferreira, F.N.; da Silva, J.M.C.; Clemente, M.I.B.; et al. Growth, Solute Accumulation, and Ion Distribution in Sweet Sorghum under Salt and Drought Stresses in a Brazilian Potiguar Semiarid Area. *Agriculture* **2023**, *13*, 803. [CrossRef]
44. Marschner, P. *Nutrição Mineral de Plantas Superiores de Marschner*, 3rd ed.; Academic Press: Cambridge, UK, 2012.
45. Lucini, L.; Borgognone, D.; Rouphael, Y.; Cardarelli, M.; Bernardi, J.; Colla, G. Mild Potassium Chloride Stress Alters the Mineral Composition, Hormone Network, and Phenolic Profile in Artichoke Leaves. *Front. Plant Sci.* **2016**, *7*, 948. [CrossRef]
46. De Sousa, C.H.C.; De Lacerda, C.F.; Bezerra, F.M.L.; Filho, G.E.; Sousa, A.E.C.; Sousa, G.G. De Respostas MORfofisiológicas de Plantas de Sorgo, Feijão-de-Corda e Algodão Sob Estresse Salino. *Agropecuária Técnica* **2010**, *31*, 29–36.
47. Du, T.; Kang, S.; Zhang, J.; Davies, W.J. Deficit Irrigation and Sustainable Water-Resource Strategies in Agriculture for China's Food Security. *J. Exp. Bot.* **2015**, *66*, 2253–2269. [CrossRef]
48. dos Santos, R.; Guimarães, V.F.; Klein, J.; Fioreze, S.L.; Macedo Júnior, E.K. Cultivares de Trigo Submetidas a Déficit Hídrico No Início Do Florescimento, Em Casa de Vegetação. *Rev. Bras. Eng. Agrícola Ambient.* **2012**, *16*, 836–842. [CrossRef]
49. Freitas, C.A.S.D.; Silva, A.R.A.D.; Bezerra, F.M.L.; Lacerda, C.F.D.; Pereira, F.J.V.; De Sousa, G.G. Produção de Matéria Seca e Trocas Gasosas Em Cultivares de Mamoneira Sob Níveis de Irrigação. *Rev. Bras. Eng. Agrícola Ambient.* **2011**, *15*, 1168–1174. [CrossRef]
50. Silva, S.M.L.D.S. Avaliação da Tolerância à Salinidade em Quatro Genótipos de Sorgo Sacarino. *CNR-ISTI Tech. Rep.* **2015**, *3*, 356–369.
51. Endres, L.; De Souza, J.L.; Teodoro, I.; Marroquim, P.M.G.; Santos, C.M.; De Brito, J.E.D. Gas Exchange Alteration Caused by Water Deficit during the Bean Reproductive Stage Alterações Das Trocas Gasosas Na Fase Reprodutiva e Produtividade Do Feijão Sob Déficit Hídrico. *Rev. Bras. Eng. Agrícola Ambient.* **2010**, *14*, 11–16. [CrossRef]
52. Taiz, L.; Zeiger, E.; Moller, I.M.; Murphy, A. *Fisiologia e Desenvolvimento Vegetal*, 6th ed.; Artimed: Porto Alegre, Brazil, 2017; ISBN 978-85-8271-366-2.
53. Carroll, A.D.; Moyen, C.; Van Kesteren, P.; Tooke, F.; Battey, N.H.; Brownlee, C. Ca^{2+}, Annexins, and GTP Modulate Exocytosis from Maize Root Cap Protoplasts. *Plant Cell* **1998**, *10*, 1267–1276. [CrossRef] [PubMed]

Disclaimer/Publisher's Note: The statements, opinions and data contained in all publications are solely those of the individual author(s) and contributor(s) and not of MDPI and/or the editor(s). MDPI and/or the editor(s) disclaim responsibility for any injury to people or property resulting from any ideas, methods, instructions or products referred to in the content.

Article

Potential of Brackish Groundwater for Different Biosaline Agriculture Systems in the Brazilian Semi-Arid Region

Carla Ingryd Nojosa Lessa [1], Claudivan Feitosa de Lacerda [1,*], Cláudio Cesar de Aguiar Cajazeiras [2], Antonia Leila Rocha Neves [1], Fernando Bezerra Lopes [1], Alexsandro Oliveira da Silva [1], Henderson Castelo Sousa [1], Hans Raj Gheyi [3], Rafaela da Silva Nogueira [4], Silvio Carlos Ribeiro Vieira Lima [5], Raimundo Nonato Távora Costa [1] and Geocleber Gomes de Sousa [4]

[1] Agricultural Engineering Department, Federal University of Ceará, Fortaleza 60455-760, Brazil
[2] Geological Survey of Brazil-CPRM, Fortaleza 60135-101, Brazil
[3] Academic Unit of Agricultural Engineering, Federal University of Campina Grande, Campina Grande 58840-000, Brazil
[4] Institute of Rural Development, University of International Integration of Afro-Brazilian Lusofonia, Redenção 62790-000, Brazil
[5] Secretariat of Economic Development and Labor of the State of Ceará, Fortaleza 60160-230, Brazil
* Correspondence: cfeitosa@ufc.br

Abstract: The objective of this research was to define the potential of brackish groundwater for 15 systems of biosaline agriculture in a representative area of the Brazilian semi-arid region. The study was conducted using a database of the State of Ceará, with 6284 wells having brackish water (EC $\geq$ 0.8 dS m^{-1} and discharge rate $\geq$ 0.5 m^3 h^{-1}). Our results show that the potential of brackish groundwater resources depends on the set of data: (i) production system (crop salt tolerance and water demand) and (ii) water source (salinity and well discharge rate). The joint analysis of these data shows that plant production systems with lesser water requirements, even with moderate tolerance levels to salt stress, present better results than more tolerant species, including halophytes and coconut orchards. About 41, 43, 58, 69, and 82% of wells have enough discharge rates to irrigate forage cactus (1.0 ha), sorghum (1.0 ha with supplemental irrigation), hydroponic cultivation, cashew seedlings, and coconut seedlings, respectively, without restrictions in terms of salinity. Otherwise, 65.8 and 71.2% of wells do not have enough water yield to irrigate an area of 1.0 ha with halophytes and coconut palm trees, respectively, but more than 98.3 and 90.7% do not reach the water salinity threshold for these crops. Our study also indicates the need for diversification and use of multiple systems on farms (intercropping, association of fish/shrimp with plants), to reach the sustainability of biosaline agriculture in tropical drylands, especially for family farming.

Keywords: drylands; aquifers; salinity; biosalinity; sustainability

1. Introduction

The semi-arid regions of the world suffer from water shortage and are increasingly vulnerable to extreme events, imposed by climate variability and enhanced by climate change on a global scale [1,2]. In particular, the tropical semi-arid regions are faced with several constraints that compromise sustainability [3–5] and the expansion of the agricultural sectors, such as high temperature, water shortage, poorly developed soils, and high salinity of groundwater sources [6–8]. Prolonged droughts, as observed in the Brazilian semi-arid region between 2012 and 2016 [9,10], cause severe losses in agricultural and livestock production, as well as impact other sectors of the economy [11].

Irrigated agriculture is responsible for the highest consumption of available water in arid and semi-arid regions [12,13]. However, farmers in these regions suffer from problems related to the availability and quality of water for agricultural production [14]. According to [15], most surface reservoirs present in the Brazilian semi-arid region have a capacity ranging from 1 to 1000 hm^3, with small (less than 10 hm^3) and medium (10 to 50 hm^3) sizes prevailing. Given the high evaporation rate and the scarcity of surface water in this region, the use of groundwater is a viable alternative [16].

According to data from the National Water Agency (ANA), in 2017 there were about 1.2 million wells drilled in the aquifers of Brazil [17]. In the Brazilian semi-arid region, there are about 160,000 wells [18], and a significant part of these water sources have a high salt concentration, with electrical conductivity of most water sources between 1.0 and 6.0 dS m^{-1} and an average discharge rate less than 3.0 m^3 h^{-1} [6,19–21]. According to [6], the largest proportion of these wells with brackish water is found in the fractured crystalline aquifers, followed by alluvial and sedimentary areas.

With the scarcity of low-salinity water for use in irrigated agriculture, the use of brackish water appears to be an alternative [21,22]. Although these brackish sources can meet the water needs of certain production systems, the high salt concentration is a constraint for the growth and productivity of most crops [23–25]. Plants under salt stress may present changes in their metabolic and biochemical activities due to the osmotic and ionic effects of excess salts in the root zone, with direct impacts on stomatal conductance and photosynthetic rate, inhibition of protein synthesis and enzymatic activities, and increased degradation of chlorophyll [26,27].

To partially circumvent salinity problems in agriculture, there is a vast literature on crop salt tolerance [28], as well as management strategies to reduce the impacts of excess salts on crop development [20]. In addition, there is a lot of information on the water needs of annual and perennial crops, and studies on the water potential of aquifers, which allow the elaboration of well-sized projects with water sustainability. However, the potential of these brackish water sources has only been assessed based on qualitative (water salinity) or quantitative (water yield) assessment [6,21,29]. According to quality indices, for example, a high percentage of the brackish groundwater (51%) in the Brazilian semi-arid region has been classified as poor quality for plant cultivation, while 87% integrated the best quality classes (excellent and good) for animal production [21]. Otherwise, studies that simultaneously evaluate the quantitative and qualitative potential of brackish groundwater for agricultural purposes have not been carried out to date. Therefore, this innovative approach needs to be developed to ensure the expansion of sustainable biosaline production under arid and semi-arid climates. The results of this type of research can give a more realistic guide for the use of brackish groundwater by farmers as well as for the improvement of public policies related to the agriculture sector.

Considering this new approach, our study tested the hypothesis that the potential of wells with brackish water in the Brazilian semi-arid region depends on the water salinity level, water discharge rates, and inherent characteristics of biosaline production systems (salt tolerance and water demand). Thus, the objective of this study was to define, based on qualitative and quantitative indicators, the potential of wells with brackish water for 15 agricultural production systems in a representative area of the Brazilian semi-arid region.

2. Material and Methods

2.1. Characterization of the Study Area

The Brazilian semi-arid region covers areas of nine States in the Northeast Region and one in the Southeast Region, covering 1262 municipalities. It has a total area of approximately 1.12 million km^2, and about 27 million inhabitants [30]. This study was carried out using a database from the State of Ceará (Figure 1). The State is made up of 184 municipalities, with a total area of 148,886 km^2, and a population of 8.9 million inhabitants. According to the Köppen classification, the State of Ceará has two types of climates: BSh (tropical semi-arid climate) and Aw (tropical climate with dry winter). However, the semi-arid tropical climate is predominant [31] in approximately 95% of the area of the state, with an average annual temperature of 27 °C, potential evapotranspiration of 1700 mm, and average annual rainfall from 500 to 800 mm, with around 80% concentrated during February to May [32].

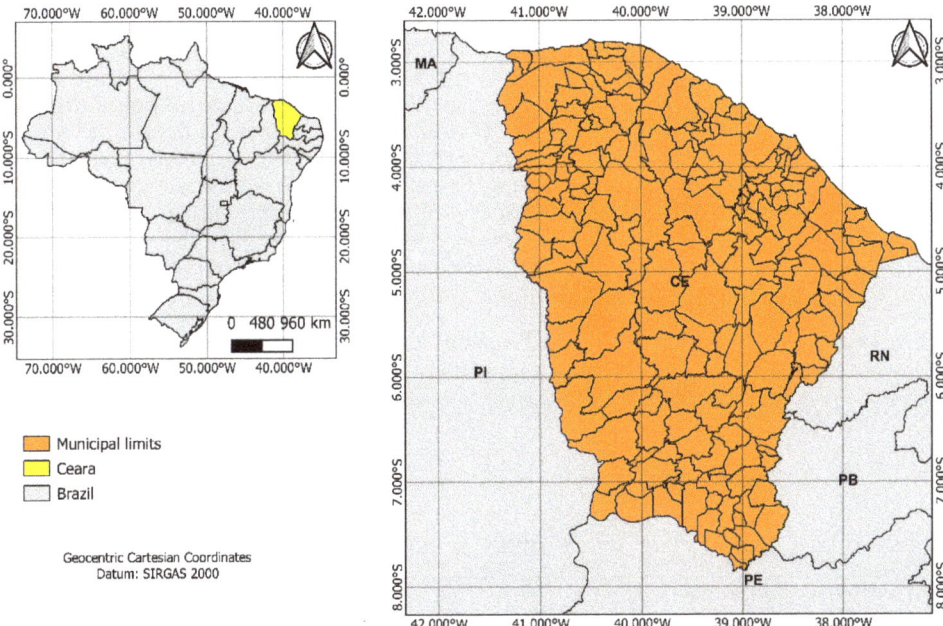

Figure 1. Geographic location of the study area. The acronyms indicate the states of the Northeast Region of Brazil that border the State of Ceará—CE: MA = Maranhão; PI = Piauí; RN = Rio Grande do Norte; PB = Paraiba; PE = Pernambuco.

2.2. Database Characterization

The research was carried out using a database of chemical analyses of water from wells in 179 municipalities in the State of Ceará, provided by the Superintendência de Obras Hidráulicas do Ceará (SOHIDRA) and by the Serviço Geológico do Brasil (CPRM). The selection of wells with brackish water was based on the electrical conductivity of the water (EC) ≥ 0.8 dS m^{-1} (quality criterion) and discharge rate (Q) ≥ 0.5 m^3 h^{-1} (water availability criterion).

The database consists of wells drilled from 1987 to 2021, totaling 25,497 wells, with 6284 wells (about 25% of the total) meeting the quality and water availability criteria established for the study (EC ≥ 0.8 dS m^{-1} and Q ≥ 0.5 m^3 h^{-1}). The database also includes relevant information, such as: city, geographic coordinates, drilling year, and depth. Table 1 presents the minimum, average, maximum, and median for electrical conductivity, total dissolved solids, discharge rate, and depth.

Table 1. Minimum, average, maximum, and median values for electrical conductivity of water (EC), total dissolved solids (TDS), discharge rate, and depth of the 6284 wells with brackish water from the Ceará State database.

Parameters	Minimum	Average	Maximum	Median
EC (dS m^{-1})	0.80	2.89	29.40	1.97
TDS (mg L^{-1})	508.8	1980.6	23,520	1260.8
Discharge (m^3 h^{-1})	0.50	4.10	180	2.48
Depth (m)	11.5	69.38	233	70

2.3. Selected Production Systems

To assess the potential use of brackish water, plant production systems and two fish and plant associations were considered, characterized as follows: (i) full irrigation (halophytes and maize); (ii) supplemental irrigation of annual crops (maize, sorghum, and cotton); (iii) irrigation of perennial crops (forage cactus, cashew, and coconut trees); (iv) hydroponics (vegetables); (v) production of seedlings in nurseries (coconut, cashew, and tree species from the Caatinga biome); (vi) ornamental plants; and (vii) association of fish and plants: (Tilapia—*Oreochromis niloticus plus* glycophytes and Tilapia *plus* halophytes). The selection of these production systems considered their adaptability to the semi-arid tropical climate, being commonly used in small and medium farms in the semi-arid region of Brazil.

2.4. Definition of Salinity Thresholds

The water salinity threshold for the 15 production systems was defined based on results published in the scientific literature (Table 2), considering a production loss of up to 10% (EC_{90}).

Table 2. Water salinity threshold for 15 production systems, assuming a maximum production loss of up to 10% (EC_{90}).

Production systems	EC_{90} (dS m^{-1}) *	References
1. Full irrigation of halophytes (Atriplex, Sarcorcórnia, etc.)	11.4	[33,34]
2. Full irrigation of maize	1.7	[28,35]
3. Supplemental irrigation of maize	3.2	[35–37]
4. Supplemental irrigation of cotton	5.1	[28,38,39]
5. Supplemental irrigation of sorghum	5.0	[28,40,41]
6. Irrigation of forage cactus	3.0	[42–45]
7. Irrigation of cashew orchards in sandy soils	5.0	[46–48]
8. Irrigation of coconut orchards in sandy soils	6.0	[49–51]
9. Hydroponics cultivation of vegetables	3.0	[52–55]
10. Coconut seedlings	4.5	[56,57]
11. Cashew seedlings	3.0	[58–60]
12. Seedlings of tree species native to Caatinga biome	2.5	[61]
13. Herbaceous ornamental plants	2.5	[62–64]
14. Tilapia *plus* glycophytes	3.0	**
15. Tilapia *plus* halophytes	9.0	***

* Electrical conductivity measured at 25 °C; ** Threshold for the glycophyte systems that will be used in association with fish (supplemental irrigation, hydroponics, forage cactus, and seedling production in nurseries); *** Threshold value for tilapia cultivation, according to [65–67].

In general, less restrictive water salinity thresholds were chosen, considering the existence of poorly developed soils in areas with a high occurrence of brackish groundwater in the Brazilian semi-arid region. A minimum leaching fraction of 15% is recommended to avoid excessive salt build-up, especially when full irrigation is used. The sodium adsorption ratio (SAR) should be previously evaluated in the case of soils with medium texture or with high clay content, although SAR values are not high in most water sources in the region studied [6]. In some cases, the potential for incrustation or corrosion must also be evaluated to avoid clogging problems in the irrigation system, especially in waters with a high concentration of carbonates and sulphate [28].

For supplemental irrigation of annual crops, the salinity thresholds are higher than for full irrigation, because of the possibility of leaching of part of the salts by rainwater [68]. However, for supplemental irrigation of cotton, it is recommended to use the threshold salinity value indicated for full irrigation, given the types of soils in which cotton is cultivated, which normally have high clay content. For perennial crops, both possibilities of full and supplemental irrigation were considered. Full irrigation of perennial crops such as coconut and cashew nut should be carried out in deep sandy soils and orchards in production. The implantation of new orchards must be performed at the beginning of the rainy season, with the use of supplemental irrigation with brackish water, if necessary.

For most cases, localized irrigation is recommended, as this method reduces the direct impact of salinity on the leaves. However, sprinkler irrigation may be used for annual crops, especially for supplemental irrigation of forage with water of moderate salinity. Supplemental irrigation can be practiced during the rainy season in the Brazilian semi-arid region. Forage cactus should only be irrigated in the dry season.

The fish *plus* glycophytes system combines tilapia cultivation with plant production systems (supplemental irrigation of annual crops or forage cactus irrigation or seedling production in nurseries or hydroponic cultivation), and the average salinity threshold for plant cultivation was considered. For the fish *plus* halophyte system, the threshold value for tilapia cultivation (9.0 dS m^{-1}) was considered [65–67].

2.5. Definition of the Minimum Required Discharge Rates

The minimum discharge rate (Table 3) was defined according to the size of the enterprise, the water demand of each production system, and the time of functioning of the deep well (6 h per day).

Table 3. Minimum discharge rates (Q) required of wells for each production system *.

Production Systems/Enterprise Size	Required Water Discharge Rate ($m^3 h^{-1}$)
Full irrigation of halophytes (Atriplex, Sarcorcórnia)—0.5 ha	2.0
Full irrigation of halophytes (Atriplex, Sarcorcórnia)—1.0 ha	4.0
Full irrigation of maize—0.5 ha	2.5
Full irrigation of maize—1.0 ha	5.0
Supplemental irrigation of annual crops (maize, cotton, sorghum)—0.5 ha	1.25
Supplemental irrigation of annual crops (maize, cotton, sorghum)—1.0 ha	2.50
Irrigation of forage cactus—0.5 ha	1.0
Irrigation of forage cactus—1.0 ha	2.0
Irrigation of cashew orchards in sandy soils—0.5 ha	1.6
Irrigation of cashew orchards in sandy soils—1.0 ha	3.2
Irrigation of coconut orchards in sandy soils—0.5 ha	3.0
Irrigation of coconut orchards in sandy soils—1.0 ha	6.0
Hydroponic cultivation of vegetables (100 m^2)	1.0
Cashew seedlings in nurseries (2000 seedlings)	0.5
Coconut seedlings in nurseries (2000 seedlings)	0.5
Seedlings of tree species native to Caatinga biome (2000 seedlings)	0.5
Herbaceous ornamental plants (2000 plants)	0.5
Tilapia *plus* glycophytes	1.5
Tilapia *plus* halophytes	2.5

For full irrigation of maize, a daily water depth of 5.0 mm was considered [69], with localized irrigation and a wetted area equal to 50% of the total area. This value represents the water consumption at the stage of maximum crop demand, with lower consumption at the other stages. For longer irrigation intervals, a 50–100 m^3 cistern/tank will be required to store water from the well on days without an irrigation event. For supplemental irrigation of annual crops, the adequate discharge rate will be half of that required for full irrigation [69–71], with the possibility of storing rainwater or well water during the dry spells.

For forage cactus irrigation, 40,000 plants ha^{-1} and an irrigation depth of 1.0 mm per day were considered [72,73], with weekly irrigation during the dry season, which requires the storage of well water. For the dwarf cashew, 204 plants ha^{-1} were considered with an average water application of 80 L $plant^{-1}$ day^{-1} [74]. For dwarf green coconut, 200 plants ha^{-1} were considered with an average water application rate of 150 L $plant^{-1}$ day^{-1} [75,76].

For the fish *plus* glycophytes system, the discharge rate required for 0.4 ha of supplemental irrigation of annual crops, or 0.5 ha of forage cactus was considered, which is also sufficient for hydroponics or seedling production. For the fish *plus* halophytes system, the discharge rate required for 0.5 ha of halophytes was considered. The total water required by each system also includes evaporation losses in fish farming. A total volume of 100 m^3 was considered for the cultivation of fish, which can be stored in one or more tanks.

2.6. Criteria for Defining the Potential of the Wells

The criteria for water quality (threshold salinity, Table 2) and water availability (water discharge required, Table 3) were adopted, and their respective adequacy and non-adequacy values, as described in Table 4:

Table 4. Water quality criteria and well productivity.

Electrical Conductivity (EC$_{90}$)	Discharge (Q)	Symbol
Adequate (ad)	Adequate (ad)	EC$_{ad}$ and Q$_{ad}$
Inadequate (nad)	Adequate (ad)	EC$_{nad}$ and Q$_{ad}$
Adequate (ad)	Inadequate (nad)	EC$_{ad}$ and Q$_{nad}$
Inadequate (nad)	Inadequate (nad)	EC$_{nad}$ and Q$_{nad}$

For the irrigation of halophytes, annual crops, and perennial crops, the irrigable area in hectares was estimated. For this calculation, the following data were considered: The volume of water produced by the well without salinity restriction (m^3) and the volume required for each cultivation system (m^3 ha^{-1}). The daily water volume produced by the well was obtained by multiplying the discharge rate by 6 (time of functioning).

2.7. Data Analysis

Data of electrical conductivity and discharge rate of the wells were organized in spreadsheets in the computer program Microsoft Excel in a file containing the respective geographic coordinates to facilitate the manipulation of the data. Then, georeferenced maps were made for each biosaline production system, using the Quantum GIS 3.22 program [77].

3. Results

3.1. Irrigation of Halophytes

Data analysis showed that 3637 wells (57.9%) have adequate electrical conductivity and discharge for full irrigation of halophytes for an area of 0.5 ha (Figure 2A). However, 70 wells (1.1%) do not have adequate electrical conductivity, but their discharge rate is adequate, 2539 wells (40.4%) have adequate electrical conductivity, but the water availability is insufficient, and 38 wells (0.6%) do not have an adequate discharge rate and the electrical conductivity is higher than the salinity threshold. For an area of 1.0 ha (Figure 2B), 2113 wells (33.7%) are adequate both from the point of view of electrical conductivity and water availability, 29 wells (0.5%) only have adequate discharge rates, 4063 wells (64.6%) only have adequate electrical conductivity, and 79 wells (1.2%) do not have suitable salinity or water availability.

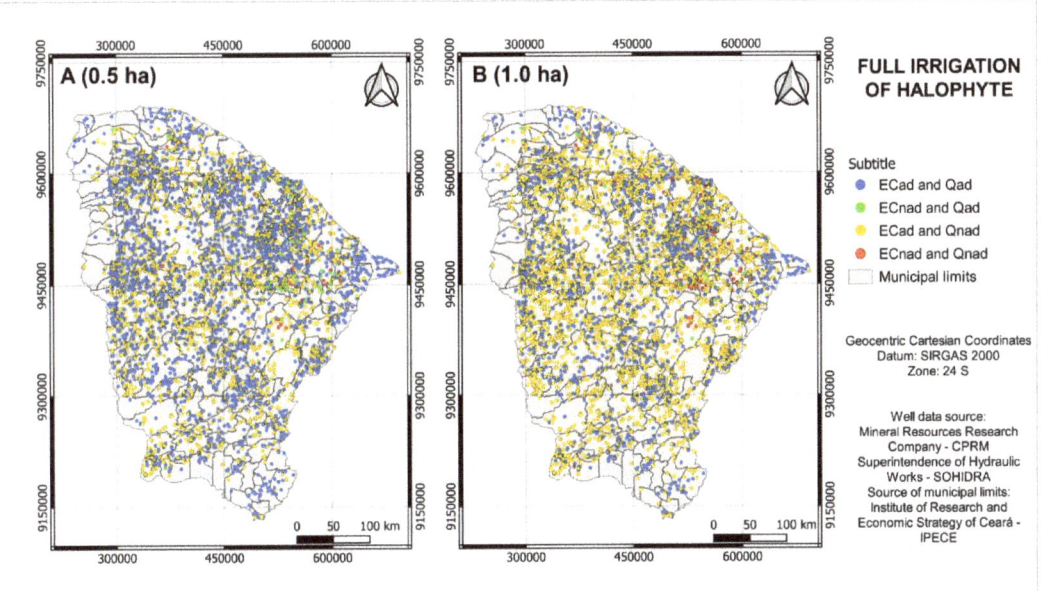

Figure 2. Potential of brackish groundwater in the State of Ceará (Northeast Brazil) for full irrigation of halophytes in 0.5 ha (**A**) and 1.0 ha (**B**), based on the threshold electrical conductivity of the irrigation water (EC) and discharge rate (Q). EC_{ad}: adequate electrical conductivity; EC_{nad}: inadequate electrical conductivity; Q_{ad}: adequate discharge; Q_{nad}: inadequate discharge.

3.2. Full and Supplemental Irrigation of Annual Crops

For full irrigation of maize, with an area of 0.5 ha (Figure 3A), about 1410 wells (22.4%) show EC and Q suitability, 1729 wells (27.5%) show inadequate EC and adequate Q, 1263 wells (20.1%) have adequate EC and inadequate Q, and 1882 wells (30.0%) do not have adequate EC or adequate Q. For 1.0 ha (Figure 3B), only 794 wells (12.6%) have adequate EC and Q, 839 wells (13.3%) only have adequate Q, 1879 wells (30.0%) only have adequate EC, and 2772 wells (44.1%) have salinity above the threshold and do not reach the minimum discharge required for the crop.

For supplemental irrigation of maize, for 0.5 ha (Figure 3C), 3278 wells (52.2%) meet the water demand of the crop and have an EC within the salinity threshold of the crop. However, 1248 wells (19.8%) only have adequate Q, 1211 wells (19.3%) only have adequate electrical conductivity of water, and 547 wells (8.7%) did not present either adequate electrical conductivity or water availability. For the area of 1.0 ha (Figure 3D), 2314 wells (36.8%) fit according to EC and Q, 825 wells (13.1%) only have adequate discharge, 2175 wells (34.6%) only have adequate electrical conductivity, and 970 wells (15.5%) are not suitable due to the high salinity and low discharge rate.

Of the 6284 wells evaluated for supplemental irrigation of sorghum in an area of 0.5 ha (Figure 4A), 62.0% (3898 wells) have adequate EC and Q, 10.0% (628 wells) only have adequate Q, 23.3% (1462 wells) have adequate EC and inadequate Q, and 4.7% (296 wells) do not have adequate EC and Q. For an area of 1.0 ha (Figure 4B), 43.5% (2732 wells) of the wells have adequate EC and Q, 6.5% (407 wells) have inadequate EC and adequate Q, 41.8% (2628 wells) have adequate EC and inadequate Q, and 8.2% (517 wells) do not have adequate EC or Q.

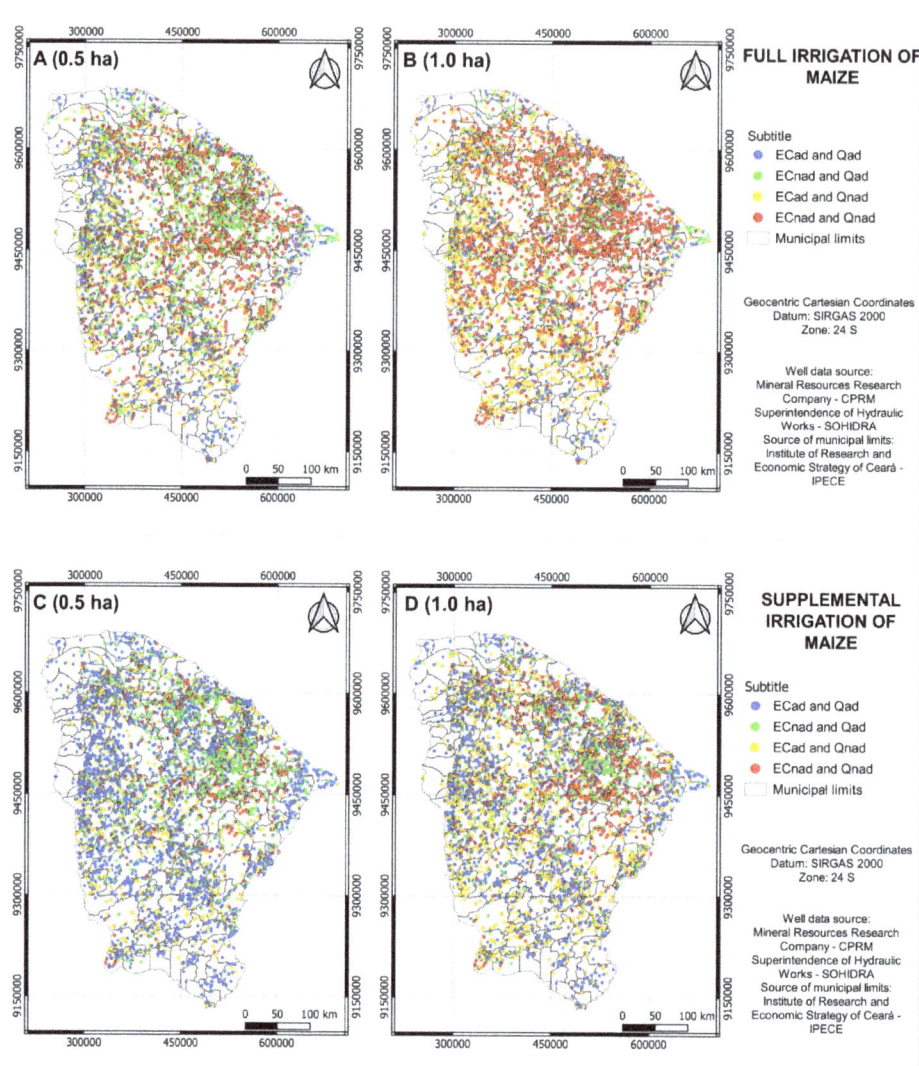

Figure 3. Potential of brackish groundwater in the State of Ceará (Northeast Brazil) for full irrigation of maize in 0.5 ha (**A**) and 1.0 ha (**B**), and supplemental irrigation of maize in 0.5 ha (**C**) and 1.0 ha (**D**), based on the threshold electrical conductivity of the irrigation water (EC) and discharge rate (Q). EC_{ad}: adequate electrical conductivity; EC_{nad}: inadequate electrical conductivity; Q_{ad}: adequate discharge; Q_{nad}: inadequate discharge.

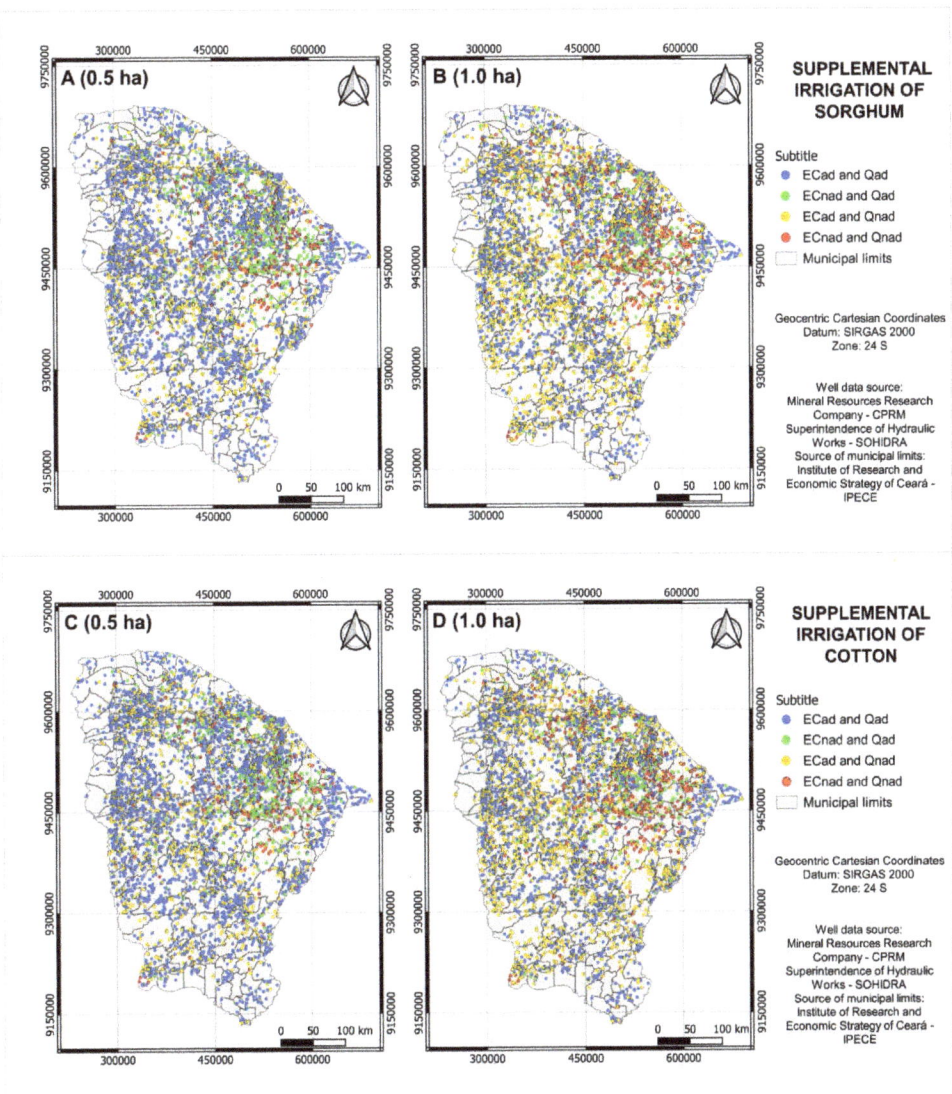

Figure 4. Potential of brackish groundwater in the State of Ceará (Northeast Brazil) for supplemental irrigation of sorghum in 0.5 ha (**A**) and 1.0 ha (**B**), and supplementary irrigation of cotton in 0.5 ha (**C**) and 1.0 ha (**D**), based on the threshold electrical conductivity of the irrigation water (EC) and discharge rate (Q). EC_{ad}: adequate electrical conductivity; EC_{nad}: inadequate electrical conductivity; Q_{ad}: adequate discharge; Q_{nad}: inadequate discharge.

For the practice of supplemental irrigation of cotton in an area of 0.5 ha (Figure 4C), 62.6% of wells (3938 wells) achieve adequacy in terms of salinity and water availability, 9.3% (588 wells) only have adequate Q, 23.5% (1477 wells) only have adequate EC, and 4.5% (281 wells) do not have either adequate water electrical conductivity or well discharge. Considering an area of 1.0 ha (Figure 4D), 44.0% of wells (2760 wells) have adequate water electrical conductivity and discharge, 6.0% (379 wells) only have adequate Q, 42.2% (2655 wells) only have adequate EC, and 7.8% (490 wells) are limited by both high salinity and low water availability.

3.3. Irrigation of Perennial Crops

In the irrigation of forage cactus for an area of 0.5 ha (Figure 5A), 3624 wells (57.7%) are adequate in terms of salinity and water availability, 1603 wells (25.5%) only have adequate Q, 703 wells (11.2%) only have adequate EC, and 354 wells (5.6%) have inadequate EC and Q. For an area of 1.0 ha (Figure 5B), 2594 wells (41.3%) present adequate EC and Q, 1108 wells (17.6%) only have adequate Q, 1733 wells (27.6%) only have adequate EC, and 849 wells (13.5%) do not have adequate EC or Q.

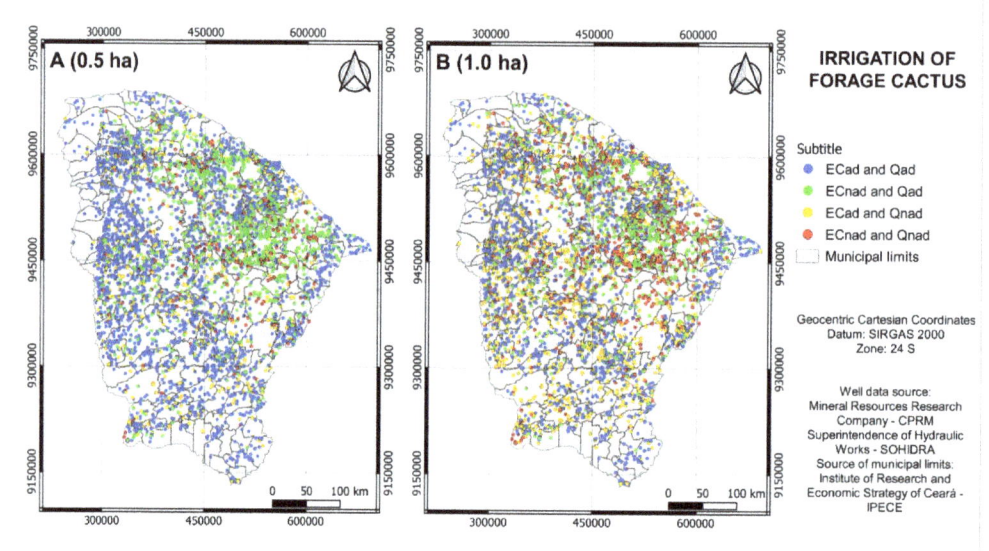

Figure 5. Potential of brackish groundwater in the State of Ceará (Northeast Brazil) for full irrigation of forage cactus in 0.5 ha (**A**) and 1.0 ha (**B**), based on the threshold electrical conductivity of the irrigation water (EC) and discharge rate (Q). EC_{ad}: adequate electrical conductivity; EC_{nad}: inadequate electrical conductivity; Q_{ad}: adequate discharge; Q_{nad}: inadequate discharge.

For the management of coconut irrigation in an area of 0.5 ha (Figure 6A), 41.2% of wells (2592 wells) have adequate electrical conductivity and discharge, 3.5% (217 wells) only have adequate Q, 49.4% (3105 wells) are limited only by low discharge, and 5.9% (370 wells) have problems of high salinity and low water availability. For an area of 1.0 ha, 19.5% (1225 wells) have adequate EC and Q, 1.4% (88 wells) have adequate Q only, 7.9% (499 wells) do not have adequate water electrical conductivity or discharge, and 71.2% (4472 wells) are limited by low water availability.

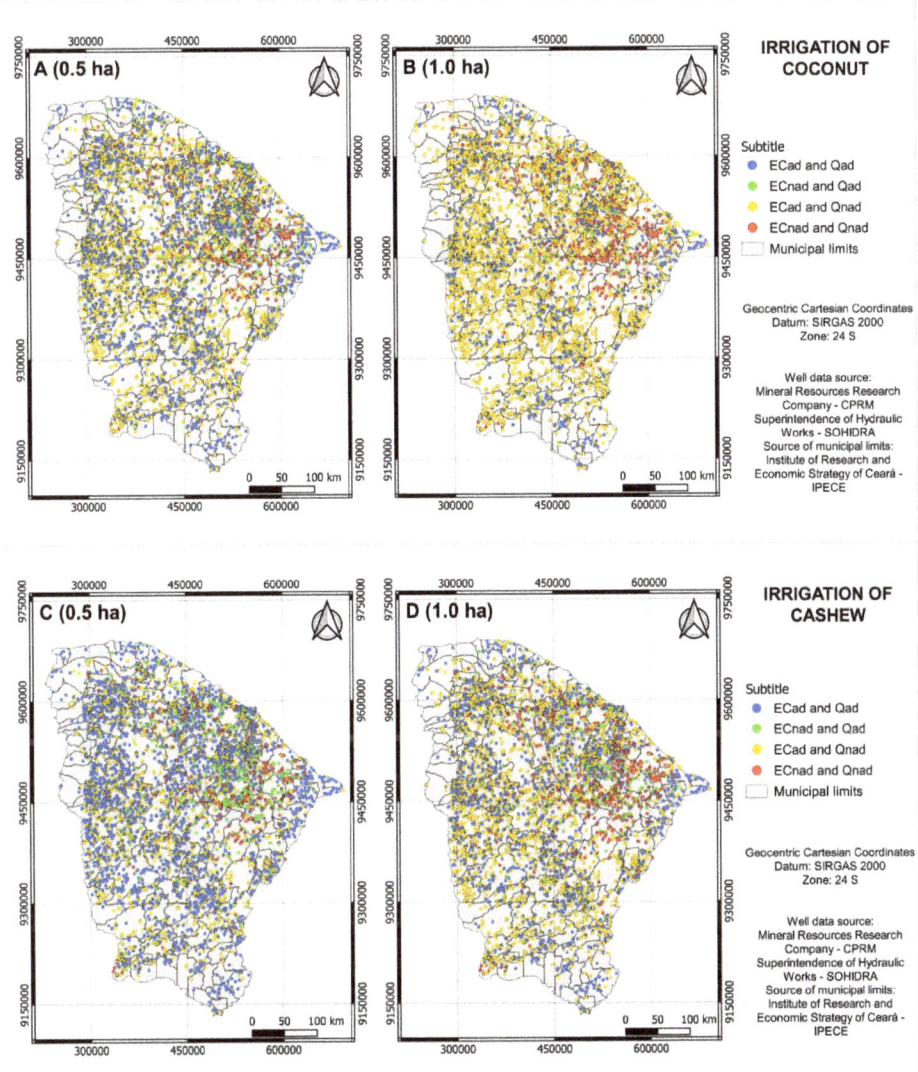

Figure 6. Potential of brackish groundwater in the State of Ceará (Northeast Brazil) for coconut irrigation in 0.5 ha (**A**) and 1.0 ha (**B**), and cashew irrigation in 0.5 ha (**C**) and 1.0 ha (**D**), based on the threshold electrical conductivity of the irrigation water (EC) and discharge rate (Q). EC_{ad}: adequate electrical conductivity; EC_{nad}: inadequate electrical conductivity; Q_{ad}: adequate discharge; Q_{nad}: inadequate discharge.

For irrigation of cashew in an area of 0.5 ha, it is observed that 55.7% (3501) of the wells have adequate EC and Q, 8.8% (554 wells) only have adequate Q, 29.6% (1859 wells) only have adequate EC, and 5.9% (370 wells) do not have adequate EC and Q (Figure 6C). For an area of 1.0 ha, 35.1% (2208) of the wells have adequate EC and Q, 5.0% (316 wells) do not have adequate EC, but have adequate Q, 9.7% (608 wells) do not have adequate EC or Q, while 50.2% (3152 wells) are unproductive due to low water availability (Figure 6D).

3.4. Seedlings, Hydroponics and Multiple Systems

Figure 7 highlights seedling production systems in nurseries and hydroponic cultivation. To produce 2000 coconut seedlings, 5195 wells (82.7%) achieve adequacy in terms of threshold water salinity and well discharge, while 1089 (17.3%) are limited by high salinity. To produce 2000 cashew seedlings (Figure 7B), 4327 wells (68.8%) have adequate water electrical conductivity and discharge, and 1957 (31.2%) are limited by high salinity. As for the production of 2000 seedlings of trees from the Caatinga biome or ornamental plants (Figure 7C), 3853 wells (61.3%) have adequate electrical conductivity and discharge, and 2431 wells (38.7%) have electrical conductivity above the water salinity threshold. For the hydroponic cultivation of vegetables in an area of 100 m² (Figure 7D), 3624 wells (57.7%) have adequate water electrical conductivity and discharge, 1603 wells (25.5%) only have adequate Q, 703 wells (11.2%) only have adequate EC, and 354 (5.6%) are limited by both high salinity and low water availability.

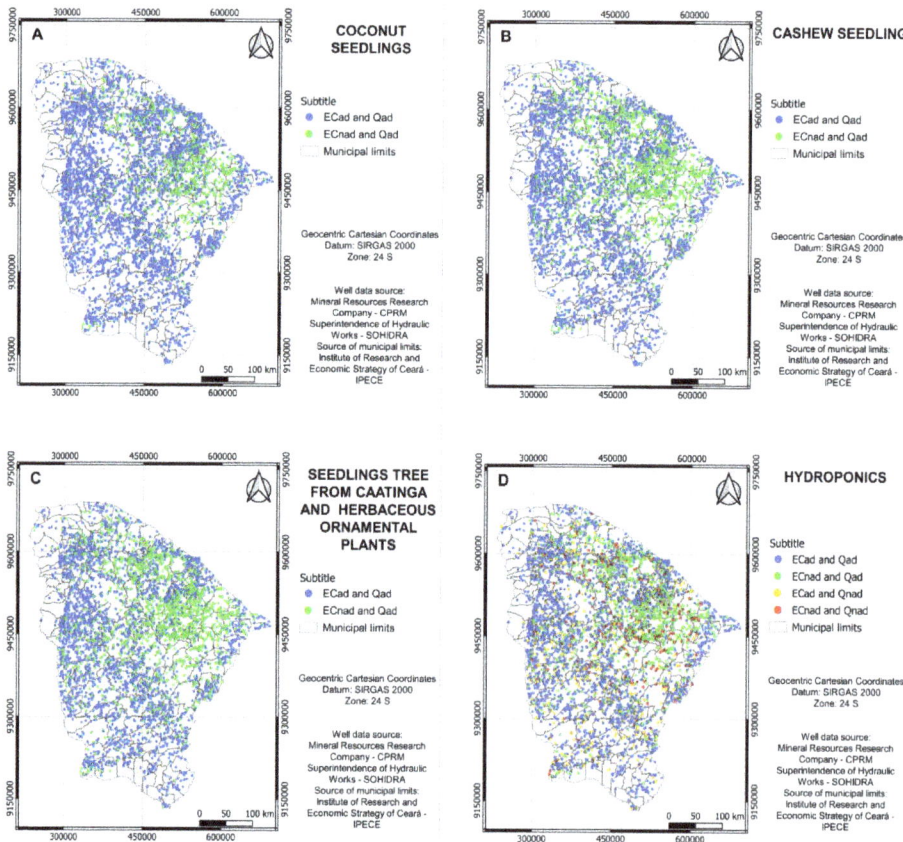

Figure 7. Potential of brackish groundwater in the State of Ceará (Northeast Brazil) to produce 2000 seedlings in nurseries of coconut (**A**), cashew (**B**), and trees native to the Caatinga biome/ornamental plants (**C**)), and to produce vegetables in hydroponic cultivation of 100 m² (**D**), based on the threshold electrical conductivity of the irrigation water (EC) and discharge rate (Q). EC_{ad}: adequate electrical conductivity; EC_{nad}: inadequate electrical conductivity; Q_{ad}: adequate discharge; Q_{nad}: inadequate discharge.

For the association of tilapia *plus* glycophytes (0.4 ha of supplemental irrigation of annual crops, or 0.5 ha of forage cactus, or 100 m² hydroponic cultivation, or production of 2000 seedlings in nurseries), 2982 wells (47.5%) have adequate EC and Q, 1272 wells (20.2%) have adequate Q but the electrical conductivity is not adequate, 1345 wells (21.4%) only have adequate EC, and 685 wells (10.9%) do not have an adequate discharge rate or electrical conductivity (Figure 8A). For the association tilapia *plus* 0.5 ha of halophytes (Figure 8B) 3042 wells (48.4%) have EC and Q adequacy, 97 wells (1.5%) have inadequate EC and adequate Q, 3016 wells (48.0%) have adequate EC and inadequate Q, and 129 (2.1%) do not have adequate EC or adequate Q.

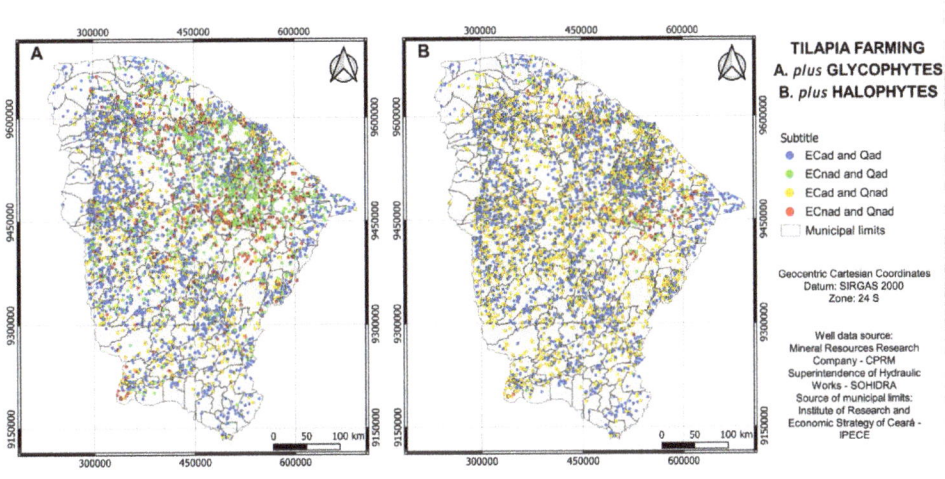

Figure 8. Potential of brackish groundwater in the State of Ceará (Northeast Brazil) for associations of fish (tilapia) *plus* glycophytes (**A**) and fish (tilapia) *plus* halophytes (**B**), based on the threshold electrical conductivity of the irrigation water (EC) and discharge rate (Q). EC_{ad}: adequate electrical conductivity; EC_{nad}: inadequate electrical conductivity; Q_{ad}: adequate discharge; Q_{nad}: inadequate discharge.

However, when only tilapia production was considered, the numbers were much more positive (Figure 9), since this production system in tanks requires less water and tilapia has a high tolerance to water salinity. For tilapia production in tanks with a total volume of up to 100 m³, 5047 wells (80.3%) are adequate in terms of EC and Q, 180 wells (2.9%) only have adequate Q, 1011 wells (16.1%) only have adequate EC, and 46 wells (0.7%) do not have adequate conditions for both criteria (Figure 9).

Table 5 summarizes all the data on the number of wells with respective suitability categories for the different biosaline systems evaluated.

Table 5. Number and percentage of wells for the productive systems tested and considering the different suitability criteria.

Production Systems	EC_{ad} and Q_{ad}*		EC_{nad} and Q_{ad}		EC_{ad} and Q_{nad}		EC_{nad} and Q_{nad}	
	Number of Wells	(%)	Number of Wells	(%)	Number of Wells	(%)	Number of Wells	(%)
Full irrigation of halophyte (0.5 ha)	3637	57.9	70	1.1	2539	40.4	38	0.6
Full irrigation of halophyte (1.0 ha)	2113	33.7	29	0.5	4063	64.6	79	1.2
Full irrigation of maize (0.5 ha)	1410	22.4	1729	27.5	1263	20.1	1882	30.0
Full irrigation of maize (1.0 ha)	794	12.6	839	13.3	1879	30.0	2772	44.1

Table 5. Cont.

Production Systems	ECad and Qad*		ECnad and Qad		ECad and Qnad		ECnad and Qnad	
	Number of Wells	(%)	Number of Wells	(%)	Number of Wells	(%)	Number of Wells	(%)
Supplemental irrigation of maize (0.5 ha)	3278	52.2	1248	19.8	1211	19.3	547	8.7
Supplemental irrigation of maize (1.0 ha)	2314	36.8	825	13.1	2175	34.6	970	15.5
Supplemental irrigation of sorghum (0.5 ha)	3898	62.0	628	10.0	1462	23.3	296	4.7
Supplemental irrigation of sorghum (1.0 ha)	2732	43.5	407	6.5	2628	41.8	517	8.2
Supplemental irrigation of cotton (0.5 ha)	3938	62.6	588	9.3	1477	23.5	281	4.5
Supplemental irrigation of cotton (1.0 ha)	2760	44.0	379	6.0	2655	42.2	490	7.8
Irrigation of forage cactus (0.5 ha)	3624	57.7	1603	25.5	703	11.2	354	5.6
Irrigation of forage cactus (1.0 ha)	2594	41.3	1108	17.6	1733	27.6	849	13.5
Irrigation of coconut (0.5 ha)	2592	41.2	217	3.5	3105	49.4	370	5.9
Irrigation of coconut (1.0 ha)	1225	19.5	88	1.4	499	7.9	4472	71.2
Irrigation of cashew (0.5 ha)	3501	55.7	554	8.8	1859	29.6	370	5.9
Irrigation of cashew (1.0 ha)	2208	35.1	316	5.0	608	9.7	3152	50.2
Coconut seedlings (2000 seedlings)	5195	82.7	1089	17.3	-	-	-	-
Cashew seedlings (2000 seedlings)	4327	68.8	1957	31.2	-	-	-	-
Caatinga Seedlings/ornamental (2000 seedlings)	3853	61.3	2431	38.7	-	-	-	-
Hydroponic cultivation (100 m²)	3624	57.7	1603	25.5	703	11.2	354	5.6
Tilapia farming *plus* glycophytes	2982	47.5	1272	20.2	1345	21.4	685	10.9
Tilapia farming *plus* halophytes	3042	48,4	97	1.5	3016	48.0	129	2.1

* EC_{ad}: adequate electrical conductivity; EC_{nad}: inadequate electrical conductivity; Q_{ad}: adequate discharge; Q_{nad}: inadequate discharge.

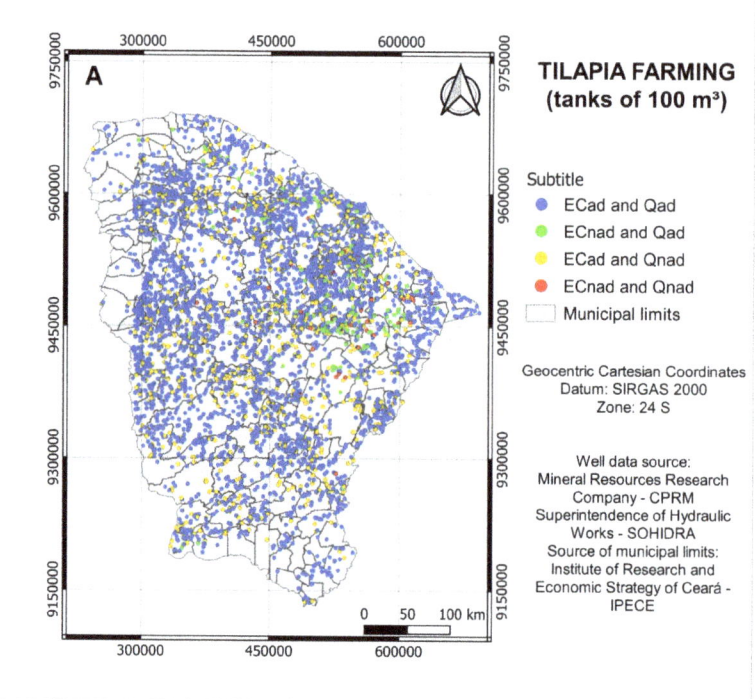

Figure 9. Potential of brackish groundwater in the State of Ceará (Northeast Brazil) for raising tilapia in tanks with a total volume of 100 m³, based on the threshold electrical conductivity of the irrigation water (EC) and discharge rate (Q). EC_{ad}: adequate electrical conductivity; EC_{nad}: inadequate electrical conductivity; Q_{ad}: adequate discharge; Q_{nad}: inadequate discharge.

3.5. Estimated Irrigated Area for Cropping Systems

Considering the sampling of wells with brackish water (6284 wells distributed throughout the State of Ceará), the discharge required by each production system, and the water productivity of each well, it was possible to estimate the potential irrigable area for each system, without restrictions in terms of salinity (Table 6). Forage cactus has the largest potential irrigable area of the evaluated systems, followed by systems with supplemental irrigation (cotton, sorghum, and maize), almost all of which has more than 9000 ha. The smallest irrigable areas were found in the irrigation systems for coconut and full irrigation for maize.

Table 6. Potential of unrestricted irrigable area with brackish water from wells in the Brazilian semi-arid region.

Production Systems	Irrigable Area (ha)
1. Full irrigation of halophytes (Atriplex, Sarcorcórnia)	6426
2. Full irrigation of maize	2566
3. Supplemental irrigation of maize	7970
4. Supplemental irrigation of sorghum	9213
5. Supplemental irrigation of cotton	9287
6. Irrigation of forage cactus	9660
7. Irrigation of coconut in sandy soils	4026
8. Irrigation of cashew in sandy soils	7197

4. Discussion

In this study, we sought to define the potential of brackish groundwater sources in the State of Ceará, a representative area of the Brazilian semi-arid region, based on the water productivity of groundwater wells, the salinity level of the water, and the water demand/salt tolerance of 15 production systems. The analyses showed that low discharge rates of brackish groundwater wells predominate in the Brazilian semi-arid region, a factor that reduces water availability and restricts enterprises to family farmers. However, it is possible to identify more promising production systems, due to their high capacity to tolerate water salinity levels, lower water demand, or both. Therefore, the expansion of biosaline agriculture in the tropical semi-arid region is not defined only by the salt tolerance of the crop.

Halophytes constitute the plant system with the highest tolerance to salinity among those analyzed, and can be irrigated with high salinity water, notably in soils with good natural drainage [33,34]. Halophytes can withstand high levels of salinity, while almost 99% of other plant species (glycophytes) have some level of sensitivity to salts [78,79]. Halophytes also show good adaptability to arid and semi-arid regions and can be cultivated under rainfed farming or in saline soils, and act as alternative forage plants, such as *Atriplex* spp., *Salicornia* spp. and *Distichlis palmeri* [80–82]. However, our data showed that about 65% of the wells with brackish water do not have enough water volume to irrigate an area of 1.0 ha with halophyte plants, although more than 95% have no salinity restriction (Figure 2). For the coconut, a salinity-tolerant glycophyte [83], the low discharge limitation affects 71% of the wells due to the high-water demand of this palm species, even though more than 80% of the water sources do not reach the salinity threshold for the crop (Figure 6). These results demonstrate that tolerance of crops to salinity is important, but it is not the only option to cope with salinity problems in semi-arid regions [84,85].

Annual crops are also characterized by relatively high-water demand and varying degrees of salt tolerance [28,86,87]. Among the annual species studied, maize is the most salt-sensitive [28,36] and has the highest water requirement. For this crop, it was observed that only 13% of the wells are adequate for the cultivation of 1.0 ha under full irrigation, while 87% are limited by low discharge rate, high salinity, or both (Figure 3). However, when supplemental irrigation is used, the numbers are much more positive, reaching a level of adequacy of about 37%. For cotton and sorghum crops, levels of adequacy with the use of supplemental irrigation were even higher (Figure 4), given the high salt tolerance of these two crops [39,88], compared to maize. The beneficial effects of supplemental irrigation with brackish water have been demonstrated in the Brazilian semi-arid region, with high efficiency in productive and economic terms [37]. Supplemental irrigation also results in low accumulation of salts in the soil, and increases photosynthesis rates and water use efficiency, demonstrating that this practice reduces the water deficit without causing significant salt damage to the crop [68]. In the context of biosaline agriculture, supplemental irrigation can reduce the losses of rainfed agriculture, especially in periods of dry spells, and is a decisive tool to deal with the limitations of the water availability in semi-arid zones both currently and when considering the future risks associated with global climate change [89,90]. In our study, the systems with supplemental irrigation of maize with brackish water resulted in larger irrigable areas than for the cultivation of halophytes (Table 6).

Plant production systems with lower water requirements presented the best results (Figure 5), even with moderate tolerance to salt stress. The forage cactus, a species with CAM metabolism [91,92], contrasted with the coconut palm trees, presenting adequacy of salinity and water availability of 41%, while coconut reached only 19.5%. Therefore, for perennial crops, it is recommended that species with low water demand should be used. For forage cactus (*Opuntia ficus*, *Nopalea cochenillifera*, *Opuntia* sp., *Opuntia stricta*) the total annual irrigation does not exceed 200 mm, with a low salt load applied to the soil during the dry season. Forage cactus is an important energy source for animal feed, as it has a high content of carbohydrates, total digestible nutrients, and water [93–95], and can also be component of multiple systems involving the use of brackish water and fish or in systems intercropped with gliricidia and other plant species [44,96,97]. The use of mulch and other water and soil conservation techniques are also recommended for the cultivation of forage cactus and other species when irrigated with brackish water [20,98,99], considering the need to increase water use efficiency in the tropical semi-arid region [13].

The cultivation of forage cactus and supplemental irrigation of annual crops also stand out in terms of the potential irrigable area without salinity restrictions (Table 6), surpassing more salinity-tolerant crops such as halophytes and coconut palm trees, and with lower risks of soil degradation. For forage cactus, the total irrigable area would be more than 9000 ha, obtained from 6284 wells with brackish water, which represent about 25% of the total number of wells contained in the database (25,497 wells). If we consider the total of 160,000 wells in the Brazilian semi-arid region [20], the total number of wells with adequate brackish water would reach around 44,000, resulting in greater irrigable areas for forage cactus and other productive systems. However, the size of the enterprise is limited by the discharge rate of each well, and effort must be made towards having a diversification of biosaline agriculture systems in areas of family farming, which can produce food to boost local businesses and fodder to feed the animals on farms. It should be noted that 65% of rural establishments in the Northeast Region of Brazil have an area of less than 10 ha, and 79% of establishments are classified as family farming [100], which strengthens the need for diversification of production systems as a way of achieving social and economic sustainability of family production.

Plant systems in nurseries reach high levels of adequacy (Figure 7), including coconut seedlings (83%), cashew seedlings (69%), and tree species native to the Caatinga biome/ornamental plants (61%). This decreasing suitability is in accordance with the salt tolerance of the species in the initial growth stage, which decreases in the same order [62,101–103]. For the hydroponic production of leafy vegetables, the adequacy level reached about 58%; this system has aroused the interest of researchers and farmers in the Brazilian semi-arid region [104–106]. Hydroponics can also be included in solar energy production systems, using semi-transparent panels [107].

The production of tilapia in tanks presents a high degree of adaptability to salinity and requires little water, reaching a degree of adequacy greater than 80% (Figure 9). According to [21], the potential for using brackish water in the Brazilian semi-arid region is higher for pisciculture than for agriculture. Fish farming can also be done in combination with plant systems, such as halophytes, supplemental irrigation, seedling production, hydroponic cultivation, and forage cactus production, with a high degree of adequacy, as biosaline systems (Figure 8). These combined systems allow the multiple use of water, as well as the use of nutrients contained in wastewater from fish farming [108–111]. Obviously, the degree of adequacy decreases for the combined crop, as demonstrated by comparing Figure 8 (associations of tilapia with vegetables) and Figure 9 (only tilapia cultivation). These differences are explained by the limitation of salinity in glycophytes or by the greater water demand of plant systems. The shrimp *Litopenaeus vannamei* is also tolerant of high salinity [112] and its cultivation has grown in Brazilian inland areas [82]. The option exists for it to form associations with glycophytes and halophytes using brackish water, despite requiring greater investment and more technology than tilapia farming. Therefore, associations between fish (or shrimp) and vegetables increase the opportunities for productive, economic, and environmental sustainability in semi-arid environments.

It is important to point out that the sustainability of production systems with brackish water requires adequate management and monitoring, especially in irrigated systems, considering the edaphoclimatic conditions in the Brazilian semi-arid region. Soil and water management techniques must be implemented, such as in situ water collection and soil cover (mulch), to increase the indicators of efficient use of water. Seasonal rains will favor the leaching of excess salts from the soil, avoiding environmental damage and losses to the farmer. However, soil fertility maintenance strategies are also recommended, such as the incorporation of crop residues, manure application, and use of wastewater sources present on farm, including treated wastewater. These techniques will contribute to the environmental sustainability of biosaline systems of family farming in semi-arid regions.

5. Conclusions

Our data show that the potential of brackish groundwater resources in the Brazilian semi-arid region depends on the set of data from the production system (salt tolerance and water demand) and the water source (water salinity and discharge rate). The joint analysis of these data shows that plant production systems with lesser water requirements (forage cactus, supplemental irrigation, hydroponic cultivation, and seedling production in nurseries) presented better results than more salt-tolerant species, including halophytes and coconut orchards.

About 41, 43, 58, 69, and 82% of wells have enough discharge rates to irrigate forage cactus (1.0 ha), sorghum (1.0 ha of supplemental irrigation), hydroponic cultivation, cashew seedlings, and coconut seedlings, respectively, without restrictions in terms of salinity. Otherwise, 65.8 and 71.2% of wells do not have enough water productivity to irrigate an area of 1.0 ha with halophytes and coconut palm trees, respectively, although more than 98.3 and 90.7% do not reach the salinity threshold for these crops. These results demonstrate that the use of quantitative and qualitative data generates more realistic information related to the potential of brackish water for agricultural purposes, and this type of evaluation should be recommended to other semi-arid regions worldwide.

Our data also indicate the need for diversification and for use of multiple systems on farms (intercropping with different plant species, association of fish/shrimp with plants, hydroponics/solar farming), to guarantee the sustainability of biosaline agriculture in the semi-arid regions, especially for family farming. However, an economic analysis of different systems should also be investigated in the future, indicating those that may result in greater net income for farmers. All these data will serve as a basis for formulating public policies aimed at the economic and social sustainability of family farming in tropical drylands.

Author Contributions: Conceptualization, C.I.N.L., C.F.d.L., A.L.R.N., C.C.d.A.C., A.O.d.S. and F.B.L.; methodology, C.I.N.L., C.F.d.L., A.L.R.N., A.O.d.S., C.C.d.A.C. and H.R.G.; investigation, C.I.N.L., H.C.S., R.d.S.N., C.F.d.L., A.L.R.N. and A.O.d.S.; writing—original draft preparation, C.I.N.L., C.F.d.L., H.C.S., R.d.S.N. and H.R.G.; writing—review and editing, C.I.N.L., C.C.d.A.C., G.G.d.S., R.N.T.C., F.B.L. and S.C.R.V.L.; project administration, C.F.d.L.; funding acquisition, C.F.d.L. and S.C.R.V.L. All authors have read and agreed to the published version of the manuscript.

Funding: This research was funded by Scientist Chief Program (Convênio 14/2022, ADECE-SEDET-FUNCAP), Fundação Cearense de Apoio ao Desenvolvimento Científico e Tecnológico–Funcap (Process number 08126425/2020), and Conselho Nacional de Desenvolvimento Científico e Tecnológico–CNPq (Process number 309174/2019-8).

Institutional Review Board Statement: Not applicable.

Data Availability Statement: Not applicable.

Acknowledgments: Acknowledgments are due to the Conselho Nacional de Desenvolvimento Científico e Tecnológico (CNPq), Agência de Desenvolvimento do Ceará (ADECE), Secretaria de Desenvolvimento Econômico e Trabalho do Ceará (SEDET), Fundação Cearense de Apoio ao Desenvovlimento Cinetífico e Tecnológico (FUNCAP), and Programa Cientista-Chefe, Brazil, for the financial support provided for this research and award of fellowship to the first author. The authors also thank the Superintendência de Obras Hidráulicas do Ceará (SOHIDRA) and the Serviço Geológico do Brasil (CPRM) that provided the database for this study.

Conflicts of Interest: The authors declare no conflict of interest.

References

1. Marengo, J.A.; Bernasconi, M. Regional differences in aridity/drought conditions over Northeast Brazil: Present state and future projections. *Clim. Chang.* **2015**, *129*, 103–115. [CrossRef]
2. Muralikrishnan, L.; Padaria, R.N.; Choudhary, A.K.; Dass, A.; Shokralla, S.; El-Abedin, T.K.Z.; Abdelmohsen, S.A.M.; Mahmoud, E.A.; Elansary, H.O. Climate change-induced drought impacts, adaptation and mitigation measures in semi-arid pastoral and agricultural watersheds. *Sustainability* **2022**, *14*, 6. [CrossRef]
3. Lopes, F.B.; Andrade, E.M.; Aquino, D.N.; Lopes, J.F.B. Proposta de um índice de sustentabilidade do Perímetro Irrigado Baixo Acaraú, Ceará, Brasil. *Rev. Ciênc. Agron.* **2009**, *40*, 185–193.
4. Cavalcante Júnior, R.G.; Freitas, M.A.V.; Silva, N.F.; Azevedo Filho, F.R. Sustainable groundwater exploitation aiming at the reduction of water vulnerability in the Brazilian semi-arid region. *Energies* **2019**, *12*, 904. [CrossRef]
5. Andrade, E.M.; Guerreiro, M.J.S.; Palácio, H.A.Q.; Campos, D.A. Ecohydrology in a Brazilian tropical dry forest: Thinned vegetation impact on hydrological functions and ecosystem services. *J. Hydrol. Reg. Stud.* **2020**, *27*, 100649. [CrossRef]
6. Silva, F.J.A.; Araújo, A.L.; Souza, R.O. Águas subterrâneas no Ceará—poços instalados e salinidade. *Rev. Tecnol.* **2007**, *28*, 136–159.
7. Batistão, A.C.; Holthusen, D.; Reichert, J.M.; Portela, J.C. Soil solution composition affects microstructure of tropical saline alluvial soils in semi-arid environment. *Soil Tillage Res.* **2020**, *203*, 104662. [CrossRef]
8. Santacruz-de León, G.; Moran-Ramírez, J.; Ramos-Leal, J.A. Impact of drought and groundwater quality on agriculture in a semi-arid zone of Mexico. *Agriculture* **2022**, *12*, 1379. [CrossRef]
9. Alvalá, R.C.S.; Cunha, A.P.M.A.; Brito, S.S.B.; Seluchi, M.E.; Marengo, J.A.; Moraes, O.L.L.; Carvalho, M.A. Drought monitoring in the Brazilian Semiarid region. *An. Acad Bras. Ciênc.* **2019**, *91*, e20170209. [CrossRef]
10. Melati, M.D.; Fleischmann, A.S.; Fan, F.M.; Paiva, R.C.D.; Athayde, G.B. Estimates of groundwater depletion under extreme drought in the Brazilian semi-arid region using GRACE satellite data: Application for a small-scale aquifer. *Hydrogeol. J.* **2019**, *27*, 2789–2802. [CrossRef]
11. Marengo, J.A.; Torres, R.R.; Alves, L.M. Drought in Northeast Brazil—past, present, and future. *Theor. Appl Climatol.* **2017**, *129*, 1189–1200. [CrossRef]
12. Gao, X.; Qu, Z.; Huo, Z.; Tang, P.; Qiao, S. Understanding the role of shallow groundwater in improving field water productivity in arid areas. *Water* **2020**, *12*, 3519. [CrossRef]
13. Frizzone, J.A.; Lima, S.C.R.V.; Lacerda, C.F.; Mateos, L. Socio-economic indexes for water use in irrigation in a representative basin of the tropical semiarid region. *Water* **2021**, *13*, 2643. [CrossRef]
14. Andrade, E.M.; Lopes, F.B.; Palácio, H.A.Q.; Aquino, D.N.; Alexandre, D.M.B. Land use and groundwater quality: The case of Baixo Acaraú Irrigated Perimeter, Brazil. *Rev. Ciênc. Agron.* **2010**, *41*, 208–215. [CrossRef]
15. Peter, S.; de Araújo, J.C.; Araújo, N.; Herrmann, H.J. Flood avalanches in a semiarid basin with a dense reservoir network. *J. Hydrol.* **2014**, *512*, 408–420. [CrossRef]
16. Nunes, K.G.; Costa, R.N.T.; Cavalcante, I.N.; Gondim, R.S.; Lima, S.C.R.V.; Mateos, L. Groundwater resources for agricultural purposes in the Brazilian semi-arid region. *Rev. Bras. Eng. Agrícola Ambient.* **2022**, *26*, 915–923. [CrossRef]

17. ANA—Agência Nacional de Águas e Saneamento Básico. 2021. Available online: https://www.gov.br/ana/pt-br (accessed on 3 October 2022).
18. CPRM—Companhia de Pesquisa de Recursos Minerais. SIAGAS—Sistema de Informações de Águas Subterrâneas. November 2022. Available online: http://siagasweb.cprm.gov.br/layout/index.php (accessed on 20 November 2020).
19. Silva Júnior, L.G.A.; Gheyi, H.R.; Medeiros, J.F. Composição química de águas do cristalino do nordeste brasileiro. *Rev. Bras. Eng. Agrícola Ambient.* **1999**, *3*, 11–17. [CrossRef]
20. Lacerda, C.F.; Gheyi, H.R.; Medeiros, J.F.; Costa, R.N.T.; Sousa, G.G.; Lima, G.S. Strategies for the use of brackish water for crop production in Northeastern Brazil. In *Saline and Alkaline Soils in Latin America*; Taleisnik, E., Lavado, R.S., Eds.; Springer: New York, NY, USA, 2021; pp. 71–99.
21. Amaral, K.D.S.; Navoni, J.A. Desalination in rural communities of the Brazilian semi-arid region: Potential use of brackish concentrate in local productive activities. *Proces. Saf. Environ. Prot.* **2023**, *169*, 61–70. [CrossRef]
22. Silva, J.E.S.B.; Matias, J.R.; Guirra, K.S.; Aragão, C.A.; Araújo, G.G.L.; Dantas, B.F. Development of seedlings of watermelon cv. Crimson Sweet irrigated with biosaline water. *Rev. Bras. Eng. Agrícola Ambient.* **2015**, *19*, 835–840. [CrossRef]
23. Munns, R. Comparative physiology of salt and water stress. *Plant Cell Environ.* **2002**, *25*, 239–250. [CrossRef]
24. Zörb, C.; Geilfus, C.M.; Dietz, K.J. Salinity and crop yield. *Plant Biol.* **2018**, *21*, 31–38. [CrossRef] [PubMed]
25. Ferreira, J.F.S.; Liu, X.; Suddarth, S.R.P.; Nguyen, C.; Sandhu, D. NaCl accumulation, shoot biomass, antioxidant capacity, and gene expression of *Passiflora edulis* f. Flavicarpa Deg. in response to irrigation waters of moderate to high salinity. *Agriculture* **2022**, *12*, 1856. [CrossRef]
26. Liang, W.; Ma, X.; Wan, P.; Liu, L. Plant salt-tolerance mechanism: A review. *Biochem. Biophys. Res. Commun.* **2018**, *495*, 286–291. [CrossRef] [PubMed]
27. Torres Mendonça, A.J.; de Silva, A.A.R.; Lima, G.S.D.; de Soares, L.A.A.; Nunes Oliveira, V.K.; Gheyi, H.R.; de Lacerda, C.F.; de Azevedo, C.A.V.; de Lima, V.L.A.; Fernandes, P.D. Salicylic acid modulates okra tolerance to salt stress in hydroponic system. *Agriculture* **2022**, *12*, 1687. [CrossRef]
28. Ayers, R.S.; Westcot, D.W. *Water Quality for Agriculture*; Food and Agriculture Organization of the United Nations (FAO): Rome, Italy, 1985; p. 174.
29. FAO; AWC. *Guidelines for Brackish Water Use for Agricultural Production in the Near East and North Africa Region*; FAO: Cairo, Egypt, 2023. [CrossRef]
30. SUDENE—Superintendência do Desenvolvimento do Nordeste. Nova Delimitação do Semiárido. 2017. Available online: http://antigo.sudene.gov.br/images/arquivos/semiarido/arquivos/Rela%C3%A7%C3%A3o_de_Munic%C3%ADpios_Semi%C3%A1rido.pdf (accessed on 17 November 2020).
31. Alvares, C.A.; Stape, J.L.; Sentelhas, P.C.; Gonçalves, J.L.M.; Sparovek, G. Köppen's climate classification map for Brazil. *Meteorol. Z.* **2013**, *22*, 711–728. [CrossRef] [PubMed]
32. CODEVASF—Companhia de Desenvolvimento dos Vales do São Francisco e do Parnaíba. Caderno de caracterização: Estado do Ceará. 2022. Available online: https://www.codevasf.gov.br/acesso-a-informacao/institucional/biblioteca-geraldo-rocha/publicacoes/outras-publicacoes/caderno-de-caracterizacao-estado-do-ceara.pdf (accessed on 18 October 2022).
33. Porto, E.R.; Amorim, M.C.C.; Dutra, M.T.; Paulino, R.V.; Brito, L.T.L.; Matos, A.N.B. Rendimento da *Atriplex nummularia* irrigada com efluentes da criação de tilápia em rejeito da dessalinização da água. *Rev. Bras. Eng. Agrícola Ambient.* **2006**, *10*, 97–100. [CrossRef]
34. Costa, C.S.B.; Bonilla, O.H. Halófitas brasileiras: Formas de cultivo e uso. In *Manejo da Salinidade na Agricultura: Estudos Básicos e Aplicados*; Gheyi, H.R., Dias, N.S., Lacerda, C.F., Gomes Filho, É., Eds.; INCTSal: Fortaleza, Brazil, 2016; pp. 243–258.
35. Lacerda, C.F.; Sousa, G.G.; Silva, F.L.B.; Guimarães, F.V.A.; Silva, G.L.; Cavalcante, L.F. Soil salinization and maize and cowpea yield in the crop rotation system using saline waters. *Eng. Agríc.* **2011**, *31*, 663–675. [CrossRef]
36. Barbosa, F.S.; Lacerda, C.F.; Gheyi, H.R.; Farias, G.C.; Silva Júnior, R.J.C.; Lage, Y.A.; Hernandez, F.F.F. Productivity and íon content in maize irrigated with saline water in a continuous or alternating system. *Ciên. Rural* **2012**, *45*, 1731–1737. [CrossRef]
37. Cavalcante, E.S.; Lacerda, C.F.; Costa, R.N.T.; Gheyi, H.R.; Pinho, L.L.; Bezerra, F.M.S.; Oliveira, A.C.; Canjá, J.F. Supplemental irrigation using brackish water on maize in tropical semi-arid regions of Brazil: Yield and economic analysis. *Sci. Agric.* **2021**, *78*, e20200151. [CrossRef]
38. Cavalcante, Í.H.L.; Oliveira, F.A.; Cavalcante, L.F.; Beckmann, M.Z.; Campos, M.C.C.; Gondim, S.C. Crescimento e produção de duas cultivares de algodão irrigadas com água salinizadas. *Rev. Bras. Eng. Agrícola Ambient.* **2005**, *9*, 108–111. [CrossRef]
39. Soares, L.A.A.; Fernandes, P.D.; Lima, G.S.; Gheyi, H.R.; Nobre, R.G.; Sá, F.V.S.; Moreira, R.C.L. Saline water irrigation strategies in two production cycles of naturally colored cotton. *Irrig. Sci.* **2020**, *38*, 401–413. [CrossRef]
40. Vieira, M.R.; Lacerda, C.F.; Cândido, M.J.D.; Carvalho, P.L.; Costa, R.N.T.; Silva, J.N. Produtividade e qualidade da forragem de sorgo irrigadas com água salina. *Rev. Bras. Eng. Agrícola Ambient.* **2005**, *9*, 42–46. [CrossRef]
41. Guimarães, M.J.M.; Simões, W.L.; Salviano, A.M.; Oliveira, A.R.; Silva, J.S.; Barros, J.R.A.; Willadino, L. Management for grain sorghum cultivation under saline water irigation. *Rev. Bras. Eng. Agrícola Ambient.* **2022**, *26*, 755–762. [CrossRef]
42. Fonseca, V.A.; Santos, M.R.; Silva, J.A.; Donato, S.L.R.; Rodrigues, C.S.; Brito, C.F.B. Morpho-phusiology, yield, and water-use efficiency of Opuntia fícus-indica irrigated with saline water. *Acta Sci. Agron.* **2019**, *41*, e42631. [CrossRef]
43. Santos, N.S.; Silva, J.C.S.; Pereira, W.S.; Melo, J.L.R.; Lima, K.V.; Lima, D.O.; Lima, K.F.; Almeida, R.S. Crescimento da palma forrageira sob estresse salino e diferentes lâminas de irrigação. *Rev. Craibeiras Agroecol.* **2020**, *5*, e9452.

44. Araújo, G.G.L.; Silva, T.G.F.; Campos, F.S. Agricultura biossalina e uso de águas salobras na produção de forragem. In *Agricultura Irrigada em Ambientes Salinos*; Cerqueira, P.R.S., Lacerda, C.F., Araújo, G.G.L., Gheyi, H.R., Simões, W.L., Eds.; CODEVASF: Brasília, Brazil, 2021; Volume 1, pp. 174–211.
45. Pereira, M.C.A.; Azevedo, C.A.V.; Dantas Neto, J.; Pereira, M.O.; Ramos, J.G.; Nunes, K.G.; Lyra, G.B.; Saboya, L.M.F. Production of forage palm cultivars (Orelha de Elefante Mexicana, IPA-Sertânia and Miúda) under different salinity levels in irrigation water. *Aust. J. Crop Sci.* **2021**, *15*, 977–982. [CrossRef]
46. Carneiro, P.T.; Fernandes, P.D.; Gheyi, H.R.; Soares, F.A.L. Germinação e crescimento inicial de genótipos de cajueiro anão-precoce em condições de salinidade. *Rev. Bras. Eng. Agrícola Ambient.* **2002**, *6*, 199–206. [CrossRef]
47. Guilherme, E.A.; Lacerda, C.F.; Bezerra, M.A.; Prisco, J.T.; Gomes Filho, E. Desenvolvimento de plantas adultas de cajueiro anão precoce irrigadas com águas salinas. *Rev. Bras. Eng. Agrícola Ambient.* **2005**, *9*, 253–257. [CrossRef]
48. Araújo, L.F.; Lima, R.E.M.; Costa, L.O.; Silveira, Ê.M.C.; Bezerra, M.A. Alocação de íons e crescimento de plantas de cajueiro anão-precoce irrigadas com água salina no campo. *Rev. Bras. Eng. Agrícola Ambient.* **2014**, *18*, S34–S38. [CrossRef]
49. Ferreira Neto, M.; Gheyi, H.R.; Holanda, J.S.; Medeiros, J.F.; Fernandes, P.D. Qualidade do fruto verde de coqueiro em função da irrigação com água salina. *Rev. Bras. Eng. Agrícola Ambient.* **2002**, *6*, 69–75. [CrossRef]
50. Marinho, F.J.L.; Ferreira Neto, M.; Gheyi, H.R.; Fernandes, P.D.; Viana, S.B.A. Uso de água salina na irrigação do coqueiro (*Cocus nucifera* L.). *Rev. Bras. Eng. Agrícola Ambient.* **2005**, *9*, 359–364. [CrossRef]
51. Gheyi, H.R.; Ferreira Neto, M.; Marinho, F.J.L.; Fernandes, P.D.; Holanda, J.S.; Soares, F.A.L. Response of green coconut Anão under saline water irrigation. In Proceedings of the 20th Congress on Irrigation and Drainage, Lahore, Pakistan, 13–18 October 2008; Volume 1, pp. 675–692.
52. Silva, M.G.; Oliveira, I.S.; Soares, T.M.; Gheyi, H.R.; Santana, G.O.; Pinho, J.S. Growth, production and water consumption of coriander in hydroponic system using brackish waters. *Rev. Bras. Eng. Agrícola Ambient.* **2018**, *22*, 547–552. [CrossRef]
53. Soares, H.R.; Silva, Ê.F.F.; Silva, G.F.; Cruz, A.F.S.; Santos Júnior, J.A.; Rolim, M.M. Salinity and flow rates of nutrients solution on cauliflower biometrics in NFT hydroponic system. *Rev. Bras. Eng. Agrícola Ambient.* **2020**, *24*, 258–265. [CrossRef]
54. Sousa, C.A.; Silva, A.O.; Santos, J.S.G.; Lacerda, C.F.; Silva, G.F. Production of watercress with brackish water and different circulation times for the nutrient solution. *Rev. Ciênc. Agron.* **2020**, *51*, e20196775. [CrossRef]
55. Silva, M.G.; Silva, P.C.C.; Cova, A.M.W.; Gheyi, H.R.; Soares, T.M. Experiências com o uso de águas salobras em hidroponia no nordeste brasileiro. In *Agricultura Irrigada em Ambientes Salinos*; Cerqueira, P.R.S., Lacerda, C.F., Araújo, G.G.L., Gheyi, H.R., Simões, W.L., Eds.; CODEVASF: Brasília, Brazil, 2021; Volume 1, pp. 290–310.
56. Marinho, F.J.L.; Gheyi, H.R.; Fernandes, P.D. Germinação e formação de mudas de coqueiro irrigadas com água salina. *Rev. Bras. Eng. Agrícola Ambient.* **2005**, *9*, 334–340. [CrossRef]
57. Lima, B.L.C. Respostas Fisiológicas e Morfométricas na Produção de Mudas de Coqueiro Anão Irrigado com Água Salina. Dissertação (Mestrado em Engenharia Agrícola)—Universidade Federal do Ceará, Fortaleza. 2014. Available online: https://repositorio.ufc.br/bitstream/riufc/10566/1/2014_dis_blclima.pdf (accessed on 10 September 2022).
58. Sousa, A.B.O.; Bezerra, M.A.; Farias, F.C. Germinação e desenvolvimento inicial de clones de cajueiro comum sob irrigação com água salina. *Rev. Bras. Eng. Agrícola Ambient.* **2011**, *15*, 390–394. [CrossRef]
59. Freitas, V.S.; Marques, E.C.; Bezerra, M.A.; Prisco, J.T.; Gomes-Filho, E. Crescimento e acúmulo de íons em plantas de cajueiro anão precoce em diferentes tempos de exposição à salinidade. *Semin. Cienc. Agrar.* **2013**, *34*, 3341–3352. [CrossRef]
60. Sousa, V.F.O.; Santos, G.L.; Maia, J.M.; Maia Júnior, S.O.; Santos, J.P.O.; Costa, J.E.; Silva, A.F.; Dias, T.J.; Ferreira-Silva, S.L.; Taniguchi, C.A.K. Salinity-tolerant dwarf cashew rootstock has better ionic homeostasis and morphophysiological performance of seedlings. *Rev. Bras. Eng. Agrícola Ambient.* **2023**, *27*, 92–100. [CrossRef]
61. Sousa Neto, O.N.; Dias, N.S.; Ferreira Neto, M.; Lira, R.B.; Rebouças, J.R.L. Utilização do rejeito da dessalinização da água na produção de mudas de espécies da Caatinga. *Rev. Caatinga* **2011**, *24*, 123–129.
62. Oliveira, E.V.; De Lacerda, C.F.; Neves, A.L.R.; Gheyi, H.R.; Oliveira, D.R.; De Oliveira, F.Í.F.; De Araújo Viana, T.V. A new method to evaluate salt tolerance of ornamental plants. *Theor. Exp. Plant Physiol.* **2018**, *30*, 173–180. [CrossRef]
63. Lacerda, C.F.; Oliveira, E.V.; Neves, A.L.R.; Gheyi, H.R.; Bezerra, M.A.; Costa, C.A.G. Morphophysiological responses and mechanisms of salt tolerance in four ornamental perennial species under tropical climate. *Rev. Bras. Eng. Agrícola Ambient.* **2020**, *24*, 656–663. [CrossRef]
64. Santos, J.W.G.; Lacerda, C.F.; Oliveira, A.C.; Mesquita, R.O.; Bezerra, A.M.E.; Marques, E.S.; Neves, A.L.R. Quantitative and qualitative responses of *Euphorbia milii* and *Zamioculcas zamiifolia* exposed to different levels of salinity and luminosity. *Rev. Ciênc. Agron.* **2022**, *53*, e20218070. [CrossRef]
65. Likongwe, J.S.; Stecko, T.D.; Stauffer Junior, J.R.; Carline, R.F. Combined effects of water temperature and salinity on growth and feed utilization of juvenile Nile tilapia *Oreochromis niloticus* (Linneaus). *Aquaculture* **1996**, *146*, 37–46. [CrossRef]
66. Kamal, A.H.M.M.; Mair, G.C. Salinity tolerance in superior genotypes of tilapia, *Oreochromis niloticus*, *Oreochromis mossambicus* and their hybrids. *Aquaculture* **2005**, *247*, 189–201. [CrossRef]
67. Souza, A.C.M.; Dias, N.S.; Arruda, M.V.M.; Fernandes, C.S.; Alves, H.R.; Nobre, G.T.N.; Peixoto, M.L.L.F.; Sousa Neto, O.N.; Silva, M.R.F.; Silva, F.V.; et al. Economic analysis and development of the nile tilapia cultivated in the nursery using reject brine as water support. *Water Air Soil Pollut.* **2022**, *233*, 8. [CrossRef]

68. Cavalcante, E.S.; Lacerda, C.F.; Mesquita, R.O.; de Melo, A.S.; da Silva Ferreira, J.F.; dos Santos Teixeira, A.; Lima, S.C.R.V.; da Silva Sales, J.R.; de Souza Silva, J.; Gheyi, H.R. Supplemental irrigation with brackish water improves carbon assimilation and water use efficiency in maize under tropical dryland conditions. *Agriculture* **2022**, *12*, 544. [CrossRef]
69. EMBRAPA—Empresa Brasileira de Pesquisa Agropecuária. Cultivo do Milho. 2010. Available online: https://ainfo.cnptia.embrapa.br/digital/bitstream/item/81707/1/Manejo-irrigacao.pdf (accessed on 8 October 2022).
70. Bastos, E.A.; Ferreira, V.M.; Silva, C.R.; Andrade Júnior, A.S. Evapotranspiração e coeficiente de cultivo do feijão-caupi no vale do Gurguéia, Piauí. *Irriga* **2008**, *13*, 182–190. [CrossRef]
71. EMBRAPA—Empresa Brasileira de Pesquisa Agropecuária. Sorgo: Oprodutor Pergunta, a Embrapa Responde. 2015. Available online: https://ainfo.cnptia.embrapa.br/digital/bitstream/item/215310/1/500-perguntas-sorgo.pdf (accessed on 2 October 2022).
72. Pereira, M.O. Desempenho Agronômico da Palma Forrageira sob Lâminas de Irrigação e Níveis de Salinidade da Água. Ph.D. Thesis, Universidade Federal de Campina Grande, Campina Grande, Brazil, 2020. Available online: http://dspace.sti.ufcg.edu.br:8080/xmlui/bitstream/handle/riufcg/12665/MARIANA%20DE%20OLIVEIRA%20PEREIRA%20-%20TESE%20%28PPGEA%29%202020.pdf?sequence=3&isAllowed=y (accessed on 13 September 2022).
73. Freitas, W.F.R. *Palma Forrageira Opuntia Fícus-Indica (L.) Mill e Nopalea Cochenilifera (L.) Salm-Dyck*; Universidade Federal de Campina Grande: Campina Grande, Brazil, 2021.
74. EMBRAPA—Empresa Brasileira de Pesquisa Agropecuária. Sistemas de Produção—Cultivo do Cajueiro Anão Precoce. 2008; 44p, Available online: https://www.infoteca.cnptia.embrapa.br/infoteca/bitstream/doc/421404/1/Sp012aed.pdf (accessed on 5 October 2022).
75. Nogueira, L.C.; Nogueira, L.R.Q.; Miranda, F.R. Irrigação do coqueiro. In *A Cultura do Coqueiro no Brasil*; Ferreira, J.M.S., Warwick, D.R.N., Siqueira, L.A., Eds.; EMBRAPA/CPATC: Brasília, Brazil, 2018; pp. 159–187.
76. EMBRAPA—Empresa Brasileira de Pesquisa Agropecuária. Adubando para Alta Produtividade e Qualidade: Fruteiras Tropicais do Brasil. 2009. Available online: https://www.embrapa.br/busca-de-publicacoes/-/publicacao/658334/adubando-para-alta-produtividade-e-qualidade-fruteiras-tropicais-do-brasil (accessed on 1 October 2022).
77. QGIS Development Team. *QGIS Geographic Information System*; Open Source Geospatial Foundation: Girona, Spain, 2022.
78. Flowers, T.J.; Galal, H.K.; Bromham, L. Evolution of halophytes: Multiple origins of salt tolerance in land plants. *Funct. Plant Biol.* **2010**, *37*, 604–612. [CrossRef]
79. Holguin Peña, R.J.; Medina-Hernández, D.; Ghasemi, M.; Rueda Puente, E.O. Salt tolerant plants as a valuable resource for sustainable food production in arid and saline coastal zones. *Acta Bio. Colomb.* **2021**, *26*, 116–126. [CrossRef]
80. Masters, D.G.; Benes, S.E.; Norman, C. Biosaline agriculture for forage and livestock production. *Agric. Ecosyst. Environ.* **2007**, *119*, 234–248. [CrossRef]
81. Panta, S.; Flowers, T.; Lane, P.; Doyle, R.; Haros, G.; Shabala, S. Halophyte agriculture: Success stories. *Environ. Exp. Bot.* **2014**, *107*, 71–83. [CrossRef]
82. Gheyi, H.R.; Lacerda, C.F.; Freire, M.B.G.S.; Costa, R.N.T.; Souza, E.R.; Silva, A.O.; Fracetto, G.G.M.; Cavalcante, L.F. Management and reclamation of salt-affected soils: General assessment and experiences in the Brazilian semiarid region. *Rev. Ciênc. Agron.* **2022**, *53*, e20217917. [CrossRef]
83. Santos, M.M.S.; Lacerda, C.F.; Neves, A.L.R.; Sousa, C.H.C.; Ribeiro, A.A.; Bezerra, M.A.; Araújo, I.C.S.; Gheyi, H.R. Ecophysiology of the tall coconut growing under different coastal areas of northeastern Brazil. *Agric. Water Manag.* **2020**, *232*, 106047. [CrossRef]
84. Medeiros, W.J.F.; Oliveira, F.Í.F.; Lacerda, C.F.; Sousa, C.H.C.; Cavalcante, L.F.; Silva, A.R.A.; Ferreira, J.F.S. Isolated and combined effects of soil salinity and waterlogging in seedlings of 'Green Dwarf' coconut. *Semin. Cienc. Agrar.* **2018**, *39*, 1459–1468. [CrossRef]
85. Rengasamy, P.; Lacerda, C.F.; Gheyi, H.R. Salinity, Sodicity and Alkalinity. In *Subsoil Constraints for Crop Production*; Oliveira, T.S., Bell, R.W., Eds.; Springer: Cham, Switzerland, 2022; pp. 83–107.
86. de Ribeiro, A.A.; de Lacerda, C.F.; Neves, A.L.R.; Sousa, C.H.C.; dos Braz, R.S.; de Oliveira, A.C.; Pereira, J.M.G.; de Ferreira, J.F.S. Uses and losses of nitrogen by maize and cotton plants under salt stress. *Arch. Agron. Soil Sci.* **2020**, *67*, 1119–1133. [CrossRef]
87. Dourado, P.R.M.; de Souza, E.R.; Santos, M.A.; Lins, C.M.T.; Monteiro, D.R.; Paulino, M.K.S.S.; Schaffer, B. Stomatal regulation and osmotic adjustment in sorghum in response to salinity. *Agriculture* **2022**, *12*, 658. [CrossRef]
88. Hostetler, A.N.; Govindarajulu, R.; Hawkins, J.S. QTL mapping in an interspecific sorghum population uncovers candidate regulators of salinity tolerance. *Plant Stress* **2021**, *2*, 100024. [CrossRef]
89. Chauhan, C.P.S.; Singh, R.B.; Gupta, S.K. Supplemental irrigation of wheat with saline water. *Agric. Water Manag.* **2008**, *95*, 253–258. [CrossRef]
90. Nangia, V.; Oweis, T.; Kemeze, F.H.; Schnetzer, J. Supplemental Irrigation: A Promising Climate-Smart Practice for Dryland Agriculture. Wageningen. CGIAR/CCAFS. 2018. Available online: https://cgspace.cgiar.org/bitstream/handle/10568/92142/GACSA%20Practice%20Brief%20Supplemental%20Irrigation.pdf (accessed on 10 November 2020).
91. Borland, A.M.; Griffiths, H.; Harwell, J.; Smith, J.A.C. Exploiting the potential of plants with crassulacean acid metabolism for bioenergy production on marginal lands. *J. Exp. Bot.* **2009**, *60*, 2879–2896. [CrossRef]
92. Cushman, J.C.; Davis, S.C.; Yang, X.H.; Borland, A.M. Development and use of bioenergy feedstocks for semi-arid and arid lands. *J. Exp. Bot.* **2015**, *66*, 4177–4193. [CrossRef]
93. Costa, R.G.; Trevino, I.H.; Medeiros, G.R.; Medeiros, A.N.; Pinto, T.F.; Oliveira, R.L. Effects of replacing corn with cactus pear (*Opuntia ficus indica* Mill) on the performance of Santa Inês lambs. *Small Rumin. Res.* **2012**, *102*, 13–17. [CrossRef]

94. Rocha Filho, R.R.; Santos, D.C.; Véras, A.S.C.; Siqueira, M.C.B.; Novaes, L.P.; Mora-Luna, R.; Monteiro, C.C.F.; Ferreira, M.A. Can spineless forage cactus be the queen of forage crops in dryland areas? *J. Arid Environ.* **2021**, *186*, 104426. [CrossRef]
95. Ravari, F.N.; Tahmasbi, R.; Dayani, O.; Khezri, A. Cactus-alfalfa blend silage as an alternative feedstuff for Saanen dairy goats: Effect on feed intake, milk yield and components, blood and rumen parameters. *Small Rumin. Res.* **2022**, *216*, 106811. [CrossRef]
96. Miranda, K.R.; Dubeux Junior, J.C.B.; Mello, A.C.L.; Silva, M.C.; Santos, M.V.F.; Santos, D.C. Forage production and mineral composition of cactus intercropped with legumes and fertilized with different sources of manure. *Ciên. Rural* **2019**, *49*, e20180324. [CrossRef]
97. Silva Brito, G.S.M.; Santos, E.M.; de Araújo, G.G.L.; de Oliveira, J.S.; Zanine, A. de M.; Perazzo, A.F.; Campos, F.S.; Lima, A.G.V. de O.; Cavalcanti, H.S. Mixed silages of cactus pear and gliricidia: Chemical composition, fermentation characteristics, microbial population and aerobic stability. *Sci. Rep.* **2020**, *10*, 6834. [CrossRef]
98. Souza, M.S.; Freire da Silva, T.G.; De Souza, L.S.B.; Alves, H.K.M.N.; Leite, R.M.C.; De Souza, C.A.A.; Araújo, G.G.L.; Campos, F.S.; Silva, M.J.; De Souza, P.J.O.P. Growth, phenology and harvesting time of cactus-millet intercropping system under biotic mulching. *Arch. Agro Soil Sci.* **2020**, *67*, 764–778. [CrossRef]
99. Alves, H.K.M.N.; Silva, T.G.F.; Jardim, A.M.R.F.; Souza, L.S.B.; Araújo Júnior, J.N.; Souza, C.A.A.; Moura, M.S.B.; Araújo, G.G.L.; CAMPOS, F.S.; Cruz Neto, J.F. The use of mulch in cultivating the forage cactus optimizes yield in less time and increases the water use efficiency of the crop. *Irrig. Drain.* **2022**, *4*, 75–89. [CrossRef]
100. IPEA—Instituto de Pesquisa Econômica Aplicada. Agricultura Nordestina: Análise Comparativa Entre os Censos Agropecuários de 2006 e 2017. 2021; 42p. Available online: http://repositorio.ipea.gov.br/handle/11058/10758 (accessed on 29 September 2022).
101. Abreu, C.E.B.; Prisco, J.T.; Nogueira, A.R.C.; Lacerda, C.F.; Gomes-Filho, E.G. Physiological and biochemical changes occurring in dwarf-cashew seedlings subjected to salt stress. *Braz. J. Plant Physiol.* **2008**, *20*, 105–108. [CrossRef]
102. Lima, B.L.C.; Lacerda, C.F.; Ferreira Neto, M.; Ferreira, J.F.S.; Bezerra, A.M.E.; Marques, E.C. Physiological and ionic changes in dwarf coconut seedlings irrigated with saline water. *Rev. Bras. Eng. Agrícola Ambient.* **2017**, *21*, 122–127. [CrossRef]
103. Bessa, M.C.; Lacerda, C.F.; Amorim, A.V.; Bezerra, A.M.E.; Lima, A.D. Mechanisms of salt tolerance in seedlings of six woody native species of the Brazilian semi-arid. *Rev. Ciênc. Agron.* **2017**, *48*, 157–165. [CrossRef]
104. Leal, L.Y.C.; Souza, E.R.; Santos Júnior, J.A.; Santos, M.A. Comparison of soil and hydroponic cultivation systems for spinach irrigated with brackish water. *Sci. Hortic.* **2020**, *274*, 109616. [CrossRef]
105. Bione, M.A.A.; Soares, T.M.; Cova, A.M.W.; Paz, V.P.S.; Gheyi, H.R.; Rafael, M.R.S.; Modesto, F.J.N.; Santana, J.A.; Neves, B.S.L. Hydroponic production of 'Biquinho' pepper with brackish water. *Agric. Water Manag.* **2021**, *245*, 106607. [CrossRef]
106. Navarro, F.E.C.; Santos Junior, J.A.; Martins, J.B.; Cruz, R.I.F.; Silva, M.M.; Medeiros, S.S. Physiological aspects and production of coriander using nutrient solutions prepared in different brackish waters. *Rev. Bras. Eng. Agrícola Ambient.* **2022**, *26*, 831–839. [CrossRef]
107. Xu, Z.; Elomri, A.; Al-Ansari, T.; Kerbache, L.; El Mekkawy, T. Decisions on design and planning of solar-assisted hydroponic farms under various subsidy schemes. *Renew. Sustain. Energy Rev.* **2022**, *156*, 111958. [CrossRef]
108. Graber, A.; Junge, R. Aquaponic systems: Nutrient recycling from fish wastewater by vegetable production. *Desalination* **2009**, *246*, 147–156. [CrossRef]
109. Kaburagi, E.; Yamada, M.; Baba, T.; Fujiyama, H.; Murillo-Amador, B.; Yamada, S. Aquaponics using saline groundwater: Effect of adding microelements to fish wastewater on the growth of Swiss chard (*Beta vulgaris* L. spp. cicla). *Agric. Water Manag.* **2020**, *227*, 105851. [CrossRef]
110. Kien, T.T.; Thao, N.T.P.; Thanh, T.V.; Hieu, T.T.; Son, L.T.; Schnitzer, H.; Luu, T.L.; Hai, L.T. Nitrogen conversion efficiency in the integrated catfish farming system toward closed ecosystem in Mekong delta, Vietnam. *Process Saf. Environ. Prot.* **2022**, *168*, 180–188. [CrossRef]
111. Andrade, M.S.; Sousa, J.F.; Morais, M.B.; Albuquerque, C.C. Saline pisciculture effluent as an alternative for irrigation of *Croton blanchetianus* (Euphorbiaceae). *Rev. Bras. Eng. Agrícola e Ambient.* **2023**, *27*, 256–263. [CrossRef]
112. Maicá, P.F.; Borba, M.R.; Martins, T.G.; Wasielesky Junior, W. Effect of salinity on performance and body composition of Pacific white shrimp juveniles reared in a super-intensive system. *R. Bras. Zootec.* **2014**, *43*, 343–350. [CrossRef]

Disclaimer/Publisher's Note: The statements, opinions and data contained in all publications are solely those of the individual author(s) and contributor(s) and not of MDPI and/or the editor(s). MDPI and/or the editor(s) disclaim responsibility for any injury to people or property resulting from any ideas, methods, instructions or products referred to in the content.

Article

Foliar Application of Salicylic Acid Mitigates Saline Stress on Physiology, Production, and Post-Harvest Quality of Hydroponic Japanese Cucumber

Valeska Karolini Nunes Oliveira [1], André Alisson Rodrigues da Silva [1], Geovani Soares de Lima [1,*], Lauriane Almeida dos Anjos Soares [2], Hans Raj Gheyi [1], Claudivan Feitosa de Lacerda [3], Carlos Alberto Vieira de Azevedo [1], Reginaldo Gomes Nobre [4], Lúcia Helena Garófalo Chaves [1], Pedro Dantas Fernandes [1] and Vera Lúcia Antunes de Lima [1]

1. Academic Unit of Agricultural Engineering, Federal University of Campina Grande, Campina Grande 58430-380, PB, Brazil
2. Academic Unit of Agrarian Sciences, Federal University of Campina Grande, Pombal 58840-000, PB, Brazil
3. Department of Agricultural Engineering, Federal University of Ceará, Fortaleza 60455-760, CE, Brazil
4. Department of Science and Technology, Federal Rural University of the Semi-Arid, Caraúbas 59780-000, RN, Brazil
* Correspondence: geovani.soares@professor.ufcg.edu.br; Tel.: +55-83-99945-9864

Abstract: Salicylic acid (SA) is a phenolic compound capable of inducing physiological and metabolic changes that enhance the tolerance of plants to saline stress associated with using a hydroponic system and enable the use of saline water in semi-arid regions. In this context, this assay aimed to evaluate the impact of the foliar application of SA on mitigating salt stress effects on Japanese cucumber cultivated in a hydroponic system. The experiment was carried out in a protected ambient (greenhouse), using the Nutrient Film Technique—NFT hydroponic system. A completely randomized design was performed in a 4 × 4 split-plot scheme, with four levels of electrical conductivity of the nutrient solution—ECns (2.1, 3.6, 5.1, and 6.6 dS m^{-1})—considered as plots and four SA concentrations (0, 1.8, 3.6, and 5.4 mM), regarded as subplots, with four replicates and two plants per plot. An increase in the ECns negatively affected the physiology, production components, and post-harvest quality of cucumber. However, the application of SA to leaves at concentrations between 1.4 and 2.0 mM reduced the deleterious effects of saline stress and promoted an increase in the production of and improvement in the post-harvest quality of cucumber fruits.

Keywords: *Cucumis sativus* L.; brackish water; soilless cultivation; phytohormone

1. Introduction

Freshwater scarcity is common in both arid and semi-arid zones worldwide, due to irregular distribution of water resources, degradation of water quality by anthropic activities, and increased consumption of water resulting from population growth [1,2]. Thus, the use of brackish water becomes necessary to ensure agricultural production in these regions and meet the need for food [3].

However, excess salts in water and/or soil are becoming an increasing threat to global agricultural production and affect nearly 20% of irrigated land of the world, causing severe losses in food production and quality [4]. The high concentrations of salts inhibit water absorption by roots, resulting in water deficit, a decrease in leaf area, and stomatal conductance, which reduces photosynthesis and plant growth [5]. Salt stress also modifies electron transport and alters the activity of photosystem II, which is responsible for oxidizing water molecules to produce electrons [6]. In addition, the absorption and excessive accumulation of Na$^+$ in cells cause ionic imbalance, lipid peroxidation, and damage to the cell membrane [7]. The water content in the fruit can also be affected by salt stress, causing changes in the concentrations of soluble solids, ascorbic acid, and titratable acidity.

Phytohormones play a fundamental role in signaling and attenuating biotic and abiotic stresses [4]. Among the phytohormones, salicylic acid (SA), a phenolic compound, stands out as a regulator of plant growth and development [8]. SA is considered a key component of the plant's antioxidant system and plays an important role in regulating the metabolism of reactive oxygen species (ROS) and in the balance of the redox system [9]. However, its effect depends on the concentration, plant species, stage of crop development, and mode of application [10,11].

The beneficial effect of salicylic acid as an attenuator of salt stress has been reported in recent years in several studies with vegetables, such as bell pepper [12], tomato [13], eggplant [14], melon [15], okra [3], and basil [16]. However, information on its use in Japanese cucumber crops, particularly in hydroponic cultivation, is still incipient.

The use of hydroponic systems is becoming more widespread, due to greater control over the rhizosphere conditions compared to cultivation in soil, which results in gains in the quantity, quality, and safety of production [17]. Hydroponic cultivation also promotes greater efficiency in the use of water and nutrientsa and the absence of matric potential minimizes the effects of salinity on plants, which enables the use of brackish water [18]. Among the hydroponics systems, the NFT (Nutrient Film Technique), a closed system with recirculation of the nutrient solution, stands out as the most used in the cultivation of fast-growing vegetables [3,19].

Cucumber (*Cucumis sativus* L.) is one of the most popular vegetables. It has a moderate sensitivity to salinity [20,21], with a threshold salinity of 2.5 dS m^{-1} in irrigation water and 3.5 dS m^{-1} in the saturation extract of soil [22,23]. Its fruits have a high nutritional value and are rich in proteins, carbohydrates, vitamin C, and minerals [24,25]. In addition, cucumber has anti-inflammatory, antioxidant, and anticancer properties that help in the treatment of several diseases [26].

Previous studies demonstrate that salt stress inhibits germination, growth, biomass production, and cucumber yield [21]. In research conducted by Brengi et al. [27], limitations in growth and chlorophyll synthesis were verified due to salt stress. Chen et al. [28] reported reductions in cucumber yield corresponding to a 13% per unit increment of electrical conductivity above 2.5 dS m^{-1}, demonstrating its sensitivity to salt stress.

This study hypothesizes that the application of salicylic acid on leaves induces tolerance to salt stress in Japanese cucumbers cultivated in a hydroponic system through the regulation of physiological and biochemical processes, which result in gains in production and post-harvest fruit quality. In this context, this study was conducted to evaluate the influence of the foliar application of salicylic acid on mitigating salt stress effects on the physiology, production, and post-harvest quality of Japanese cucumbers in the NFT hydroponic system.

2. Materials and Methods

2.1. Experiment Site

The study was carried out in a protected ambient (greenhouse) belonging to the Center of Science and Agri-Food Technology (CCTA) of the Federal University of Campina Grande (UFCG), in Pombal (6°46'13" S, 37°48'6" W, 184 m a.s.1), Paraíba, Brazil. The daily temperature (maximum, mean, and minimum) and average relative humidity during the experimental period (from May to June 2022) are shown in Figure 1.

2.2. Cultivar Studied

'Hiroshi' Japanese cucumber seeds from Isla® were used in this assay. This variety has a cycle of approximately 60 days, with vigorous and highly productive plants, as well as adaptability to different growing regions in Brazil. It produces cylindrical and uniform fruits of bright dark green color, with from 18 to 22 cm length and diameters between 30 and 40 mm [29].

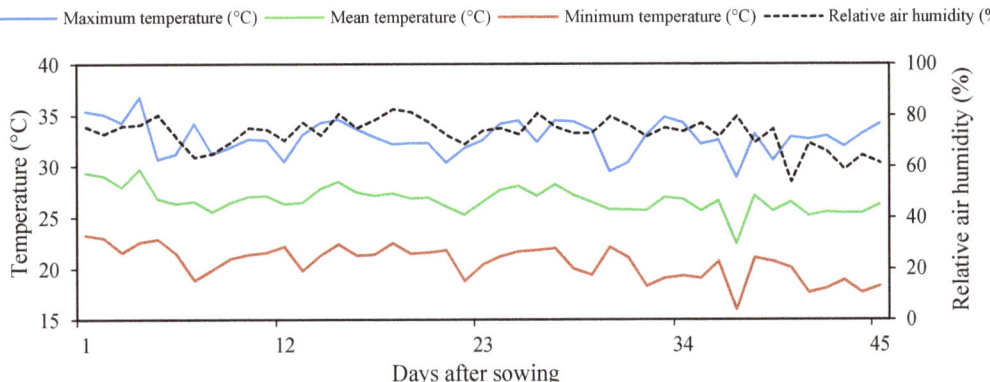

Figure 1. Daily maximum, mean, and minimum temperatures and average relative air humidity observed in the internal area of the greenhouse during the experimental period.

2.3. Experimental Design and Treatments

The treatments included four levels of electrical conductivity of the nutrient solution—ECns (2.1, 3.6, 5.1, and 6.6 dS m^{-1}) and four concentrations of salicylic acid—SA (0, 1.8, 3.6, and 5.4 mM). The treatments were distributed in a completely randomized design in a split-plot scheme, with the ECns levels as plots and salicylic acid concentrations as subplots with four replicates and two plants per plot. Salicylic acid was applied using foliar spraying.

The concentrations of SA used in this study were based on a study conducted with melon [15], while the salinity levels of the nutrient solution were adapted from the assay carried out by [30] with cucumber cv. 'Hokushin'.

2.4. Setting Up and Management of Experiment

The hydroponic system used was the Nutrient Film Technique—NFT type, performed using six-meter-long polyvinyl chloride (PVC) tubes of 100 mm diameter spaced 0.40 m apart. In the hydroponic profile, the circular planting cells had a diameter of 54.17 mm at a distance of 0.50 m and the spacing between treatments (subsystems) was 1.0 m. The hydroponic profiles were supported by 0.60 m high sawhorses with 4% slopes to permit the flow of nutrient solution (Figure 2). At the end of each subsystem, a 150-L polyethylene recipient was placed to collect the excess returning nutrient solution and to recirculate it into the system. The nutrient solution was injected into the hydroponic profile at the top of each channel with a pump of 35 W at a flow rate of 3 L min^{-1}. A timer was used to program the circulation of the nutrient solution into the system, with an intermittent flow of 15 minutes at hourly intervals during the day and 30 minutes at night.

In the study, the nutrient solution recommended by Hoagland and Arnon [31] was used containing N, P, K, Ca, Mg, S, B, Mn, Zn, Cu, Mo, and Fe in concentrations of 210, 31, 234, 200, 48, 64, 0.5, 0.5, 0.05, 0.02, 0.01, and 5 mg L^{-1}, respectively. The fertilizers used as sources of macronutrients in the preparation of the solution were monobasic potassium phosphate (KH_2PO_4), potassium nitrate (KNO_3), calcium nitrate ($Ca(NO_3)_2 \cdot 4H_2O$), and magnesium sulfate ($MgSO_4 \cdot 7H_2O$). As sources of micronutrients, boric acid (H_3BO_3), manganese sulfate ($MnSO_4 \cdot 4H_2O$), zinc sulfate ($ZnSO_4 \cdot 7H_2O$), copper sulfate ($CuSO_4 \cdot 5H_2O$), ammonium molybdate ($(NH_4)_6Mo_7O_{24} \cdot 4H_2O$), ferrous sulfate ($FeSO_4$), and EDTA-Na, respectively, were employed.

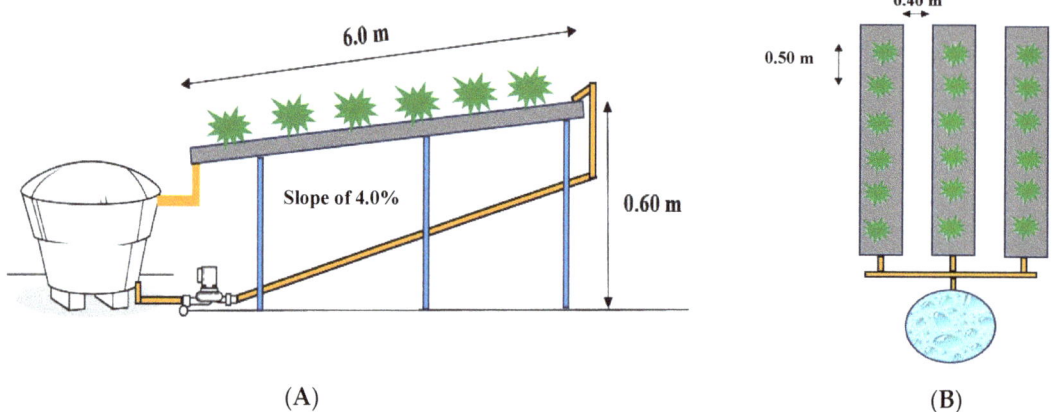

Figure 2. Side view (**A**) and top view (**B**) of the hydroponic system used in the study.

The sowing was conducted in polyethylene cups of 50 mL capacity containing vegetable sponges of the genus Luffa (*Luffa aegytiaca*) arranged in trays. The sponges were sanitized using sodium hypochlorite (2.5%) before sowing. From germination to the emergence of the first true leaf (on average ten days after sowing), a half-strength (50%) nutrient solution was used. The vegetable sponge was removed after the emergence of the first true leaf, the seedlings were inserted directly into the hydroponic channels, and a full-strength nutrient solution was employed.

The saline solutions used in the experiment were obtained from the addition of sodium chloride (NaCl), calcium chloride ($CaCl_2 \cdot 2H_2O$), and magnesium chloride ($MgCl_2 \cdot 6H_2O$) in the equivalent proportion of 7:2:1, respectively, to the nutrient solution prepared in the municipal supply water as described above. The proportion of Na, Ca, and Mg added is commonly found in the waters used for irrigation in the semi-arid region of northeastern Brazil [32]. The amounts of the individual salts added to the nutrient solution are shown in Table 1.

Table 1. Quantities of salts incorporated per 100 L of nutrient solution.

ECns (dS m^{-1})	NaCl	$CaCl_2 \cdot 2H_2O$	$MgCl_2 \cdot 6H_2O$
		g	
2.1	-	-	-
3.6	61.43	20.55	15.23
5.1	122.85	41.10	30.45
6.6	184.28	61.65	45.68

At intervals of eight days, the nutrient solution was completely replaced. The electrical conductivity and pH were monitored daily and, when necessary, the nutrient solution was adjusted by the addition of local-supply water or nutrient solution, maintaining the ECns as per treatments established initially. The pH was maintained between 5.5 and 6.5 by adding either 0.1 M potassium hydroxide (KOH) or hydrochloric acid (HCl). The aplants were cultivated using a vertical support fixed with a plastic string (number 10) (Figure 3).

The salicylic acid solution (100 mM) was prepared by dissolution in 30% ethyl alcohol. The solutions of adequate concentrations, as per treatment, were prepared by diluting this solution in water before each application event. To reduce the surface tension of the drops on the leaf surface, a Wil-fix adjuvant was added (0.5 mL L^{-1}) to the solution. The first application was performed 48 h after transferring the seedlings and 72 h before the application of the saline nutrient solution. The other applications were carried out at 10-day intervals until the beginning of the flowering stage, spraying the abaxial and adaxial leaf surfaces. The sprayings were performed between 17:00 and 18:00 h and the average volume of solution applied per plant was 80 mL. During the spraying of salicylic acid, a plastic tarpaulin structure was used to prevent the solution from drifting onto neighboring plants.

Figure 3. Cultivation of Japanese cucumber in NFT—Nutrient Film Technique hydroponic system—at different stages of development (Vegetative stage—(**A**), flowering stage—(**B**), and fruiting stage—(**C**)).

2.5. Traits Analyzed

At 40 days after transplanting (DAT), the relative water content, percentage of electrolyte leakage in the leaf blade, leaf gas exchange, photosynthetic pigments, and chlorophyll *a* fluorescence were evaluated. Harvest began at 43 DAT; the following production components were obtained: number of fruits per plant, average fruit weight, total production, fruit length, and fruit diameter. In addition, the following post-harvest characteristics were determined in the fruit pulp: pH, titratable acidity, ascorbic acid, and soluble solids contents.

2.5.1. Relative Water Content

For the determination of the relative water content (RWC), two leaves were removed from the middle third of the main branch obtaining ten discs 12 mm in diameter. The disks were immediately weighed for the fresh mass (FM); then, the discs were transferred to a beaker and immersed in 50 mL of distilled water for 24 h. After this period, with a paper towel, the excess water was removed from the disks and the turgid mass (TM) was determined. The discs were dried at a temperature of $\approx 65 \pm 3\ °C$ in an oven until constant weight to obtain the dry mass (DM). The relative water content was determined using Equation (1) as recommended by [33].

$$RWC = ((FM - DM)/(TM - MS)) \times 100 \qquad (1)$$

where:

RWC—relative water content (%);
FM—fresh mass of leaves (g);
TM—turgid mass (g);
DM—dry mass (g).

2.5.2. Percentage of Electrolyte Leakage

The electrolyte leakage (EL) was obtained using a copper hole puncher to produce five leaf discs with an area of 1.54 cm² each. The discs were washed and placed in an Erlenmeyer® flask containing 50 mL of distilled water. After closing with aluminum foil, the Erlenmeyer® flasks were kept at a temperature of 25 °C for 24 h and then the initial electrical conductivity of the medium (Xi) was measured using a benchtop conductivity meter (MB11, MS Techonopon®, Piracicaba—SP, Brazil). Later, the Erlenmeyer® flasks were subjected to a temperature of 80 °C for 120 minutes in an oven (SL100/336, SOLAB®) and, after cooling, the final electrical conductivity (Xf) was determined. In turn, the percentage of electrolyte leakage (% EL) was obtained through Equation (2), as recommended by [34].

$$\% \text{ EL} = (Xi/Xf) \times 100 \quad (2)$$

where:

% EL—percentage of electrolyte leakage (%);
Xi—initial electrical conductivity;
Xf—final electrical conductivity.

2.5.3. Photosynthetic Pigments

The quantification of the photosynthetic pigments (chlorophyll *a*, chlorophyll *b*, chlorophyll *total*, and carotenoids) was carried out according to [35], with extracts from disc samples of the third mature leaf blade from the apex. In each sample, 6.0 mL of 80% acetone PA was added. Through these extracts, the concentrations of chlorophyll and carotenoids were determined using a spectrophotometer (UV/VIS—UV17030, AKSO®, São Leopoldo—RS, Brazil) at the absorbance wavelength (470, 647, and 663 nm), using Equations (3)–(6), with results expressed in μg mL⁻¹.

$$\text{Chl } a = (12.25 \times \text{ABS}663) - (2.79 \times \text{ABS}647) \quad (3)$$

$$\text{Chl } b = (21.5 \times \text{ABS}647) - (5.10 \times \text{ABS}647) \quad (4)$$

$$\text{Chl } t = (7.15 \times \text{ABS}663) + (18.71 \times \text{ABS}647) \quad (5)$$

$$\text{Car} = \frac{|(1000 \times \text{ABS}470) - (1.82 \times \text{Cl } a) - (85.02 \times \text{Cl } b)|}{198} \quad (6)$$

where:

Chl *a*—chlorophyll *a*;
Chl *b*—chlorophyll *b*;
Chl *t*—chlorophyll *total*;
Car—carotenoids.

2.5.4. Leaf Gas Exchange Parameters

The internal CO_2 concentration (C_i, μmol CO_2 m⁻² s⁻¹), stomatal conductance (g_s, mol H_2O m⁻² s⁻¹), transpiration (E, mmol H_2O m⁻² s⁻¹), and CO_2 assimilation rate (A, μmol CO_2 m⁻² s⁻¹) were determined on the third mature leaf counted from the apex of the main branch of the plant, using an irradiation of 1200 μmol m⁻² s⁻¹, obtained from the photosynthetic light saturation curve, and airflow of 200 mL min⁻¹, using the portable photosynthesis meter "LCPro+" from ADC BioScientific Ltd. The leaf gas exchange determinations were performed between 08:00 and 10:00 a.m., under ambient conditions of temperature and CO_2 concentration.

2.5.5. Chlorophyll Fluorescence

The chlorophyll fluorescence was performed on the third mature leaf, counted from the apex of the main branch of the plant between 08:00 and 10:00 a.m., using an OS5p pulse-modulated fluorimeter from Opti Science, employing the Fv/Fm protocol to obtain the variables—initial fluorescence (F0), maximum fluorescence (Fm), variable fluorescence (Fv = Fm − F0), and quantum efficiency of photosystem II (Fv/Fm). This protocol was performed after adaptation of the leaves to the dark for 30 minutes, using a clip of the device, to ensure that all the acceptors were oxidized, i.e., with the reaction centers open. Subsequently, the evaluations were determined under light conditions, using an actinic light source with a multi-flash saturating pulse coupled to a clip to determine the initial fluorescence before the saturation pulse (Fs), maximum fluorescence after adaptation to saturating light (Fms), electron transport rate (ETR), and quantum efficiency of photosystem II (Y_{II}).

2.5.6. Production Components

The fruits were harvested from each plant based on their degree of maturation when they were from 18 to 22 cm in length [36]. The following production components were obtained: number of fruits per plant, average fruit weight (g per fruit), total production per plant (g per plant), average fruit length (cm), and average fruit diameter (mm). The average diameter of the cucumber fruit was measured in the center of the fruit.

2.5.7. Post-Harvest

The pH of the pulp was determined immediately after harvest, using a digital pH meter (COMBO5, AKSO®, São Leopoldo—RS, Brazil) previously calibrated at pH 4.0 and 7.0 with buffer solutions; the soluble solids (°Brix) were determined by direct reading using a digital refractometer (MA871, AKSO®, São Leopoldo—RS, Brazil); and the ascorbic acid content (mg per 100g of pulp) was obtained using titration. The determinations were created using the methodologies recommended by [37]. The titratable acidity was expressed as a percentage of citric acid.

2.6. Data Analysis

For the data collected in this study regarding relative water content, the percentage of electrolyte leakage, photosynthetic pigments, leaf gas exchange parameters, chlorophyll fluorescence, production components and yield, and post-harvest quality, the levels of sources of variation, electrical conductivity of the nutrient solution (ECns), and salicylic acid (SA) concentrations were explored and submitted to the distribution normality test (Shapiro–Wilk) to verify if the data obeyed normal distribution. Then, the analysis of variance (ANOVA) was realized and, in cases of significance, being the quantitative factors, linear and quadratic regression analyzes were performed using the SISVAR-ESAL statistical program [38]. The definition of the model (linear or quadratic) was based on the values of the coefficient of determination (R^2) and the biological significance of the phenomenon. Further, the effects of the interaction (ECns × SA) were analyzed using response surface curves, prepared with the SigmaPlot v.12.5 software.

3. Results

The relative water content (RWC), the percentage of electrolyte leakage in the leaf blade (% EL), and all the variables of leaf gas exchange were affected significantly ($p \leq 0.01$) by the interaction between the electrical conductivity of the nutrient solution and the concentrations of salicylic acid (ECns × SA) (Table 2).

Table 2. Summary of the analysis of variance (ANOVA) for relative water content (RWC), percentage of electrolyte leakage (% EL), internal CO_2 concentration (Ci), stomatal conductance (gs), transpiration (E), and CO_2 assimilation rate (A) of Japanese cucumber grown in a hydroponic system with saline nutrient solution and foliar application of salicylic acid, 40 days after transplanting.

Source of Variation	DF	Mean Squares					
		RWC	% EL	Ci	gs	E	A
Saline nutrient solution (ECns)	3	1607.72 **	106.41 **	19,058.13 **	0.012 **	2.93 **	237.04 **
Linear regression	1	4218.93 **	307.07 **	51,606.70 **	0.03 **	7.64 **	685.88 **
Quadratic regression	1	601.59 ns	8.19 *	3835.94 *	0.001 ns	0.79 ns	25.11 ns
Residual 1	9	24.42	0.06	180.69	2.0×10^{-6}	0.01	0.05
Salicylic acid (SA)	3	11.26 ns	136.83 **	4900.61 **	0.002 **	0.38 **	130.82 **
Linear regression	1	-	60.31 *	1070.47 ns	0.001 *	0.07 *	141.21 *
Quadratic regression	1	-	223.59 **	11,184.12 **	0.004 **	1.04 **	165.86 **
Interaction (ECns × SA)	9	107.06 *	0.33 **	1300.44 *	6.0×10^{-6} *	3.52 **	10.86 **
Residual 2	36	29.07	3.40	415.59	5.7×10^{-4}	0.28	0.64
CV 1 (%)		6.62	3.66	5.93	10.55	6.32	5.81
CV 2 (%)		7.23	5.10	8.99	13.89	12.32	9.83

DF: degree of freedom; CV: Coefficient of variation; ns, *, and **, respectively, not significant, significant at a $p \leq 0.05$ and $p \leq 0.01$.

The foliar application of salicylic acid (SA) up to concentrations of 2.0 mM promoted an increase in the RWC, even when plants were cultivated with the highest ECns (6.6 dS m^{-1}) (Figure 4A). The highest RWC (85.05%) was observed in plants subjected to ECns of 2.1 dS m^{-1} and SA concentration of 2.0 mM, corresponding to an increase of 3.34% compared to plants cultivated with the same ECns (2.1 dS m^{-1}) and without SA application (0 mM). However, the application of SA on leaves at concentrations greater than 2.0 mM intensified the harmful effects of salt stress on RWC, and the lowest value (59.03%) was obtained in plants that received ECns of 6.6 dS m^{-1} and SA concentration of 5.4 mM.

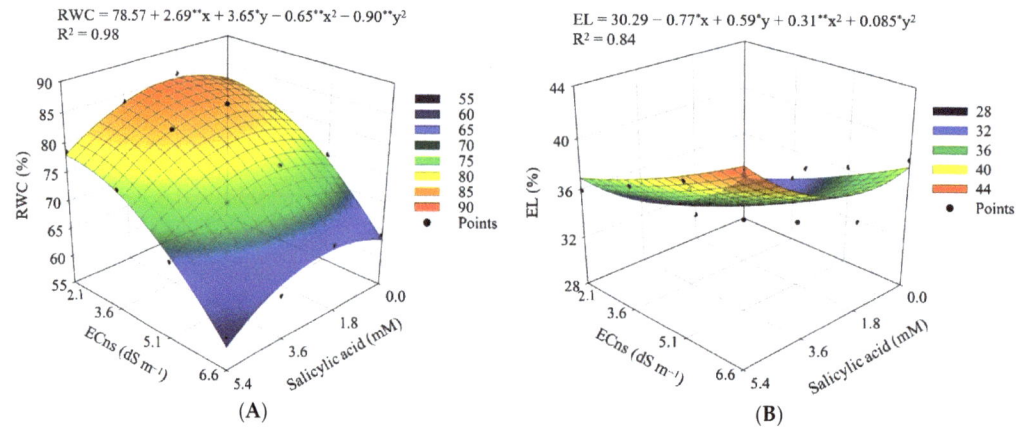

Figure 4. Response surface for relative water content—RWC (A)—and percentage of electrolyte leakage in the leaf blade—% EL (B) of Japanese cucumber—as a function of the interaction between the electrical conductivity of the nutrient solution (ECns) and the concentrations of salicylic acid (SA), grown in a hydroponic system, 40 days after transplanting. X and Y—concentration of SA and ECns, respectively; * and ** significant at a $p \leq 0.05$ and $p \leq 0.01$, respectively.

The increase in the electrical conductivity of the nutrient solution increased electrolyte leakage in the leaf blade (% EL), regardless of the concentration of SA (Figure 4B). The application of SA at concentrations greater than 1.5 mM intensified the harmful effects of saline stress with the highest value of % EL (42.77%) obtained in plants subjected to ECns of 6.6 dS m^{-1} and SA concentration of 5.4 mM. However, cucumber plants subjected to the highest level of ECns (6.6 dS m^{-1}) and SA concentration of 1.4 mM showed an EL of 37.41%, i.e., a reduction of 12.53% compared to plants cultivated with the same ECns and SA application of 5.4 mM, demonstrating the beneficial effect of SA on the acclimatization of plants to saline stress when applied at appropriate concentrations.

The internal CO_2 concentration (Ci) was reduced by the application of SA up to the concentration of 2.0 mM, regardless of the ECns level (Figure 5A). The lowest value of the internal CO_2 concentration (164.7 μmol CO_2 m^{-2} s^{-1}) was observed in plants subjected to ECns of 2.1 dS m^{-1} and SA concentration of 2.0 mM. Cucumber plants subjected to the highest level of ECns (6.6 dS m^{-1}) and SA concentration of 2.0 mM showed a reduction of 9.1% (24.0 μmol CO_2 m^{-2} s^{-1}) in the internal CO_2 concentration compared to plants cultivated with the same ECns and without SA application (0 mM).

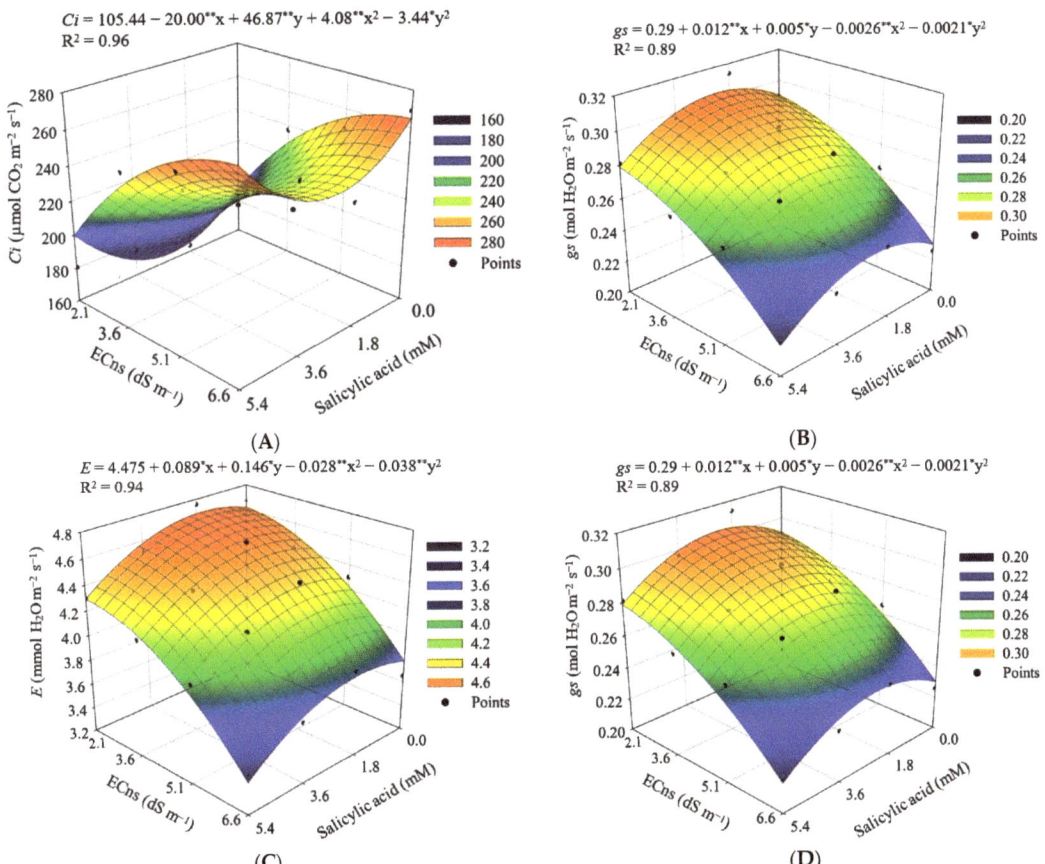

Figure 5. Response surface for internal CO_2 concentration—Ci (**A**), stomatal conductance—gs (**B**), transpiration—E (**C**), and CO_2 assimilation rate—A (**D**) of Japanese cucumber as a function of the interaction between the electrical conductivity of the nutrient solution (ECns) and the concentrations of salicylic acid (SA), grown in a hydroponic system, 40 days after transplanting. X and Y—concentration of SA and ECns, respectively; * and ** significant at a $p \leq 0.05$ and $p \leq 0.01$, respectively.

The foliar application of salicylic acid with concentrations of up to 2.0 mM promoted an increase in stomatal conductance (gs), even when plants were cultivated under ECns of

6.6 dS m^{-1} (Figure 5B). The highest value of stomatal conductance (0.305 mol H$_2$O m^{-2} s^{-1}) was verified in plants cultivated with ECns of 2.1 dS m^{-1} and an SA concentration of 2.0 mM, corresponding to an increase of 4.81% (0.014 mol H$_2$O m^{-2} s^{-1}) compared to plants cultivated with the same salinity level (2.1 dS m^{-1}) and without SA application (0 mM). However, the foliar application of SA at concentrations greater than 2.0 mM intensified the harmful effects of salt stress on stomatal conductance; the lowest value (0.221 mol H$_2$O m^{-2} s^{-1}) was obtained in plants subjected to ECns of 6.6 dS m^{-1} and SA concentration of 5.4 mM.

The transpiration (E) and CO$_2$ assimilation rate (A) of cucumber plants were also favored by the application of salicylic acid on leaves up to a concentration of 2.0 mM, regardless of the ECns (Figure 5C and 5D). The plants subjected to SA concentration of 2.0 mM and ECns of 2.1 dS m^{-1} attained the highest transpiration (4.7 mmol H$_2$O m^{-2} s^{-1}) and CO$_2$ assimilation rate (34.3 µmol CO$_2$ m^{-2} s^{-1}). In relative terms, the transpiration and CO$_2$ assimilation rate of plants cultivated with ECns of 2.1 dS m^{-1} and subjected to SA concentration of 2.0 mM compared to those cultivated under the same level of salinity and without SA application (0 mM) enabled increments of 1.52% (0.07 mmol H$_2$O m^{-2} s^{-1}) and 5.9% (1.90 µmol CO$_2$ m^{-2} s^{-1}) to be observed, respectively.

There was a significant effect ($p \leq 0.01$) of the interaction between the ECns and the concentrations of SA only for chlorophyll *a* and chlorophyll *total* (Table 3). On the other hand, the electrical conductivity of the nutrient solution significantly influenced ($p \leq 0.01$) all the variables of photosynthetic pigments, while the concentrations of salicylic acid alone did not affect the chlorophyll and carotenoid contents of Japanese cucumber, 40 days after transplanting.

Table 3. Summary of the analysis of variance (ANOVA) for chlorophyll *a* (Chl *a*), chlorophyll *b* (Chl *b*), chlorophyll *total* (Chl *t*), and carotenoids (Car) of Japanese cucumber grown in a hydroponic system with saline nutrient solution and foliar application of salicylic acid, 40 days after transplanting.

Source of Variation	DF	Mean Squares			
		Chl *a*	Chl *b*	Chl *t*	Car
Saline nutrient solution (ECns)	3	54.47 **	27.09 **	154.76 **	1.90 **
Linear regression	1	158.72 **	67.54 **	433.61 **	5.41 **
Quadratic regression	1	4.16 ns	12.06 ns	30.36 ns	0.26 ns
Residual 1	9	3.94	0.59	6.59	0.62
Salicylic acid (SA)	3	0.49 ns	1.26 ns	9.73 ns	1.01 ns
Interaction (ECns × SA)	9	5.42 *	0.76 ns	18.33 *	0.29 ns
Residual 2	36	3.24	1.20	7.52	0.38
CV 1 (%)		10.87	12.02	10.43	13.65
CV 2 (%)		9.87	17.19	11.14	10.48

DF: degree of freedom; CV: Coefficient of variation; ns, *, and **, respectively, not significant, significant at a $p \leq 0.05$ and $p \leq 0.01$.

The increase in the ECns reduced the chlorophyll *a* (Figure 6A) and chlorophyll *total* (Figure 6B) contents of cucumber plants. However, the foliar application of SA up to concentration of 2.0 mM reduced the effects of saline stress, promoting increments in Chl *a* and Chl *t*, even when plants were subjected to the highest ECns (6.6 dS m^{-1}). The highest values of chlorophyll *a* (21.17 µg mL^{-1}) and chlorophyll *total* (28.76 µg mL^{-1}) were obtained in plants cultivated with ECns of 2.1 dS m^{-1} and SA concentration of 2.0 mM, corresponding to increments of 3.61% (0.74 µg mL^{-1}) in chlorophyll *a* and 3.50% (0.97 µg mL^{-1}) in chlorophyll *total* for plants cultivated with the same salinity level (2.1 dS m^{-1}) and without the application of SA (0 mM). On the other hand, the application of SA on leaves at concentrations greater than 2.0 mM intensified the harmful effects of saline stress, with the lowest values of chlorophyll *a* (15.66 µg mL^{-1}) and chlorophyll *total* (20.23 µg mL^{-1}) in plants subjected to ECns of 6.6 dS m^{-1} and SA concentration of 5.4 mM.

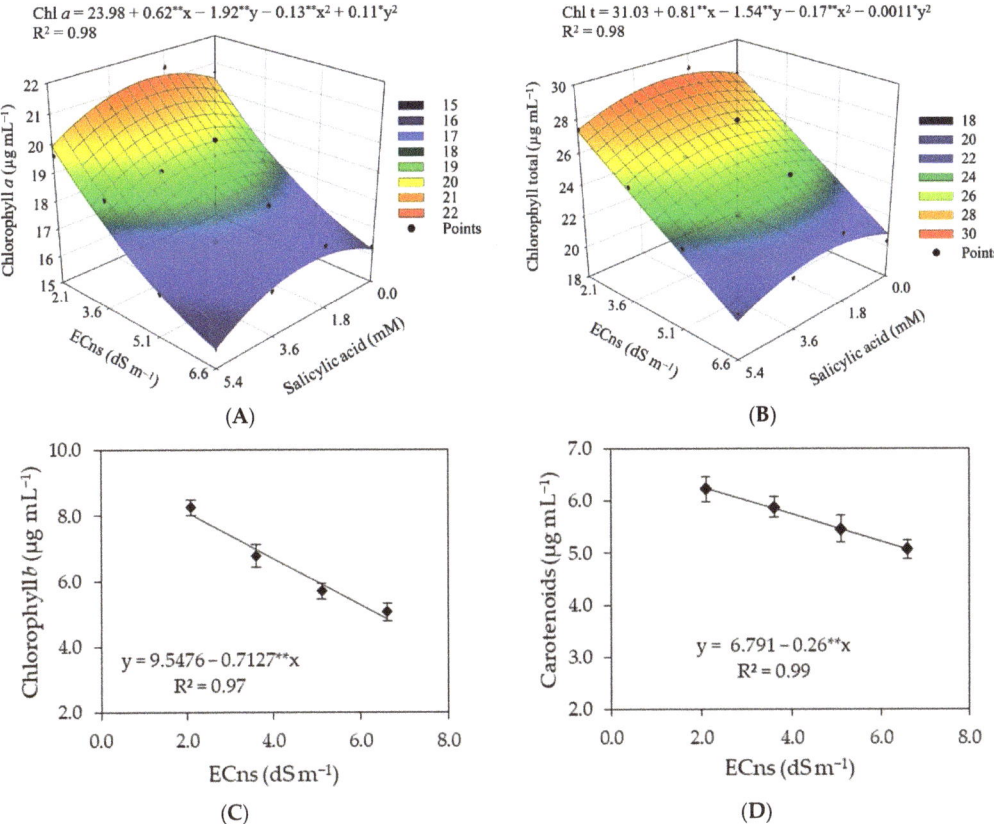

Figure 6. Response surface for chlorophyll a—Chl a (**A**), chlorophyll total—Chl total (**B**), chlorophyll b—Chl b (**C**), and carotenoids—Car (**D**) of Japanese cucumber as a function of the interaction between the electrical conductivity of the nutrient solution (ECns) and the concentrations of salicylic acid (SA), grown in a hydroponic system, 40 days after transplanting. X and Y—concentration of SA and ECns, respectively; * and ** significant at a $p \leq 0.05$ and $p \leq 0.01$, respectively. Vertical lines represent the standard error of the mean (n = 4).

The synthesis of chlorophyll b (Figure 6C) and carotenoids (Figure 6D) of cucumber plants were negatively affected by the increase in the electrical conductivity of the nutrient solution, with reductions, per unit increment in ECns, of 7.46% in chlorophyll b content and 3.82% in carotenoid content. Comparing the chlorophyll b and carotenoid contents of plants grown under ECns of 6.6 dS m^{-1} to those of plants subjected to ECns of 2.1 dS m^{-1}, reductions of 39.9% (3.21 µg mL^{-1}) and 18.17% (1.17 µg mL^{-1}) were observed, respectively.

According to the summary of the analysis of variance, there was a significant effect ($p \leq 0.01$) of the interaction between the ECns and the concentrations of SA only on the quantum efficiency of photosystem II (Fv/Fm) (Table 4). As a single factor, the levels of ECns significantly influenced ($p \leq 0.01$) all variables, except the electron transport rate (ETR). On the other hand, SA concentrations had no significant effect on any of the chlorophyll fluorescence variables.

The increase in the ECns had an increasing linear effect on the initial fluorescence (F0) of cucumber plants (Figure 7A), with an increment of 3.22% per unit increase in ECns. The plants cultivated with ECns of 6.6 dS m^{-1} had an increase of 13.57% (73.06) compared to those grown with ECns of 2.1 dS m^{-1}. Unlike the effect observed on the initial fluorescence

(Figure 7A), the maximum fluorescence was reduced as the ECns increased (Figure 7B). The plants grown under ECns of 6.6 dS m^{-1} showed a reduction of 20.68% (512.6) compared to those cultivated with ECns of 2.1 dS m^{-1}.

Table 4. Summary of the analysis of variance (ANOVA) for initial fluorescence (F0), maximum fluorescence (Fm), variable fluorescence (Fv), quantum efficiency of photosystem II (Fv/Fm), initial fluorescence before saturation pulse (Fs), quantum efficiency of photosystem II in the light phase (Y_{II}), and electron transport rate (ETR) of Japanese cucumber grown in a hydroponic system with saline nutrient solution and foliar application of salicylic acid, 40 days after transplanting.

Source of Variation	DF	Mean Squares						
		F0	Fm	Fv	Fv/Fm	Fs	Y_{II}	ETR
Saline nutrient solution (ECns)	3	16,234.27 **	878,366.35 **	889,782.38 **	0.021 **	174.32 **	0.017 **	120.39 ns
Linear regression	1	47,458.15 **	225,506.26 **	20,197.31 **	0.028 *	505.99 **	0.006 ns	-
Quadratic regression	1	669.52 ns	73,902.41 ns	178,325 ns	0.034 **	10.51 ns	0.037 **	-
Residual 1	9	1368.75	46,782.09	178.25	72.0×10^{-6}	4.73	0.001	53.01
Salicylic acid (SA)	3	1906.64 ns	44,209.09 ns	695.02 ns	25.6×10^{-5} ns	93.97 ns	0.0001 ns	19.61 ns
Interaction (ECns × SA)	9	2315.81 ns	113,099.59 ns	26,025.02 ns	0.007 **	37.07 ns	0.002 ns	62.11 ns
Residual 2	36	2395.35	64,148.96	15,211.43	11.8×10^{-5}	5.87	0.001	31.74
CV 1 (%)		6.44	9.73	9.94	10.21	5.41	5.94	12.94
CV 2 (%)		8.51	11.40	8.66	11.42	6.79	5.08	10.01

DF: degree of freedom; CV: Coefficient of variation; ns, *, and **, respectively, not significant, significant at a $p \leq 0.05$ and $p \leq 0.01$.

The variable fluorescence (Fv) of cucumber plants decreased with the increase in the electrical conductivity of the nutrient solution (Figure 7C). The plants cultivated with ECns of 2.1 dS m^{-1} had a variable fluorescence of 1697.34, while the minimum value (1152.08) was verified under ECns of 6.6 dS m^{-1}, i.e., there was a reduction of 545.26 (32.12%) under the highest salinity level. The application of salicylic acid on leaves at the concentration of 2.0 mM promoted an increase in the quantum efficiency of photosystem II (Figure 7D), even in plants cultivated with the highest ECns level (6.6 dS m^{-1}); however, the highest value of Fv/Fm (0.802) was obtained in plants cultivated with ECns of 3.8 dS m^{-1}, corresponding to an increase of 8.52% (0.063) compared to plants subjected to ECns of 2.1 dS m^{-1} and without SA application (0 mM). The lowest quantum efficiency of photosystem II (0.684) was recorded in plants cultivated with ECns of 6.6 dS m^{-1} and SA concentration of 5.4 mM.

For the initial fluorescence before the saturation pulse (Fs) (Figure 8A), it was observed that the increase in ECns promoted an increment of 12.55% in plants cultivated with ECns of 6.6 dS m^{-1} compared to those under ECns of 2.1 dS m^{-1}. On the other hand, the quantum efficiency of photosystem II in the light phase (Figure 8B) decreased when nutrient solutions with electrical conductivity above 2.1 dS m^{-1} were used, with the lowest value of Y_{II} (0.565) observed in plants cultivated with ECns of 6.6 dS m^{-1}.

The interaction between the ECns and SA concentrations significantly influenced ($p \leq 0.05$) the number of fruits, total production per plant, and average fruit weight ($p \leq 0.01$) (Table 5). The levels of ECns significantly influenced ($p \leq 0.01$) all the variables of the production components, while the concentrations of salicylic acid alone did not affect any variable of the production components of cucumber.

The increase in the ECns negatively affected the number of cucumber fruits per plant (NFP), regardless of the SA concentration (Figure 9A). It is worth pointing out that the foliar application of SA above the estimated concentration of 2.3 mM intensified the effects of saline stress, with the lowest value of NFP (0.90 fruits per plant) in plants cultivated with ECns of 6.6 dS m^{-1} and SA concentration of 5.4 mM, while the highest number of fruits per plant was 5.05, observed in plants subjected to ECns of 2.1 dS m^{-1} and an SA concentration of 2.0 mM.

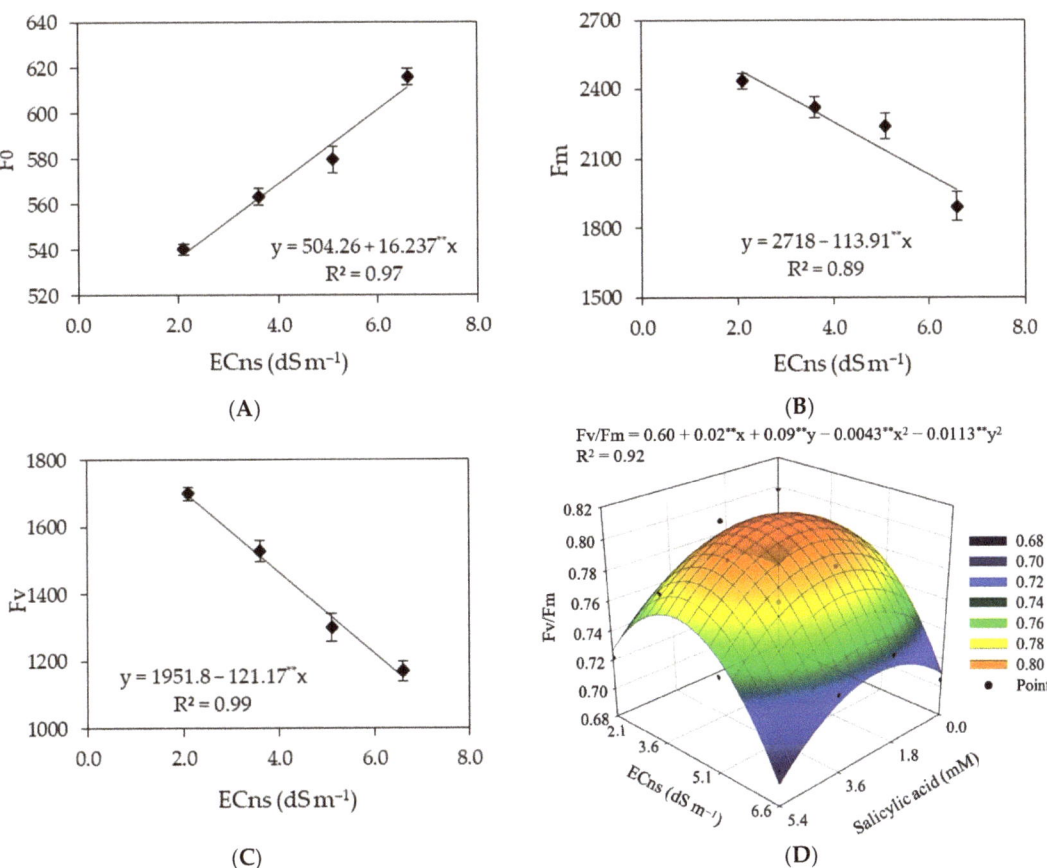

Figure 7. Initial fluorescence—F0 (**A**), maximum fluorescence—Fm (**B**), and variable fluorescence—Fv (**C**) as a function of electrical conductivity of the nutrient solution (ECns) and response surface for the quantum efficiency of photosystem II—Fv/Fm (**D**) of Japanese cucumber as a function of the interaction between ECns and the concentrations of salicylic acid (SA), grown in a hydroponic system, 40 days after transplanting. X and Y—concentration of SA and ECns, respectively; ** significant at $p \leq 0.01$. Vertical lines represent the standard error of the mean ($n = 4$).

The foliar application of SA at the estimated concentration of 2.0 mM also promoted increments in average fruit weight (Figure 9B) and total production per plant—TPP (Figure 9C). The plants cultivated with ECns of 2.1 dS m^{-1} and SA concentrations of 2.0 mM stood out with the highest values of AFW (383.40 g per fruit) and TPP (1932.59 g plant), corresponding to increments of 7.95% in AFW and 7.66% in TPP, compared to plants cultivated with the same level of ECns (2.1 dS m^{-1}) but without SA application (0 mM).

However, it is worth noting that the application of SA on leaves at concentrations greater than 2.1 mM, associated with increased ECns, reduced the AFW and TPP of cucumber (Figure 9B,C), with the lowest values of AFW (276.54 g per fruit) and TPP (263.35 g per plant) in plants cultivated with ECns of 6.6 dS m^{-1} and SA concentrations of 5.4 mM.

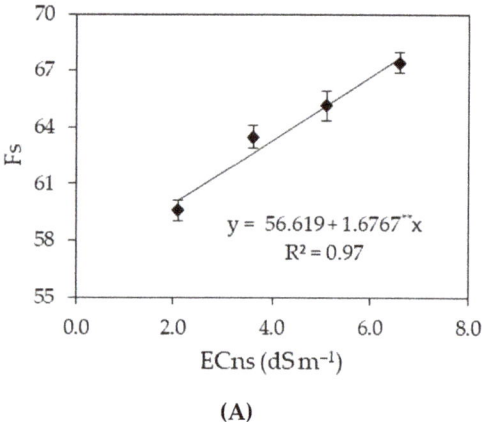
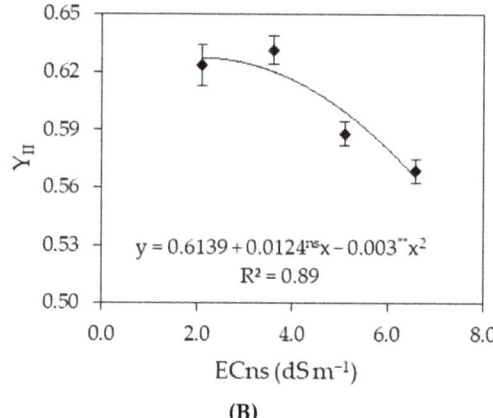

Figure 8. Initial fluorescence before the saturation pulse—Fs (**A**) and the quantum efficiency of photosystem II in the light phase—Y_{II} (**B**) of Japanese cucumber as a function of the levels of electrical conductivity of the nutrient solution (ECns), grown in a hydroponic system, 40 days after transplanting. ns and **, respectively, not significant and significant at a $p \leq 0.01$. Vertical lines represent the standard error of the mean ($n = 4$).

Table 5. Summary of the analysis of variance (ANOVA) for the number of fruits (NF), average fruit weight (AFW), total production per plant (TPP), average fruit length (AFL), and average fruit diameter (AFD) of Japanese cucumber grown in a hydroponic system with saline nutrient solution and foliar application of salicylic acid, 45 days after transplanting.

Source of Variation	DF	Mean Squares				
		NF	AFW	TPP	AFL	AFD
Saline nutrient solution (ECns)	3	40.18 **	13,910.47 **	5,869,190.94 **	49.47 **	126.64 **
Linear regression	1	117.83 **	40,951.25 **	16,913,963.71 **	143.91 **	364.10 **
Quadratic regression	1	2.71 ns	481.58 ns	409,158.52 ns	4.36 *	11.78 ns
Residual 1	9	0.47	753.89	184,044.58	0.64	2.42
Salicylic acid (SA)	3	0.17 ns	527.18 ns	66,638.93 ns	3.88 ns	2.62 ns
Interaction (ECns × SA)	9	1.67 *	5877.88 **	543,499.86 *	1.01 ns	1.60 ns
Residual 2	36	0.42	3054.67	89,665.21	0.63	2.79
CV 1 (%)		21.29	8.44	19.76	4.63	4.93
CV 2 (%)		19.42	16.98	17.75	3.89	5.29

DF: degree of freedom; CV: Coefficient of variation; ns, *, and **, respectively, not significant, significant at a $p \leq 0.05$ and $p \leq 0.01$.

The ECns negatively affected the length and diameter of cucumber fruits (Figure 10A and 10B), with reductions of 3.78% in the average fruit length—AFL—and 3.76% in the average fruit diameter—AFD—per unit increment in ECns. Comparing the AFL and AFD of plants cultivated with ECns of 6.6 dS m^{-1} to those of plants subjected to ECns of 2.1 dS m^{-1}, reductions of 18.5% (4.03 cm) and 18.4% (6.40 mm) were observed in the length and diameter of fruit, respectively.

There was a significant effect (Table 6) of the interaction between the ECns and the concentrations of SA on the hydrogen potential (pH), soluble solids (SS), ascorbic acid (AA), and titratable acidity (TA) of fruits of Japanese cucumber cultivated in a hydroponic system, at 45 days after transplanting.

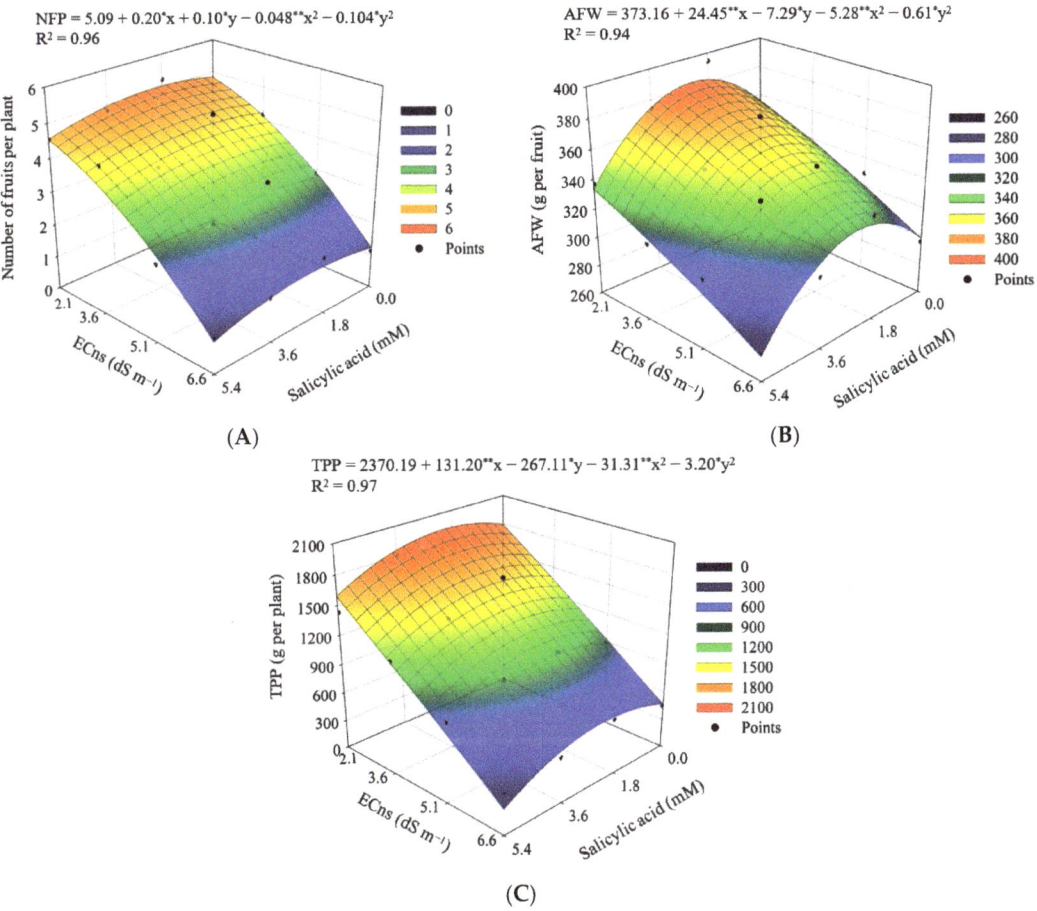

Figure 9. Response surface for the number of fruits per plant—NFP (**A**), average fruit weight—AFW (**B**), and total production per plant—TPP (**C**) of Japanese cucumber as a function of the interaction between the electrical conductivity of the nutrient solution (ECns) and the concentrations of salicylic acid (SA), grown in a hydroponic system, 45 days after transplanting. X and Y—concentration of SA and ECns, respectively; * and ** significant at a $p \leq 0.05$ and $p \leq 0.01$, respectively.

The increase in the ECns reduced the pH of the cucumber fruit pulp (Figure 11A), and the reductions were intensified with salicylic acid concentrations greater than 2.1 mM, with the lowest pH value (5.12) verified in plants cultivated with ECns of 6.6 dS m^{-1} and an SA concentration of 5.4 mM. However, it is observed that the foliar application of SA up to the concentration of 2.0 mM promoted an increase in pH, with the highest pH value (5.92) obtained in plants subjected to ECns of 2.1 dS m^{-1} and SA concentrations of 2.0 mM. The soluble solids (Figure 11B) of cucumber fruits were also reduced by the increase in the ECns, with the highest value of soluble solids (5.92 °Brix) observed in plants cultivated with ECns of 2.1 dS m^{-1} and SA concentration of 2.0 mM, while the fruits of plants subjected to the same concentration of SA (2.0 mM) and ECns of 6.6 dS m^{-1} showed a 43.9% (2.60 °Brix) reduction in soluble solids compared to the fruits of cucumber plants under ECns of 2.1 dS m^{-1}.

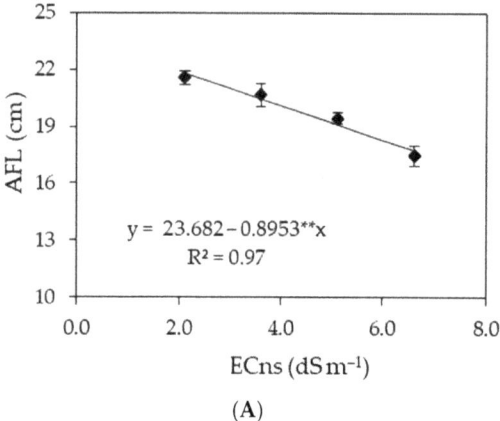
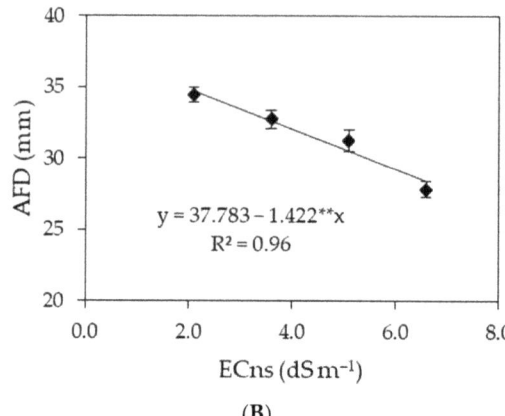

Figure 10. Average fruit length—AFL (**A**) and average fruit diameter—AFD (**B**) of Japanese cucumber cultivated in a hydroponic system as a function of the electrical conductivity of the nutrient solution (ECns), 45 days after transplanting. ** significant at a $p \leq 0.01$. Vertical lines represent the standard error of the mean ($n = 4$).

Table 6. Summary of the analysis of variance (ANOVA) for hydrogen potential (pH), soluble solids (SS), ascorbic acid (AA), and titratable acidity (TA) of fruits of Japanese cucumber grown in a hydroponic system with saline nutrient solution and foliar application of salicylic acid, 45 days after transplanting.

Source of Variation	DF	Mean Squares			
		pH	SS	AA	TA
Saline nutrient solution (ECns)	3	0.53 **	17.97 **	0.73 **	0.41 **
Linear regression	1	1.56 **	52.91 **	1.99 **	1.16 **
Quadratic regression	1	0.01 ns	0.99 ns	0.16 ns	0.005 ns
Residual 1	9	0.03	0.004	0.006	0.001
Salicylic acid (SA)	3	0.35 **	0.009 ns	0.35 **	0.65 **
Linear regression	1	0.17 ns	-	0.49 *	0.69 *
Quadratic regression	1	0.55 **	-	0.45 **	1.21 **
Interaction (ECns × SA)	9	0.13 *	0.45 **	0.006 **	0.04 *
Residual 2	36	0.02	0.05	0.002	0.008
CV 1 (%)		3.24	3.42	2.27	2.74
CV 2 (%)		3.00	4.62	4.86	6.75

DF: degree of freedom; CV: Coefficient of variation; ns, *, and **, respectively, not significant, significant at a $p \leq 0.05$ and $p \leq 0.01$.

The foliar application of SA at the estimated concentration of 2.0 mM promoted an increase in the ascorbic acid content (Figure 11C) and titratable acidity (Figure 11D) of cucumber fruits, even when plants were subjected to the highest salinity of nutrient solution (6.6 dS m^{-1}). However, the highest values of ascorbic acid (1.44 mg 100g^{-1} pulp) and titratable acidity (1.67%, Figure 11D) were obtained in plants grown under ECns of 2.1 dS m^{-1}, corresponding to an increase of 6.67% (0.09 mg 100g^{-1} pulp) in AA and 7.05% (0.11%) in TA compared to the fruits of plants cultivated with the same level of ECns but without an SA application (0 mM). The lowest value of AA (0.67 mg 100g^{-1} pulp) and TA (0.88%) were obtained in plants cultivated with ECns of 6.6 dS m^{-1} and an SA concentration of 5.4 mM.

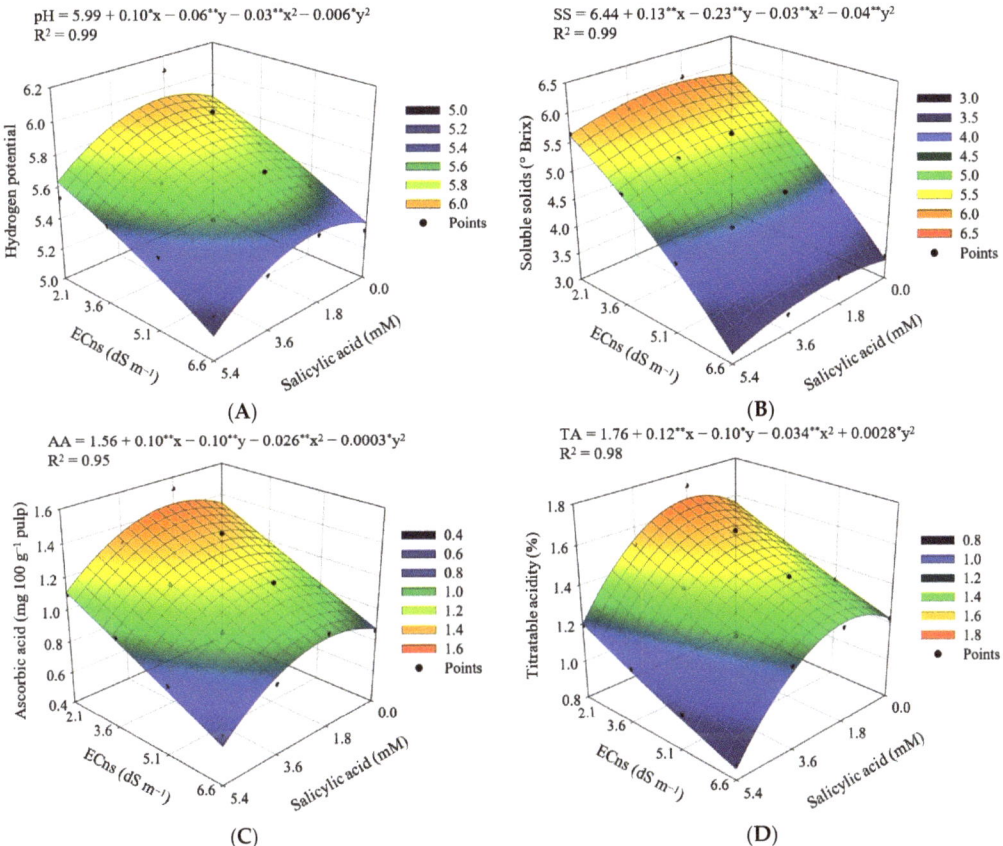

Figure 11. Response surface for hydrogen potential—pH (**A**), soluble solids—SS (**B**), ascorbic acid—AA (**C**), and titratable acidity—TA (**D**) of fruits of Japanese cucumber as a function of the interaction between the electrical conductivity of the nutrient solution (ECns) and the concentrations of salicylic acid (SA), grown in a hydroponic system, 45 days after transplanting. X and Y—concentration of SA and ECns, respectively; * and ** significant at a $p \leq 0.05$ and $p \leq 0.01$, respectively.

4. Discussion

The employment of saline water in crop production, either for supplemental irrigation or in the preparation of nutrient solutions in hydroponic systems, has long been debated [2,3]. Several studies have indicated that the use of brackish water can lead to saline stress and reduced crop yields [39,40]. This study demonstrated that saline stress caused by increased ECns has negative impacts on the physiology, production, and post-harvest quality of fruits of Japanese cucumber grown in an NFT hydroponic system. However, the harmful effects of salt stress were to some extent attenuated by the application of salicylic acid on leaves.

The root is the main organ that remains in direct contact with the nutrient solution and therefore accumulates most of the potentially toxic elements [41]. Under conditions of saline stress, an excess of toxic ions is common, especially Na^+ and Cl^-, which restrict the capacity of roots to absorb water [3,42]. The reduction in water absorption leads to reduced leaf status, as observed in this study through the relative water content (Figure 4A). Similar results were reported for different vegetable crops cultivated in the hydroponic system, such as melon [15], okra [3], and tomatoes [43].

An increase in the ECns caused an increase in the percentage of electrolyte leakage (EL) in the leaf blade (Figure 4B). Salt stress increases the production of reactive oxygen species (ROS) and damages proteins and nucleic acids, as well as cell membrane lipids, causing lipid peroxidation [44,45], and, consequently, greater EL in the leaf blade. However, the increase observed in the present study did not cause injuries to the cell membranes of leaf tissues, as damage is considered to only occur when the EL exceeds 50% [46]. Corroborating the present study, no injuries to the cell membrane of leaves were observed in hydroponic okra and melon plants under salt stress [3,47], with electrolyte leakage in the leaf blade less than 50%.

The harmful effects of salt stress on the RWC and percentage of EL in the leaf blade were attenuated by the foliar application of SA at the concentration of 2.0 mM, i.e., salicylic acid increased the RWC (Figure 4A) and reduced the percentage of EL in the leaf blade (Figure 4B). The beneficial effects of foliar application of salicylic acid have also been reported by [3], in the hydroponic cultivation of okra plants under salt stress (ECns ranging from 2.1 to 9.0 dS m^{-1}). These authors found that the foliar application of SA at a concentration of 1.5 mM was able to increase the RWC and reduce the percentage of EL in the leaf blade. Salicylic acid is one of the principal phenolic compounds and acts as a growth regulator, playing a unique role in various physiological and biochemical processes [9,48]. In addition, it acts in the reduction in lipid peroxidation and can interact with other plant hormones to increase the tolerance of plants to salt stress [49,50].

The leaf gas exchange in cucumber plants was negatively impacted by the increase in ECns (Figure 5). Under conditions of abiotic stress such as saline stress, there is an imbalance between the production of ROS and antioxidant defense, leading to oxidative stress [51]. Thus, the availability of water to plants decreases and the stomata are partially closed (Figure 5B) as a mechanism to reduce the transpiration rate and salt absorption (Figure 5C) [52]. This directly reduces the activity of the ribulose-1,5-bisphosphate carboxylase/oxygenase enzyme in the Calvin cycle and increases the production of ROS by incomplete oxygen recovery, reducing the CO_2 assimilation rate (Figure 5D) [52]. Reductions in the leaf gas exchange in cucumber plants as a function of salt stress have also been observed in other studies, such as [53,54].

On the other hand, the application of SA on leaves at the concentration of 2.0 mM alleviated the effect of salt stress on leaf gas exchange (Figure 5). Salicylic acid can increase the activity of ribulose-1,5-bisphosphate carboxylase/oxygenase as well as potassium absorption and ATP content, maintaining adequate Na^+/K^+ ratio in plants, thus favoring tolerance to saline stress [55]. The SA also increases the accumulation of osmoprotectants, improving turgor in plant cells under stress, and the activation of antioxidant enzymes, resulting in better photosynthetic activity [56].

Photosynthetic pigments play an essential role in energy assimilation in plants and their levels are significantly altered under salt stress [57]. The results of the present study reveal that the increase in ECns negatively affected the synthesis of chlorophyll and carotenoids in cucumber plants (Figure 6). However, the foliar application of SA at a concentration of 2.0 mM mitigated the effects of salt stress, promoting increments in chlorophyll *a* (Figure 6A) and chlorophyll *total* (Figure 6B).

The beneficial effect of SA is associated with its role as a signaling molecule and in the activation of the plant's defense system, which includes osmoregulation, elimination of ROS, and ionic homeostasis [58]. Salicylic acid can stimulate chlorophyll biosynthesis and/or reduce its degradation [59]. An increase in chlorophyll synthesis as a function of the foliar application of SA has also been observed in studies with bell peppers [15], cherry tomatoes [18], and strawberries [60].

Salt stress damages the chloroplast structure, promotes non-radiative heat dissipation, and inhibits electron transfer in photosystem II [2]. The results obtained in the present study demonstrate that increasing the ECns reduced the maximum fluorescence (Figure 7B), indicating damage to the light-harvesting complex of photosystem II. An increase in F0 (Figure 7A) is also an indication of damage to the reaction center of photosystem II or a reduction in the capacity to transfer excitation energy from the light-harvesting system to the reaction center of photosystem II (PSII) [61].

Maximum fluorescence (Figure 7B) and variable fluorescence (Figure 7C) were reduced by the increase in the ECns, with no influence of SA concentrations. Fm is the point at which the fluorescence of the plant reaches its maximum capacity and practically all quinone is reduced [62]. The decrease in Fm reflects a reduction in maximum light energy absorbed by photosystem II and the degradation of photosynthetic pigments [63,64], as observed in the present study (Figure 6). Variable fluorescence refers to the plant's capacity to transfer the energy of electrons ejected from pigment molecules to the formation of NADPH, ATP, and reduced ferredoxin, so its reduction may indicate that the photosynthetic apparatus was damaged by saline stress, compromising photosystem II, with negative effects on the photosynthetic process [65].

The foliar application of SA at the concentration of 2.0 mM increased the quantum efficiency of photosystem II in the dark phase (Figure 7D), regardless of the level of ECns. Thus, the results reveal that the Fv/Fm of plants sprayed with SA at a concentration of 2.0 mM was not compromised up to 5.0 dS m^{-1}, as the values of Fv/Fm ranged from 0.75 to 0.80, i.e., they were greater than or equal to 0.75. Several authors consider Fv/Fm values between 0.75 and 0.85 as normal in non-stressed plants [66,67].

Salicylic acid, in addition to signaling antioxidant genes and proteins under salt stress conditions, can lead to a greater accumulation of ions responsible for osmoregulation and membrane structure, such as K^+ and Ca^{2+}, and reduce the concentration of toxic Na^+ and Cl^- ions [16]. This may be related to increased photochemical efficiency in cucumber plants sprayed with SA at a concentration of 2.0 mM [68,69]. Corroborating the present study, Mendonça et al. [3] evaluated the effect of foliar application of SA on hydroponic okra plants under salt stress (ECns ranging from 2.1 to 9.0 dS m^{-1}) and reported that the application of SA at a concentration of 1.2 mM promoted an increase in Fv/Fm.

The reductions in the initial fluorescence before the saturation pulse and in the quantum efficiency of photosystem II in the light phase (Y_{II}) (Figure 8) due to the increase in the ECns indicate a decrease in photosynthetic activity, which corroborates the reductions observed in the CO_2 assimilation rate (Figure 5D) of plants subjected to the highest levels of ECns.

The results of this study indicated that an increase in the electrical conductivity of the nutrient solution negatively affected the production components of Japanese cucumber, with reductions in the NFP, AFW, TPP, AFL, and AFD. These results are a consequence of the osmotic and ionic effects induced by the high salinity of the nutrient solution [3]. The reduction in osmotic potential causes water stress and accumulation of Na^+ and Cl^- ions, leading to nutritional imbalance and a consequent reduction in production components [70]. Reductions in production components as a function of salt stress in hydroponic cultivation have also been observed in other studies with okra [71], cucumber [72], 'Biquinho' pepper [73], and Italian zucchini [40].

The foliar application of SA at a concentration of 2.0 mM promoted an increase in the number of fruits, average fruit weight, and total production per plant, especially in plants cultivated under ECns of 2.1 dS m^{-1}. The increase in production components observed in plants treated with SA (2.0 mM) may be associated with chlorophyll a fluorescence and gas exchange responses. In summary, the foliar application of SA in cucumber plants acted as an elicitor, increased leaf turgor (Figure 4A), and reduced lipid peroxidation (Figure 4B), consequently protecting the photosynthetic apparatus of the plants. In addition, this protection was observed through the higher quantum efficiency of photosystem II (Figure 7D). Finally, the plants treated with SA at a concentration of 2.0 mM showed the highest CO_2 assimilation rate (Figure 5D) and this photo-assimilated carbon was translocated to the fruits, resulting in a greater number of fruits, average fruit weight, and total production per plant (Figure 9).

It is worth mentioning that, in this study, SA concentrations above 2.1 mM intensified the deleterious effects of salt stress, causing reductions in the production components, the quantum efficiency of photosystem II, and CO_2 assimilation rate with lower values obtained in plants cultivated under ECns of 6.6 dS m^{-1} and sprayed with SA at a concentration of 5.4 mM. Therefore, the beneficial effect of SA depends on several factors, including concentration and mode of application [10,11]. According to Aires et al. [74], high concentrations of SA can cause high levels of oxidative stress, leading to a reduction in stress tolerance.

From the results obtained in this study, it can be suggested that the application of SA on leaves at adequate concentrations can increase the production of Japanese cucumber cultivated in a hydroponic system, probably by stimulating physiological processes involved in the active transfer of photosynthetic products from the source to the sink, which is consistent with the results obtained by [75]. This beneficial effect of SA on the production components of cucumber may be related to its role in reducing the absorption of Na$^+$ and increasing the absorption of N, P, K, Ca, and Mg by plants [76]. Increases in production components as a function of the foliar application of SA have also been observed in hydroponic okra [3] and melon [15].

Average fruit length and diameter are variables of great interest to the consumer because they define the size of the edible part and the classification of the fruit. In this study, there was no positive or negative influence of the application of SA on the diameter and length of the cucumber fruit, although the SA increased the number of fruits, their average weight, and, consequently, the total production per plant. However, the length and diameter of cucumber fruits were reduced by the increase in the ECns. These results are a consequence of the high salinity of the nutrient solution, which can cause a water deficit by lowering the osmotic potential and toxicity of specific ions such as Cl$^-$ and Na$^+$ [13].

Post-harvest variables play an important role because they are responsible for the organoleptic characteristics that provide the feeling of freshness and palatability [77]. The pH is an important characteristic in fruit quality evaluation because low values can guarantee pulp conservation without the need for heat treatment, hence avoiding the loss of nutritional quality. Thus, the reduction observed in pH with an increase in the ECns shows that the saline stress imposed on the cucumber increased the acidic character of the pulp, besides reducing the concentration of the total soluble solids. It is worth pointing out that changes in the post-harvest quality of the fruits produced under salt stress conditions occur due to the action of the osmotic effect of the nutrient solution, inhibiting the absorption of water and nutrients by plants and their photosynthetic capacity [78].

The foliar application of SA at a concentration of 2.0 mM promoted an increase in soluble solids, especially in plants grown under ECns of 2.1 dS m^{-1}. The increase in soluble solids may be related to the role of SA in reducing the rate of degradation of polysaccharides and, consequently, the greater availability of simple sugars, besides delaying fruit maturity by inhibiting the production and effects of ethylene [79]. An increase in soluble solids content due to SA application was also reported by [46] in the melon fruits cultivated in a hydroponic system with brackish water. The ascorbic acid content and titratable acidity were also increased as a function of the application of SA at a concentration of 2.0 mM. The increase in these variables is a highly valued characteristic, particularly when one intends to use cucumber for industrial processing, as it reduces the employment of acidifiers, thus improving nutritional and organoleptic quality, being considered a principal attribute for post-harvest evaluation [80].

5. Conclusions

Cucumber is a vegetable crop that is sensitive to saline stress even when grown in a hydroponic system, being negatively affected by the electrical conductivity of nutrient solution above 2.1 dS m^{-1}. However, the application of salicylic acid on leaves between concentrations of 1.4 and 2.0 mM not only promotes an increase in the relative water content in the leaf blade but also positively influences the synthesis of photosynthetic pigments, leaf gas exchange, and the quantum efficiency of photosystem II, reducing the percentage of electrolyte leakage in the leaf blade. In addition, salicylic acid increases the production and postharvest quality of cucumber fruits. These observations reinforce the hypothesis that the foliar application of salicylic acid in adequate concentrations can act as a crucial signaling molecule in the attenuation of saline stress in cucumber plants, which can enhance the use of brackish water in hydroponic cultivation, especially in regions with the scarcity of freshwater of low salinity. Further studies are needed to understand how salicylic acid acts in salt stress signaling through biochemical analysis. These studies will help in understanding the metabolic and detoxifying mechanisms that occur in plants to develop efficient strategies to mitigate the harmful effects of saline stress.

Author Contributions: Conceptualization, V.K.N.O. and G.S.d.L.; methodology, A.A.R.d.S. and V.K.N.O.; software, A.A.R.d.S. and P.D.F.; validation, V.K.N.O., G.S.d.L. and L.A.d.A.S.; formal analysis, R.G.N. and L.A.d.A.S.; investigation, G.S.d.L. and V.K.N.O.; resources, G.S.d.L. and L.A.d.A.S.; data collection, A.A.R.d.S.; writing—original draft preparation, V.K.N.O. and G.S.d.L.; writing—review and editing, A.A.R.d.S., G.S.d.L., C.A.V.d.A., V.L.A.d.L., C.F.d.L., L.H.G.C., P.D.F. and H.R.G.; supervision, H.R.G.; project administration, G.S.d.L. and L.A.d.A.S. All authors have read and agreed to the published version of the manuscript.

Funding: CNPq (National Council for Scientific and Technological Development) (Proc. 313796/2020-3), CAPES (Coordination for the Improvement of Higher Education Personnel) financial code—001, and UFCG (Universidade Federal de Campina Grande).

Data Availability Statement: Not applicable.

Acknowledgments: The authors thank the Graduate Program in Agricultural Engineering of the Federal University of Campina Grande, the National Council for Scientific and Technological Development (CNPq), and the Coordination for the Improvement of Higher Education Personnel (CAPES) for the support in carrying out this research.

Conflicts of Interest: The authors declare no conflict of interest.

References

1. Zhang, Y.; Li, X.; Simunek, J.; Shi, H.; Chen, N.; Hu, Q.; Tian, T. Evaluating soil salt dynamics in a field drip-irrigated with brackish water and leached with freshwater during different crop growth stages. *Agric. Water Manag.* **2021**, *244*, e106601. [CrossRef]
2. Zhang, W.; Du, T. Fresh/brackish watering at growth period provided a trade-off between lettuce growth and resistance to NaCl-induced damage. *Sci. Hortic.* **2022**, *304*, e111283. [CrossRef]
3. Mendonça, A.J.T.; da Silva, A.A.R.; de Lima, G.S.; Soares, L.A.A.; Oliveira, V.K.N.; Gheyi, H.R.; Fernandes, P.D. Salicylic acid modulates okra tolerance to salt stress in hydroponic system. *Agriculture* **2022**, *12*, 1687. [CrossRef]
4. Bachani, J.; Mahanty, A.; Aftab, T.; Kumar, K. Insight into calcium signalling in salt stress response. *S. Afr. J. Bot.* **2022**, *151*, 1–8. [CrossRef]
5. Sousa, H.C.; de Sousa, G.G.; Cambissa, P.B.; Lessa, C.I.; Goes, G.F.; Silva, F.D.B.; Viana, T.V.A. Gas exchange and growth of zucchini crop subjected to salt and water stress. *Rev. Bras. Eng. Agricola Ambient.* **2022**, *26*, 815–822. [CrossRef]
6. Najar, R.; Aydi, S.; Sassi-Aydi, S.; Zarai, A.; Abdelly, C. Effect of salt stress on photosynthesis and chlorophyll fluorescence in Medicago truncatula. *Plant Biosyst.* **2019**, *153*, 88–97. [CrossRef]
7. Fang, S.; Hou, X.; Liang, X. Response mechanisms of plants under saline-alkali stress. *Front. Plant Sci.* **2021**, *12*, e667458. [CrossRef]
8. Arif, Y.; Sami, F.; Siddiqui, H.; Bajguz, A.; Hayat, S. Salicylic acid in relation to other phytohormones in plant: A study towards physiology and signal transduction under challenging environment. *Environ. Exp. Bot.* **2020**, *175*, 104040. [CrossRef]
9. Poór, P. Effects of salicylic acid on the metabolism of mitochondrial reactive oxygen species in plants. *Biomolecules* **2020**, *10*, 341. [CrossRef]
10. Horváth, E.; Szalai, G.; Janda, T. Induction of abiotic stress tolerance by salicylic acid signaling. *J. Plant Growth Regul.* **2007**, *26*, 290–300. [CrossRef]
11. Poór, P.; Borbély, P.; Bódi, N.; Bagyánszki, M.; Tari, I. Effects of salicylic acid on photosynthetic activity and chloroplast morphology under light and prolonged darkness. *Photosynthetica* **2019**, *57*, 367–376. [CrossRef]
12. Veloso, L.L.S.A.; de Lima, G.S.; da Silva, A.A.R.; Souza, L.P.; de Lacerda, C.N.; Silva, I.J.; Fernandes, P.D. Attenuation of salt stress on the physiology and production of bell peppers by treatment with salicylic acid. *Semin. Ciênc. Agrar.* **2021**, *42*, 2751–2768. [CrossRef]
13. Da Silva, A.A.R.; Sousa, P.F.N.; de Lima, G.S.; Soares, L.A.A.; Gheyi, H.R.; Azevedo, C.A.V. Hydrogen peroxide reduces the effect of salt stress on growth and postharvest quality of hydroponic mini watermelon. *Water Air Soil Pollut.* **2022**, *233*, 198. [CrossRef]
14. Sousa, V.F.O.; Santos, A.S.; Sales, W.S.; Silva, A.J.; Gomes, F.A.L.; Dias, T.J.; Araújo, J.R.E.S. Exogenous application of salicylic acid induces salinity tolerance in eggplant seedlings. *Braz. J. Biol.* **2022**, *84*, e257739. [CrossRef]

15. Soares, M.D.M.; Dantas, M.V.; de Lima, G.S.; Oliveira, V.K.N.; Soares, L.A.A.; Gheyi, H.R.; Fernandes, P.D. Physiology and yield of 'Gaúcho' melon under brackish water and salicylic acid in hydroponic cultivation. *Arid. Land Res. Manag.* **2022**, *37*, 134–153. [CrossRef]
16. Silva, T.I.; Silva, J.S.; Dias, M.G.; Martins, J.V.S.; Ribeiro, W.S.; Dias, T.J. Salicylic acid attenuates the harmful effects of salt stress on basil. *Rev. Bras. Eng. Agricola Ambient.* **2022**, *26*, 399–406. [CrossRef]
17. Kaloterakis, N.; van Delden, S.H.; Hartley, S.; De Deyn, G.B. Silicon application and plant growth promoting rhizobacteria consisting of six pure *Bacillus* species alleviate salinity stress in cucumber (*Cucumis sativus*). *Sci. Hortic.* **2021**, *288*, e110383. [CrossRef]
18. Silva, A.A.R.; Veloso, L.L.S.A.; de Lima, G.S.; Soares, L.A.A.; Chaves, L.H.G.; Silva, F.A.; Fernandes, P.D. Induction of salt stress tolerance in cherry tomatoes under different salicylic acid application methods. *Semin. Cienc. Agrar.* **2022**, *43*, 1145–1166. [CrossRef]
19. Bezerra, R.R.; Júnior, J.A.S.; Pessoa, U.C.; Silva, Ê.F.d.F.e.; de Oliveira, T.F.; Nogueira, K.F.; de Souza, E.R. Water efficiency of coriander under flows of application of nutritive solutions prepared in brackish waters. *Water* **2022**, *14*, 4005. [CrossRef]
20. Yildirim, E.; Turan, M.; Guvenc, I. Effect of foliar salicylic acid applications on growth, chlorophyll, and mineral content of cucumber grown under salt stress. *J. Plant Nutr.* **2008**, *31*, 593–612. [CrossRef]
21. Das, A.; Singh, S.; Islam, Z.; Munshi, A.D.; Behera, T.K.; Dutta, S.; Dey, S. Current progress in genetic and genomics-aided breeding for stress resistance in cucumber (*Cucumis sativus* L.). *Sci. Hortic.* **2022**, *300*, e111059. [CrossRef]
22. Maas, E.V.; Hoffman, G.J. Crop Salt tolerance—Current assessment. *J. Irrig. Drain. Div. ASCE.* **1977**, *103*, 115–134.
23. Medeiros, P.R.; Duarte, S.N.; Dias, C.T. Tolerância da cultura do pepino à salinidade em ambiente protegido. *Rev. Bras. Eng. Agricola Ambient.* **2009**, *13*, 406–410. [CrossRef]
24. Sotiroudis, G.; Melliou, E.; Sotiroudis, T.G.; Chinou, I. Chemical analysis, antioxidant and antimicrobial activity of three Greek cucumber (*Cucumis sativus*) cultivars. *J. Food Biochem.* **2010**, *34*, 61–78.
25. Uthpala, T.G.G.; Marapana, R.A.U.J. Study on nutritional composition on firmness of two gherkin (*Cucumis sativus* L.) varieties (Ajax and Vlasset) on brine fermentation. *Am. J. Food Sci. Technol.* **2017**, *5*, 61–63. [CrossRef]
26. Agatemor, U.M.; Nwodo, O.F.; Anosike, C.A. Phytochemical and proximate composition of cucumber (*Cucumis sativus*) fruit from Nsukka, Nigeria *Afr. J. Biotechnol.* **2018**, *17*, 1215–1219. [CrossRef]
27. Brengi, S.H.; Abd Allah, E.M.; Abouelsaad, I.A. Effect of melatonin or cobalt on growth, yield and physiological responses of cucumber (*Cucumis sativus* L.) plants under salt stress. *J. Saudi Soc. Agric. Sci.* **2022**, *21*, 51–60. [CrossRef]
28. Chen, T.W.; Gomez Pineda, I.M.; Brand, A.M.; Stützel, H. Determining ion toxicity in cucumber under salinity stress. *Agronomy* **2020**, *10*, 677. [CrossRef]
29. Isla. Available online: https://www.isla.com.br/arquivos-para-download/catalogos (accessed on 5 November 2022).
30. Medeiros, P.R.; Duarte, S.N.; Dias, C.T.; Silva, M.F. Tolerância do pepino à salinidade em ambiente protegido: Efeitos sobre propriedades físico-químicas dos frutos. *Irriga* **2010**, *15*, 301–311. [CrossRef]
31. Hoagland, D.R.; Arnon, D.I. The water-culture method for growing plants without soil. *Circ. Calif. Agric. Exp. Stn.* **1950**, *347*, 32.
32. Medeiros, J.F.; Lisboa, R.A.; de Oliveira, M.; de Silva Júnior, M.J.; Alves, L.P. Caracterização das águas subterrâneas usadas para irrigação na área produtora de melão da Chapada do Apodi. *Rev. Bras. Eng. Agricola Ambient.* **2003**, *7*, 469–472. [CrossRef]
33. Weatherley, P.E. Studies in the water relations of the cotton plant. I. The field measurement of water deficits in leaves. *New Phytol.* **1950**, *49*, 81–97. [CrossRef]
34. Scotti-Campos, P.; Pham-Thi, A.T.; Semedo, J.N.; Pais, I.P.; Ramalho, J.C.; Matos, M.C. Physiological responses and membrane integrity in three Vigna genotypes with contrasting drought tolerance. *Emir. J. Food Agric.* **2013**, *25*, 1002–1013. [CrossRef]
35. Arnon, D.I. Copper enzymes in isolated chloroplasts: Polyphenoloxidase in *Beta vulgaris*. *Plant Physiol.* **1949**, *24*, 1–15. [CrossRef]
36. Carvalho, A.D.F.; Amaro, G.B.; Lopes, J.F.; Vilela, N.J.; Michereff Filho, M.; Andrade, R. *A Cultura do Pepino*; MAPA: Brasília, Brazil, 2013; Volume 1, 18p.
37. Instituto Adolfo Lutz—IAL. Normas analíticas do Instituto Adolfo Lutz. In *Métodos Químicos e Físicos Para Análise de Alimentos*, 3rd ed.; IMESP: São Paulo, Brazil, 2008; p. 1020.
38. Ferreira, D.F. SISVAR: A computer analysis system to fixed effects split plot type designs. *Rev. Bras. Biom.* **2019**, *37*, 529–535. [CrossRef]
39. Cavalcante, E.S.; Lacerda, C.F.D.; Costa, R.N.T.; Gheyi, H.R.; Pinho, L.L.; Bezerra, F.M.S.; Canjá, J.F. Supplemental irrigation using brackish water on maize in tropical semi-arid regions of Brazil: Yield and economic analysis. *Sci. Agric.* **2021**, *78*, e20200151. [CrossRef]
40. Dantas, M.V.; de Lima, G.S.; Gheyi, H.R.; Pinheiro, F.W.A.; Silva, P.C.C.; Soares, L.A.A. Gas exchange and hydroponic production of zucchini under salt stress and H_2O_2 application. *Rev. Caatinga.* **2022**, *35*, 436–449. [CrossRef]
41. Yadav, M.; Gupta, P.; Seth, C.S. Foliar application of α-lipoic acid attenuates cadmium toxicity on photosynthetic pigments and nitrogen metabolism in *Solanum lycopersicum* L. *Acta Physiol. Plant.* **2022**, *44*, 112. [CrossRef]
42. Roque, I.A.; Soares, L.A.A.; de Lima, G.S.; Lopes, I.A.P.; Silva, L.A.; Fernandes, P.D. Biomass, gas exchange and production of cherry tomato cultivated under saline water and nitrogen fertilization. *Rev. Caatinga* **2022**, *35*, 686–696. [CrossRef]
43. Ali, M.; Kamran, M.; Abbasi, G.H.; Saleem, M.H.; Ahmad, S.; Parveen, A.; Fahad, S. Melatonin-induced salinity tolerance by ameliorating osmotic and oxidative stress in the seedlings of two tomato (*Solanum lycopersicum* L.) cultivars. *J. Plant Growth Regul.* **2021**, *40*, 2236–2248. [CrossRef]

44. Rady, M.M.; Belal, H.E.E.; Gadallah, F.M.; Semida, W.M. Selenium application in two methods elevates drought tolerance in *Solanum lycopersicum* by increasing yield, quality, and antioxidant defense system and suppressing oxidative stress biomarkers. *Sci. Hortic.* **2020**, *266*, e109290. [CrossRef]
45. Sachdev, S.; Ansari, S.A.; Ansari, M.I.; Fujita, M.; Hasanuzzaman, M. Abiotic stress and reactive oxygen species: Generation, signaling, and defense mechanisms. *Antioxidants* **2021**, *10*, 277. [PubMed]
46. Sullivan, C.Y. Mechanisms of heat drought resistence in grain sorghum and methods of measurement. In *Sorghum in Seventies*; Rao, N.G.P., House, L.R., Eds.; Oxford and IBH Publication: New Delhi, India, 1971; Volume 1, 247p.
47. Oliveira, V.K.N.; de Lima, G.S.; Soares, M.D.M.; Soares, L.A.A.; Gheyi, H.R.; Silva, A.A.R.; Fernandes, P.D. Salicylic acid does not mitigate salt stress on the morphophysiology and production of hydroponic melon. *Braz. J. Biol.* **2022**, *82*, e262664. [CrossRef] [PubMed]
48. Xavier, A.V.; de Lima, G.S.; Gheyi, H.R.; da Silva, A.A.R.; de Lacerda, C.N.; Soares, L.A.A.; Fernandes, P.D. Salicylic acid alleviates salt stress on guava plant physiology during rootstock formation. *Rev. Bras. Eng. Agrícola e Ambient.* **2022**, *26*, 855–862. [CrossRef]
49. Gharbi, E.; Martínez, J.P.; Benahmed, H.; Fauconnier, M.L.; Lutts, S.; Quinet, M. Salicylic acid differently impacts ethylene and polyamine synthesis in the glycophyte *Solanum lycopersicum* and the wild-related halophyte *Solanum chilense* exposed to mild salt stress. *Physiol. Plant.* **2016**, *158*, 152–167. [CrossRef]
50. Jini, D.; Joseph, B. Physiological mechanism of salicylic acid for alleviation of salt stress in rice. *Rice Sci.* **2017**, *24*, 97–108. [CrossRef]
51. Roumani, A.; Biabani, A.; Karizaki, A.R.; Alamdari, E.G. Foliar salicylic acid application to mitigate the effect of drought stress on isabgol (*Plantago ovata forssk*). *Biochem. Syst. Ecol.* **2022**, *104*, e104453. [CrossRef]
52. Dias, A.S.; de Lima, G.S.; Sá, F.V.S.; Gheyi, H.R.; Soares, L.A.A.; Fernandes, P.D. Gas exchanges and photochemical efficiency of West Indian cherry cultivated with saline water and potassium fertilization. *Rev. Bras. Eng. Agricola Ambient.* **2018**, *22*, 628–633. [CrossRef]
53. Gurmani, A.R.; Khan, S.U.; Ali, A.; Rubab, T.; Schwinghamer, T.; Jilani, G.; Zhang, J. Salicylic acid and kinetin mediated stimulation of salt tolerance in cucumber (*Cucumis sativus* L.) genotypes varying in salinity tolerance. *HEB* **2018**, *59*, 461–471. [CrossRef]
54. Turan, M.; Ekinci, M.; Kul, R.; Boynueyri, F.G.; Yildirim, E. Mitigation of salinity stress in cucumber seedlings by exogenous hydrogen sulfide. *J. Plant Res.* **2022**, *135*, 517–529. [CrossRef]
55. Lee, S.Y.; Damodaran, P.N.; Roh, K.S. Influence of salicylic acid on rubisco and rubisco activase in tobacco plant grown under sodium chloride in vitro. *Saudi J. Biol. Sci.* **2014**, *21*, 417–426. [CrossRef]
56. Costa, A.A.; Paiva, E.P.; Torres, S.B.; Souza Neta, M.L.; Pereira, K.T.O.; Leite, M.S.; Benedito, C.P. Osmoprotection in *Salvia hispanica* L. seeds under water stress attenuators. *Braz. J. Biol.* **2021**, *82*, e233547. [CrossRef] [PubMed]
57. Nigam, B.; Dubey, R.S.; Rathore, D. Protective role of exogenously supplied salicylic acid and PGPB (*Stenotrophomonas* sp.) on spinach and soybean cultivars grown under salt stress. *Sci. Hortic.* **2022**, *293*, e110654. [CrossRef]
58. Kang, G.; Li, G.; Guo, T. Molecular mechanism of salicylic acid-induced abiotic stress tolerance in higher plants. *Acta Physiol. Plant.* **2014**, *36*, 2287–2297. [CrossRef]
59. Hundare, A.; Joshi, V.; Joshi, N. Salicylic acid attenuates salinity-induced growth inhibition in in vitro raised ginger (*Zingiber officinale* Roscoe) plantlets by regulating ionic balance and antioxidative system. *Plant Stress.* **2022**, *4*, e100070. [CrossRef]
60. Lamnai, K.; Anaya, F.; Fghire, R.; Zine, H.; Janah, I.; Wahbi, S.; Loutfi, K. Combined effect of salicylic acid and calcium application on salt-stressed strawberry plants. *Russ. J. Plant Physiol.* **2022**, *69*, 12. [CrossRef]
61. Da Silva, A.A.R.; de Lima, G.S.; de Azevedo, C.A.V.; Gheyi, H.R.; Soares, L.A.A.; Veloso, L.L.S.A. Salicylic acid improves physiological indicators of soursop irrigated with saline water. *Rev. Bras. Eng. Agricola Ambient.* **2022**, *26*, 412–419. [CrossRef]
62. Veloso, L.L.S.A.; de Azevedo, C.A.V.; Nobre, R.G.; de Lima, G.S.; de Capitulino, J.D.; Silva, F.A. H_2O_2 alleviates salt stress effects on photochemical efficiency and photosynthetic pigments of cotton genotypes. *Rev. Bras. Eng. Agricola Ambient.* **2023**, *27*, 34–41. [CrossRef]
63. Silva, M.M.P.; Vasquez, H.M.; Bressan-Smith, R.; da Silva, J.F.C.; Erbesdobler, E.D.A.; Andrade Junior, P.S.C. Eficiência fotoquímica de gramíneas forrageiras tropicais submetidas à deficiência hídrica. *Rev. Bras. Zootec.* **2006**, *35*, 67–74. [CrossRef]
64. Tatagiba, S.D.; Moraes, G.A.B.K.; Nascimento, K.J.T.; Peloso, A.F. Limitações fotossintéticas em folhas de plantas de tomateiro submetidas a crescentes concentrações salinas. *Eng. Agri.* **2014**, *22*, 138–149.
65. Sá, F.V.S.; Gheyi, H.R.; de Lima, G.S.; de Paiva, E.P.; Silva, L.A.; Moreira, R.C.L.; Dias, A.S. Ecophysiology of West Indian cherry irrigated with saline water under phosphorus and nitrogen doses. *Biosci. J.* **2019**, *35*, 211–221. [CrossRef]
66. Larbi, A.; Baccar, R.; Boulal, H. Response of olive tree to ammonium nitrate fertilization under saline conditions. *J. Plant Nutr.* **2020**, *44*, 1432–1445. [CrossRef]
67. da Silva, J.S.; Sá, F.V.D.S.; Dias, N.D.S.; Neto, M.F.; Jales, G.D.; Fernandes, P.D. Morphophysiology of mini watermelon in hydroponic cultivation using reject brine and substrates. *Rev. Bras. Eng. Agrícola Ambient.* **2021**, *25*, 402–408. [CrossRef]
68. Agnihotri, A.; Gupta, P.; Dwivedi, A.; Seth, C.S. Counteractive mechanism (s) of salicylic acid in response to lead toxicity in *Brassica juncea* (L.) Czern. cv. Varuna. *Planta* **2018**, *248*, 49–68. [CrossRef]
69. Gupta, S.; Seth, C.S. Salicylic acid alleviates chromium (VI) toxicity by restricting its uptake, improving photosynthesis and augmenting antioxidant defense in *Solanum lycopersicum* L. *Physiol. Mol. Biol. Plants* **2021**, *27*, 2651–2664. [CrossRef] [PubMed]

70. Ó, L.M.G.D.; Cova, A.M.W.; Neto, A.D.D.A.; da Silva, N.D.; Silva, P.C.C.; Santos, A.L.; Gheyi, H.R.; da Silva, L.L. Osmotic adjustment, production, and post-harvest quality of mini watermelon genotypes differing in salt tolerance. *Sci. Hort.* **2022**, *306*, e111463. [CrossRef]
71. Modesto, F.J.N.; dos Santos, M.Â.C.M.; Soares, T.M.; Santos, E.P.M. Crescimento, produção e consumo hídrico do quiabeiro submetido à salinidade em condições hidropônicas. *Irriga* **2019**, *24*, 2486–2497. [CrossRef]
72. Abbasi, F.; Khaleghi, A.; Khadivi, A. The effect of salicylic acid on physiological and morphological traits of cucumber (*Cucumis sativus* L. cv. Dream). *Gesunde Pflanz.* **2020**, *72*, 155–162. [CrossRef]
73. Bione, M.A.A.; Soares, T.M.; Cova, A.M.W.; Silva Paz, V.P.; Gheyi, H.R.; Rafael, M.R.S.; Neves, B.S.L. Hydroponic production of 'Biquinho' pepper with brackish water. *Agric. Water Manag.* **2021**, *245*, e106607. [CrossRef]
74. Aires, E.S.; Ferraz, A.K.L.; Carvalho, B.L.; Teixeira, F.P.; Rodrigues, J.D.; Ono, E.O. Foliar application of salicylic acid intensifies antioxidant system and photosynthetic efficiency in tomato plants. *Bragantia* **2022**, *81*, e1522. [CrossRef]
75. Khan, M.I.R.; Fatma, M.; Per, T.S.; Anjum, N.A.; Khan, N.A. Salicylic acid-induced abiotic stress tolerance and underlying mechanisms in plants. *Front. Plant Sci.* **2015**, *6*, 462. [CrossRef] [PubMed]
76. Farhangi-Abriz, S.; Ghassemi-Golezani, K. How can salicylic acid and jasmonic acid mitigate salt toxicity in soybean plants? *Ecotoxicol. Environ. Saf.* **2018**, *147*, 1010–1016. [CrossRef] [PubMed]
77. Seymen, M.; Yavuz, D.; Ercan, M.; Akbulut, M.; Çoklar, H.; Kurtar, E.S.; Türkmen, Ö. Effect of wild watermelon rootstocks and water stress on chemical properties of watermelon fruit. *HEB* **2021**, *62*, 411–422. [CrossRef]
78. De Lima, G.S.; Pinheiro, F.W.A.; Gheyi, H.R.; Soares, L.A.D.A.; Silva, S.S. Growth and post-harvest fruit quality of West Indian cherry under saline water irrigation and potassium fertilization. *Rev. Caatinga* **2020**, *33*, 775–784. [CrossRef]
79. Blankenship, S.M.; Dole, J.M. 1-Methylcyclopropene: A review. *Postharvest Biol. Technol.* **2003**, *28*, 1–25. [CrossRef]
80. Brasil, A.S.; Sigarini, K.D.S.; Pardinho, F.C.; Faria, R.; Siqueira, N.F.M.P. Avaliação da qualidade físico-química de polpas de fruta congeladas comercializadas na cidade de Cuiabá MT. *Rev. Bras. Fruticultura* **2016**, *38*, 167–175. [CrossRef]

Disclaimer/Publisher's Note: The statements, opinions and data contained in all publications are solely those of the individual author(s) and contributor(s) and not of MDPI and/or the editor(s). MDPI and/or the editor(s) disclaim responsibility for any injury to people or property resulting from any ideas, methods, instructions or products referred to in the content.

Article

NaCl Accumulation, Shoot Biomass, Antioxidant Capacity, and Gene Expression of *Passiflora edulis* f. Flavicarpa Deg. in Response to Irrigation Waters of Moderate to High Salinity

Jorge F. S. Ferreira *, Xuan Liu, Stella Ribeiro Prazeres Suddarth, Christina Nguyen [†] and Devinder Sandhu

US Salinity Laboratory (USDA-ARS), 450 W. Big Springs Rd., Riverside, CA 92507, USA
* Correspondence: jorge.ferreira@usda.gov
[†] deceased.

Abstract: *Passiflora edulis* f. flavicarpa (yellow passion fruit) is a high-value tropical crop explored for both fruit and nutraceutical markets. As the fruit production in the US rises, the crop must be investigated for the effects of salinity under semi-arid climates. We assessed the effects of irrigation-water salinity, leaf age, and drying method on leaf antioxidant capacity (LAC) and plant genetic responses. Plants were grown in outdoor lysimeter tanks for three years, with waters of electrical conductivities of 3.0, 6.0, and 12.0 dS m^{-1}. Both Na and Cl significantly increased with salinity; leaf biomass at 3.0 and 6.0 dS m^{-1} were similar but reduced significantly at 12.0 dS m^{-1}. Salinity had no effect on LAC, but new leaves had the highest LAC compared to older leaves. Low-temperature oven-dried (LTO) and freeze-dried (FD) leaves had the same LAC. The analyses of twelve transporter genes, six involved in Na$^+$ transport and six in Cl$^-$ transport, showed higher expressions in roots than in leaves, indicating a critical role of roots in ion transport and the control of leaf salt concentration. Passion fruit's moderate tolerance to salinity and its high leaf antioxidant capacity make it a potential new fruit crop for California, as well as a rich source of flavonoids for the nutraceutical market. Low-temperature oven drying is a potential alternative to lyophilization in preparation for Oxygen Radical Absorbance Capacity (ORAC) analysis of passion fruit leaves.

Keywords: *Passiflora edulis* f. flavicarpa; irrigation-water salinity; salinity response; antioxidant capacity; drying procedure; mineral status; genetic response to salinity

Citation: Ferreira, J.F.S.; Liu, X.; Suddarth, S.R.P.; Nguyen, C.; Sandhu, D. NaCl Accumulation, Shoot Biomass, Antioxidant Capacity, and Gene Expression of *Passiflora edulis* f. Flavicarpa Deg. in Response to Irrigation Waters of Moderate to High Salinity. *Agriculture* **2022**, *12*, 1856. https://doi.org/10.3390/agriculture12111856

Academic Editor: Mercè Llugany

Received: 4 October 2022
Accepted: 1 November 2022
Published: 5 November 2022

Publisher's Note: MDPI stays neutral with regard to jurisdictional claims in published maps and institutional affiliations.

Copyright: © 2022 by the authors. Licensee MDPI, Basel, Switzerland. This article is an open access article distributed under the terms and conditions of the Creative Commons Attribution (CC BY) license (https://creativecommons.org/licenses/by/4.0/).

1. Introduction

Passion fruit (*Passiflora edulis* spp., Passifloraceae) is one of the most consumed small fruits worldwide, with approximately 150 native species found in Brazil, although only a few species produce edible fruits [1]. These species have fruit peels of different colors, ranging from pale yellow to purple. Aside from the economic value given by the different colors, sizes, flavors, and aromas of the different fruits, the multicolored flowers give the species a high ornamental value. The yellow passion fruit (*Passiflora edulis* Sims. forma flavicarpa Deg.) is reported to be originally from Brazil, where its common names are based on the color of the fruit rind, used to differentiate the cultivars [2]. A recent hybrid named 'BRS Gigante Amarelo 1' is the 'Brasil Yellow Giant 1' developed by Embrapa Cerrados and collaborators and released in 2008. This cultivar produces fruits ranging in fresh weight from 120 to 350 g, and fruit yield ranging from 16–42 ton ha^{-1} when spaced at 3 m × 5 m, fertilized and irrigated according to the crop recommendation, and manually pollinated daily during flowering [3]. The global data for the trade of minor tropical fruits remain difficult to obtain, but the 2015–2017 worldwide estimated production of passion fruit was around 1.47 million metric tons (MT), the fifth highest after guava (6.7 MT), longan and lychee (both 3.4 MT), and durian (2.3 MT) [4]. Thus, there are no data for worldwide passion fruit production by country compiled by FAO. Brazil is the largest producer and consumer of passion fruit; however, the yield of *Passiflora edulis* f. flavicarpa decreased from 838,000 MT in 2013 to 593,400 MT in 2019 (https://www.statista.com/statistics/1078358/production-passion-fruit-brazil/), accessed on 27 October 2022. In contrast, the Colombian passion fruit cultivated area grew from 15,000 ha in 2016 to 21,000 ha in 2017 (https://farmfolio.net/articles/colombias-passion-fruit-production-continues-rise/), accessed 10 September 2022. Colombia reported passion fruit production of 150,000 MT in 2016, with most of its product staying home and only 6,000 MT exported to foreign markets. In 2018, Colombia's production was mostly (72%) *P. edulis* f. flavicarpa (169,000 MT), followed by *P. ligularis* (sweet granadilla) with 47,460 MT, and *P. edulis* f. edulis (purple passion fruit) with 24,800 MT (https://storymaps.arcgis.com/stories/bf57e656b0334bac83f73f5f9e5fca50), accessed on 10 September 2022. Most of the produced *P. edulis* f. edulis was used for export purposes.

The passion fruit is primarily famous for its sweet (purple fruit) or sour (yellow fruit) pulp; the nutraceutical market has been long established for the use of the leaf and flower extracts due to their anxiolytic and sedative effects [5] as well as for being generally regarded as safe [6]. Although considered safe for most users, the extract of *Passiflora incarnata*, at therapeutic doses of 500–1000 mg three times a day, was reported to cause adverse effects in one 34-year-old female patient postpartum [7]. Because there was no toxicity associated with the analysis of the extract, the authors hypothesized that the patient had a defective cytochrome P450 enzyme, and was unable to metabolize Harman alkaloids present in *P. incarnata* extracts.

Passiflora edulis is the species with the highest commercial value due to its easy adaptation to different soils and climates [8], great use for industrial products (food, nutraceutical, and cosmetics), and for having mild tolerance to salinity [9]. Tolerance to salinity is vital for the commercial feasibility of the crop in arid and semi-arid regions. Regarding the effects of salinity on the crop, the yellow passion fruit was reported to have a minimal reduction in fruit quality when irrigated with waters of electrical conductivity (EC_w) of 4 dS m^{-1} [10,11].

There is recent information on the effect of salinity on passion fruit growth and anatomical, physiological, and nutritional responses to salinity [12]. These authors evaluated the species *P. edulis*, *P. mucronata*, and their hybrid for 20 days after saline irrigation. They reported that *P. mucronata* was the most tolerant to salt due to its lowest leaf Na accumulation and highest stomatal conductance, photosynthesis, and leaf number when challenged with 150 mM NaCl. Another recent publication evaluated the antioxidant capacity and flavonoid characterization of passion fruit pulp, peel, and seeds [13]. However, there are no data on the effects of salinity on leaf antioxidant capacity and leaf biomass accumulation in adult passion fruit plants.

Our search of the literature revealed no data on the effects of different drying procedures on the antioxidant capacity of passion fruit leaves. In the United States, the food industry evaluates the antioxidant capacity of fruits, juices, and vegetables through the Oxygen Radical Absorbance Capacity (ORAC) method. For the ORAC analysis, the samples must be freeze-dried and lyophilized. Both the lyophilization process and ORAC method are expensive and time-consuming, with the ORAC analysis being unaffordable for small labs in the US and most labs in developing countries. In the past, it has been demonstrated that drying *Artemisia annua* leaves in a forced-air oven at 50 °C for 48 h preserved the highest leaf antioxidant capacity, without any significant reduction [14]. As the leaves of passion fruit vines are used to prepare medicinal remedies, it is important to evaluate practical, affordable, and rapid methods to determine the antioxidant capacity of the different species of interest. In this work, we compared low-temperature oven drying with freeze-drying and lyophilization and tested antioxidant capacity by both the ORAC and total phenolics (TP) methods.

Despite passion fruit's ecological and economic importance, molecular markers have only recently been used in genetic studies [15]. Considering that the genus is a rich source of ornamental vines (large array of flower colors), folk medicine (extract of leaves and flowers), cosmetics (hydrating creams), and food for humans (such as fresh juice, ice creams, frozen concentrated juices, pulp, and seeds), it is surprising that there are no studies on the molecular genetic response of passion fruit to salinity. Despite taxonomical disagreements on the number of genera (between 18 and 23) and species (between 520 and 700), approximately 96% of the species are distributed in the Americas, with major centers of diversity in Brazil and Colombia (both having approximately 30% of the *Passiflora* species) [15]. However, a few species have been reported from India, China, Southeast Asia, Australia, the Pacific Islands, and neighboring regions [15]. Recently, an effort has been funded by FAPESP (Research Support Foundation of São Paulo, São Paulo, Brazil) for the genetic characterization of 150 Passiflora species from Brazil as a valuable gene bank to manage conservation and biodiversity [16]. Although the main producers of passion fruit in Brazil are in the semi-arid region, afflicted by natural and anthropogenic irrigation water salinity, the study does not mention tolerance to salinity.

The objectives of the study were to evaluate the effects of irrigation-water salinity on leaf Na and Cl accumulation, mineral status, leaf biomass, antioxidant capacity, and genetic responses of yellow passion fruit plants. The study also evaluated the effect of leaf age and drying methods on leaf antioxidant capacity.

2. Materials and Methods

The United State Salinity Laboratory (USSL) is located in Riverside, southern California, at the latitude 33.9°58′24″ N, longitude 117°19′12″ E, and altitude of 311 m above sea level, with average high temperatures from June to September ranging from 30.6 °C to 34.5 °C. However, extreme summer weather in July and August elevates these temperatures to over 40 °C and brings relative humidity to 25% or lower. Riverside has an average annual precipitation of 262.6 mm, concentrated mostly from November to March. The day length ranged from 9 h:53 min:36 s on the shortest day of the year (21 December 2018, the solstice) to 14 h:25 min on the longest day of the year (21 June 2020), according to https://www.timeanddate.com/sun/usa/riverside?month=9, accessed on 10 September 2022.

2.1. Plant Material

Passiflora edulis f. flavicarpa seeds were obtained from the state of Paraíba, a state located on the semi-arid coast of northeastern Brazil, transferred to the US through APHIS-USDA, and planted in six-inch pots in April 2015 in a greenhouse at the USSL. Seedlings were kept under natural illumination and temperature until they were three months old, with an approximate height of 45 cm. Seedlings were irrigated with Riverside municipal water (EC = 0.65 dS m^{-1}) and received basic fertilization of 20-0-20 (NPK) in the pot and a leaf spray of 2% urea every two weeks until transplant. Seedlings were transplanted to outdoor sand tanks with two plants per tank and a spacing of 2.0 m between plants.

2.2. Irrigation Water Salinity and Sand Tank Cultivation

Plants were irrigated with Riverside municipal water (average EC = 0.65 dS m^{-1} and pH = 7.5) with ionic composition reported in Table 1, enriched with NO_3^- (9 mmol$_c$ L^{-1}), K$^+$ (6 mmol$_c$ L^{-1}), and P (1 mmol$_c$ L^{-1}) for the first month while plants were being established in the outdoor lysimeter tanks. After the plants were established, 30 days after transplanting (4 August 2015), they were irrigated with waters of salinity of 1.8, 3.8, and 6.8 dS m^{-1} for the first year. As plants did not show any significant difference in leaf biomass under these salinities, the irrigation water salinities were increased to 3.0, 6.0, and 12.0 dS m^{-1} for the second and third years of crop cultivation. Over time, and as nutrients were added, the final EC of the control water reached 3.0 dS m^{-1}. The treatments of 3.8 and 6.8 dS m^{-1} were increased to 6.0 and 12.0 dS m^{-1} and maintained for the last two years of the experiment. The ionic composition of the control and saline treatment waters is shown in Table 1.

Table 1. Averaged target electrical composition (EC_w), pH, and mineral ion composition of passion fruit saline irrigation waters and Riverside municipal water (RMW). Ion composition in the table reflects the concentration of ions that were added to existing ion composition in RMW as NaCl, CaCl$_2$, MgCl$_2$, MgSO$_4$, Na$_2$PO$_4$, and KNO$_3$ in order to provide the ions specified in the table and to achieve each target EC_w. No CO$_3$H$^-$ was added to treatment waters in addition to what RMW already had. Micronutrients were added through a solution based on Hoagland and Arnon [17].

Treatment	Target EC_w	Averaged Final EC_w	pH	K$^+$	Na$^+$	Ca^{2+}	Mg^{2+}	Cl$^-$	SO$_4^{2-}$	PO$_4^{3-}$	NO$_3^-$	CO$_3$H$^-$
	dS m^{-1}	dS m^{-1}						mmol$_c$ L^{-1}				
RMW		0.6	7.7	0.1	1.88	3.32	0.80	1.01	1.30	ND	0.38	3.4
T1	3.0	3.0	7.8	5.9	13.62	3.98	2.80	1.99	14.70	1.0	8.62	0.0
T2	6.0	6.0	7.7	5.9	32.12	12.88	7.40	25.99	22.70	1.0	8.62	0.0
T3	12.0	12.0	7.3	5.9	70.12	27.18	14.50	85.39	22.70	1.0	8.62	0.0

Note: The chemical composition of each target EC_w was calculated using Extract Chem V2.0 [18] based on the water composition of the Riverside municipal water (RMW).

To this basic nutrient solution, we added salts of Na, Cl, SO$_4$, Ca, Mg, and S in order to provide the target salinities (Table 1). Saline treatment solutions were pumped from 24 water reservoirs (Vol = 3605 L each) housed underneath the lysimeter facility to each of the 24 large sand tanks above, completely saturating and leaching the sand culture medium. Sand tanks were filled with coarse sand with dimensions of 1.5 m (W) × 3.0 m (L) × 2.0 m (D) (Figure 1). Plants were irrigated twice a week with treatment waters stored in underground reservoirs of 3605 L, each designated to irrigate one large sand tank above. After irrigating each tank, each saline nutrient solution returned to the reservoir through a subsurface drainage system at the bottom of the tanks (Figure 1), thus maintaining a uniform and constant salinity in the plant root zone.

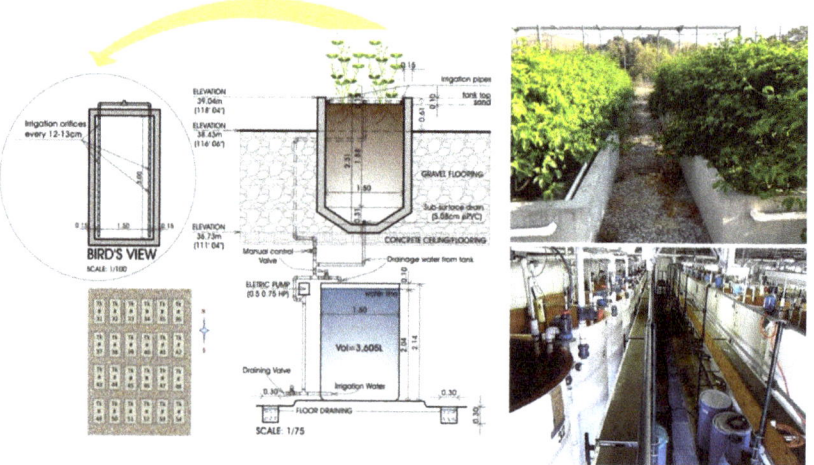

Figure 1. Schematic cross-section representation of large outdoor sand tanks (**top left**) and water reservoirs (**bottom left**) used to grow yellow passion fruit vines, with a bird's view of a single tank (**top left**) and the 25 sand tanks on a north/south orientation (**bottom left**). On the right, a representation of the front view of passion fruit in sand tanks (**top right**) and of the 24 water reservoirs (**bottom right**) housed underneath the sand tanks.

2.3. Plant Mineral Nutrition

The base nutrient solution was prepared following the nutrition recommendation for passion fruit: N as NO_3^- (9 mmol L^{-1}), P as $H_2PO_4^-$ (1.0 mmol L^{-1}), K (6.0 mmol L^{-1}). For the micronutrients, we used the Hoagland and Arnon No. 1 solution [17] with the following composition: 46 µmol L^{-1} of B; 0.3 µmol L^{-1} of Cu; 12.6 µmol L^{-1} of Mn; 0.1 µmol L^{-1} of Mo; 90 µmol L^{-1} of Fe, and 1.3 µmol L^{-1} of Zn.

2.4. Lysimeter Tanks

Each lysimeter tank had two plants spaced 2.5 m apart. The coarse sand in each tank was mixed with 10% peat moss (v/v), resulting in an average bulk density of 1380 kg m^{-3} and an average volumetric water + air content of 0.30 m^3 m^{-3}, determined by packing dry sand, weighing, saturating with water, and reweighing. Plants were assigned to tanks in a complete randomized design with eight replicates per treatment (eight tanks per treatment) with two plants per tank, totaling sixteen plants per treatment.

2.5. Sample Collection

Leaves were collected in groups according to their position in the new tertiary stem generated in 2019, then compared to one large leaf from the secondary stem that originated in 2018 (Figure 2). Leaf samples for Na, Cl, and macronutrient analyses were fully expanded and taken from positions 7–9 in the new branch, counted from the newest leaf in the apex in April 2019, with approximately 15 leaves collected per tank. For antioxidant analyses, leaves were collected from positions 1–3 (new leaves, NL), 4–7 (young leaves, YL), 8–12 (physiologically mature, ML), and one previous-year (PY) leaf (Figure 2). After collection, a portion of the leaf samples was placed in cloth bags and immediately dipped into liquid nitrogen, freeze-dried (lyophilized), and ground in order to determine the antioxidant capacity through both ORAC and total phenolic tests. The other portion of the sampled leaves was oven-dried at 50 °C for 48 h for ORAC and total phenolic analyses to evaluate if this oven temperature would preserve most of the antioxidant capacity when compared to freeze-drying, as previously determined for *Artemisia annua* [14]. Leaves from positions 1–7 were also collected from different branches and from the east or west side of the plant in order to assess whether the direction of sun exposure would affect antioxidant capacity. All leaves were kept in the dark and stored in a −80 °C freezer before being directed to either freeze-drying or low-temperature oven drying. They were then ground in a Willey Mill to a 1 mm-particle size and maintained at −80 °C before ORAC and total phenolic analyses. For fresh biomass accumulation, plants were pruned up to the secondary stem, placed in a large plastic tarp, and immediately weighed on a large scale. The vine dry weight was later estimated by taking six samples from different tanks and drying them in an oven at 70 °C until they reached a stable weight.

2.6. Antioxidant Capacity Tests

Leaves were analyzed by the ORAC test as well as by the TP (Folin–Ciocalteu) test following methodology previously validated [19,20] and currently used in our USDA laboratory in order to determine the shoot antioxidant capacity of several crops, including spinach, alfalfa, and artemisia [14,21,22].

2.7. Expression Analyses

Expression changes under salinity were studied using quantitative Reverse Transcription—Polymerase Chain Reaction (qRT-PCR). Some genes involved in Na^+ transport and Cl^- transport were selected for our expression analyses. The Arabidopsis sequence for each selected gene was used to identify the corresponding cassava (*Manihot esculenta*) gene, which was further used against the passion fruit whole genome shotgun sequence to identify the passion fruit ortholog. Primers were designed for the sequence that showed the highest homology (Table S1).

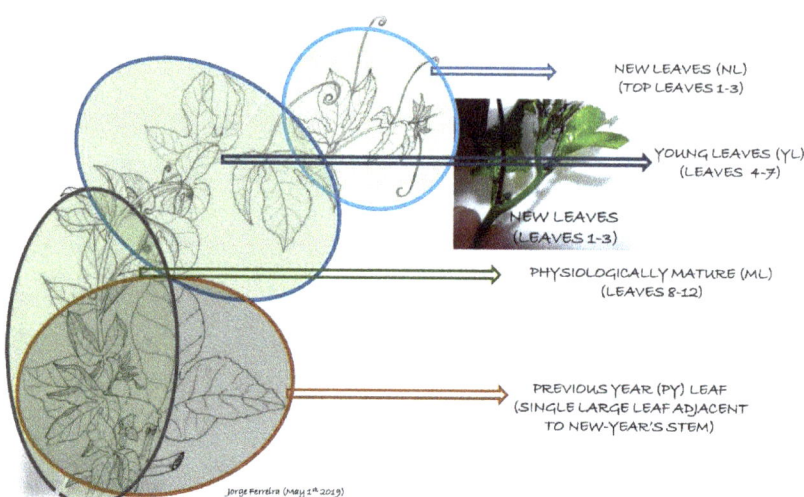

Figure 2. Schematic drawing of a new passion fruit stem arising from the previous year's stem. Elliptical lines show the group of leaves collected, starting from the new leaves at the tip of the stem to the oldest at the base. The large leaf at the base of that new stem is from the previous year's (PY) growth and was also used to compare antioxidant capacity. Schematic illustration by Jorge F.S. Ferreira.

RNA samples were collected from leaves and roots, and RNA was isolated using TRIzol® reagent (Invitrogen, Carlsbad, CA, USA). DNA contamination was removed using DNase I (Thermo Scientific, Waltham, MA, USA), and the samples were diluted to 5 ng/µL. The qRT-PCR analyses were performed in the BioRad CFX96 machine, utilizing the iTaq™ Universal SYBR® Green One-Step Kit (Bio-Rad Laboratories, Hercules, CA, USA). The PCR reaction was carried out in 10 µL volume containing 10 ng total RNA, 0.75 µM each primer, 5 µL 2X one-step SYBR® Green Reaction mix, and 0.125 µL Bio-Rad iScriptTM Reverse Transcriptase (Bio-Rad Laboratories, Hercules, CA, USA). Amplification was carried out using the following program: 50 °C for 10 min, 95 °C for 1 min, followed by 40 cycles of 95 °C for 10 s, 57 °C for 30 s, and 68 °C for 30 s. Three passion fruit housekeeping genes encoding histone (XP_002525279.1), 60S ribosomal protein (XP_002531173.1), and transcription initiation factor (XP_002299546.1) were used as reference genes for the qRT-PCR analyses [23].

3. Results

The results reflect the salt, mineral, antioxidant capacity, and gene expression in leaves of adult passion fruit plants after three years of exposure to saline waters and hot summer temperatures in Riverside, CA, USA.

3.1. Effects of Leaf Na and Cl Accumulation on Shoot Biomass

Leaf mineral analysis showed that an increase in irrigation water salinity resulted in significant increases in Na and Cl leaf concentrations. Passion fruit plants accumulated significantly more sodium and chloride with increasing salinity (Figure 3a). While leaves of plants irrigated with control water of EC_w=3.0 dS m^{-1} had, on average, 9.8 g kg^{-1} of Na and 10.2 g kg^{-1} of Cl, leaves of plants irrigated with EC_w = 6.0 dS m^{-1} had 18.4 g kg^{-1} of Na and 22.8 g kg^{-1} of Cl, and plants irrigated with EC_w = 12.0 dS m^{-1} had, on average, 29.4 g kg^{-1} of Na and 43.3 g kg^{-1} of Cl (Figure 3a, Table 2). Plants irrigated with waters of the highest salinity (12 dS m^{-1}) presented leaf edge scorching (Figure 4). Dry shoot (leaves and stems) biomass was, on average, 31.4% of fresh weight, regardless of salinity.

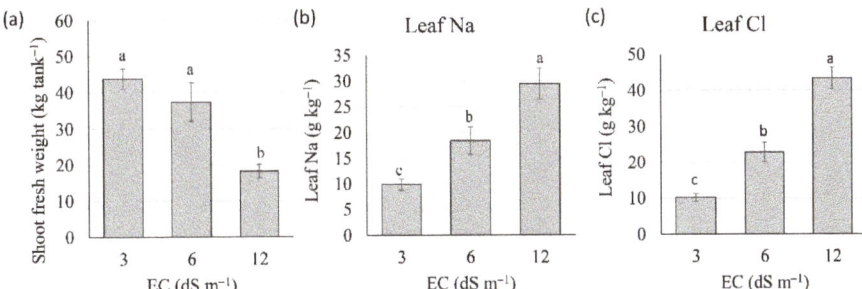

Figure 3. Biomass and ion concentrations of passion fruit plants irrigated with saline waters of three electrical conductivities (EC_w). (**a**) Shoot fresh weight. (**b**) Sodium (Na) concentrations. (**c**) Chloride (Cl) concentrations. Data are means ± 1SE, (n = 8). Different letters indicate significant differences ($p \leq 0.05$) among the three salinity treatments.

Table 2. Leaf macro- and micronutrient concentrations, including sodium and chloride, of passion fruit plants irrigated with waters of three electrical conductivities (EC_w). Data are the means of eight outdoor lysimeter tanks (n = 8), except for the highest salinity (n = 6). Different letters indicate significant differences ($p \leq 0.05$) among the three salinities. DM: dry matter. The water pH was 7.8.

EC_w	K	Na	Cl	P	Ca	Mg	S	N	Fe	Zn	Cu	B	Mn
dS m^{-1}				g kg^{-1} DM						mg kg^{-1} DM			
3.0	20.2 a	9.8 c	10.2 c	2.1 a	7.8 a	2.0 a	3.6 a	34.3 b	116 a	50.5 a	3.7 a	118 a	11.5 a
6.0	20.1 a	18.4 b	22.8 b	2.5 a	7.8 a	2.1 a	3.9 a	41.6 ab	119 a	39.9 a	3.6 a	56.0 b	13.8 a
12.0	20.5 a	29.4 a	43.3 a	2.4 a	9.8 a	2.7 a	3.2 a	43.2 a	138 a	51.3 a	3.3 a	83.3 ab	11.8 a

Figure 4. The general aspect of leaves of passion fruit plants irrigated with waters of different salinities. (**a**) 3.0 dS m^{-1}. (**b**) 6.0 dS m^{-1}. (**c**) 12.0 dS m^{-1}. Plant biomass was significantly reduced by 58.45% at the highest salinity, with leaves showing burned edges typical of excess Cl accumulation.

Although there was no significant reduction in average leaf biomass between plants irrigated with 3.0 and 6.0 dS m^{-1} (43.8 vs. 37.3 kg tank^{-1}), there was a significant reduction in leaf biomass (18.2 kg tank^{-1}) for plants irrigated with EC_w = 12.0 dS m^{-1} (Figure 3b). The leaves of plants irrigated with waters of EC_w of 3.0 and 6.0 dS m^{-1} (Figure 4a,b) had no salt-toxicity symptoms, but the leaves of plants irrigated with EC_w = 12.0 dS m^{-1} showed lateral burning of mature leaves (Figure 4c). Both plants from each of the two tanks under EC_w = 12 dS m^{-1} died during the third year.

Although leaf concentrations of Na and Cl increased with each increase in irrigation-water salinity (Figure 3), the concentrations of macro- and micronutrients remained fairly constant across salinities (Table 2). In comparison to the expected macro- and micronutrient leaf composition of field-grown *Passiflora alata* Curtis [24], our plants were low in Mn. Although Mn was provided in a mixture of essential minerals, the water pH (7.8) was above the ideal pH (>6.2) for most plants, after which manganese has a decreased availability for plant uptake.

3.2. Effects of Na and Cl Accumulation, Drying Method, and Leaf Age on Antioxidant Capacity

Salinity did not affect the antioxidant capacity of passion fruit leaves. The leaves used for this analysis were the same leaves collected for mineral analysis, and were from the middle portion of the stem. These leaves, sampled from the middle of the tertiary branch (leaves 7–9), had ORAC concentrations ranging from 620–750 µmoles TE g^{-1} DM, and TP ranging from 15–19 mg GAE/g DM (Figure 5). There was also no difference in antioxidant capacity for either ORAC and TP between leaves from the east or west side of the vine (Figure 5).

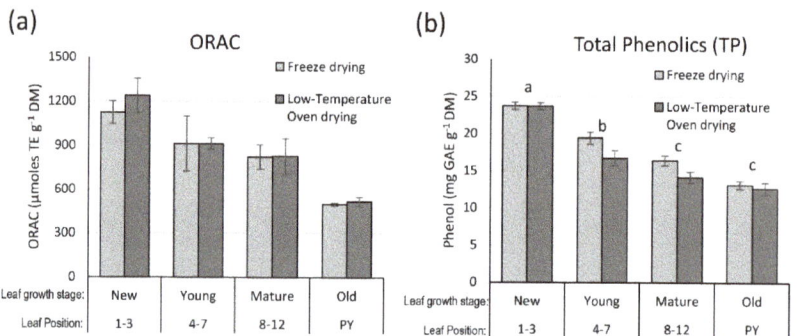

Figure 5. Comparison of drying methods and leaf developmental stages (1–12) for passion fruit leaves measured through the hydrophilic oxygen radical absorbance capacity (ORAC), and total phenolics (TP) concentration of freeze-dried and oven-dried leaves at different developmental stages. (**a**) Leaf ORAC at different leaf developmental stages. (**b**) Leaf TP at different leaf developmental stages. Vertical bars represent means plus/minus standard error (±1SE) of four replicates (n = 4) (4 tanks each with two vines). Different letters indicate significant differences ($p \leq 0.05$) among different leaf growth stages. Lack of letters between drying methods indicate lack of significant differences. TE: Trolox equivalent, GAE: gallic acid equivalent. There was no significant difference ($p > 0.05$) between the two drying methods at any leaf developmental stage. PY, previous-year leaves.

When comparing freeze-drying at −50 °C for 3–4 days with oven drying at 50 °C for 48 h, leaves at different positions of the newest branch of passion fruit vines showed no difference in antioxidant capacity (ORAC) or TP (Figure 5). However, depending on their physiological age, leaf groups had significantly different ($p < 0.05$) ORAC values, with new leaves (NL, leaves 1–3) having average ORAC values of 1200 µmoles TE g^{-1} DW, while young leaves (YL, leaves 4–7) and physiologically mature leaves (ML, leaves 8–12) of the same branch had average ORAC value of 850 µmoles TE g^{-1}, and leaves from the previous year (PY) had the lowest ORAC values of approximately 580 µmoles TE g^{-1} (Figure 6a). Leaf TP followed the same trend as ORAC, with NL having the highest values in TP, approximately 24 mg GAE g^{-1} DW, and ML and PY leaves having values of approximately 15 and 14 mg GAE g^{-1} DW, respectively (Figure 6b).

3.3. Effect of Salinity on Gene Expression

Twelve genes known to be involved in salt stress were used to study expression changes between control and salinity treatments. Of these, six genes (*AKT1*, *NHX1*, *NHX2*, *SOS1*, *SOS2*, and *SOS3*) are known to be involved in Na transport, and six genes (*ALMT12*, *ALMT9*, *CCC*, *CLCc*, *CLCg*, and *SLAH3*) are known for their roles in Cl transport. Expression analyses revealed higher expression of all 12 genes in roots than in leaves, irrespective of salinity treatment (Figure 7).

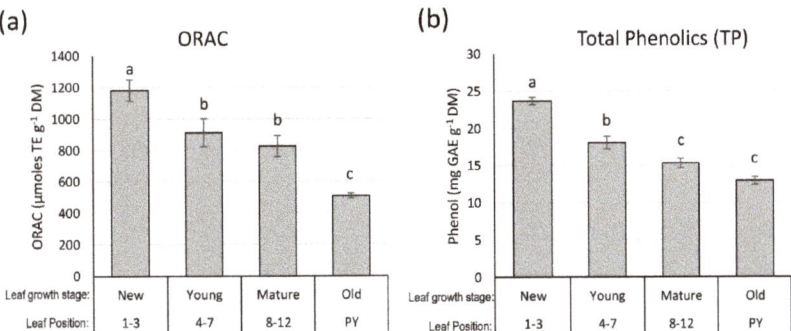

Figure 6. Effect of passion fruit leaf developmental stage on the hydrophilic oxygen radical absorbance capacity (ORAC) and total phenolics (TP) concentration. (**a**) Leaf ORAC at different leaf developmental stages. (**b**) Leaf TP at different leaf developmental stages. Vertical bars represent means plus/minus one standard error (±1SE) of eight replicated samples (n = 8) (two vines per tank, freeze-dried and oven-dried sample data were combined). TE: Trolox equivalent, GAE: gallic acid equivalent. Different letters indicate significant differences ($p \leq 0.05$) among different leaf growth stages. PY, previous-year leaves.

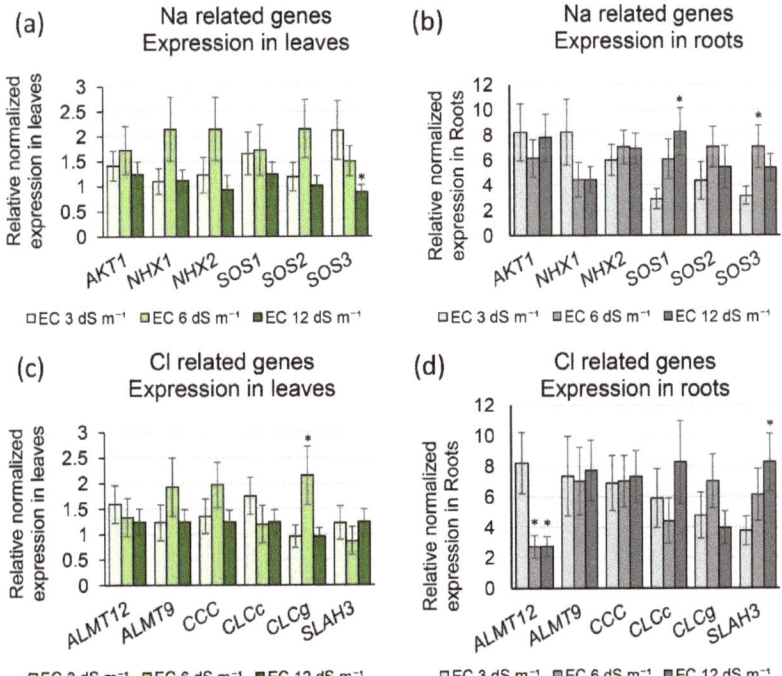

Figure 7. Expression of genes involved in Na^+ and Cl^- transport under EC = 3 dS m^{-1}, EC = 6 dS m^{-1}, and EC = 12 dS m^{-1} in leaves and roots of passion fruit. (**a**) Expression of genes involved in Na^+ transport in leaves. (**b**) Expression of genes involved in Na^+ transport in roots. (**c**) Expression of genes involved in Cl^- transport in leaves. (**d**) Expression of genes involved in Cl^- transport in roots. * indicate a significant difference ($p \leq 0.05$) between EC = 3 dS m^{-1} and EC = 6 dS m^{-1} or 3 dS m^{-1} and EC = 12 dS m^{-1}. Bars represent standard errors.

Of the genes involved in Na transport, *SOS1* and *SOS3* were significantly upregulated in roots, at 12 dS m^{-1} compared to 3 dS m^{-1}, whereas *SOS3* was significantly upregulated at 6 dS m^{-1} compared to 3 dS m^{-1} (Figure 7b). On the other hand, *SOS3* was downregulated in leaves at 12 dS m^{-1} compared to the control (Figure 7a). None of the other genes were differentially expressed under control and salinity conditions.

Of the genes regulating Cl transport, *ALMT12* was downregulated at 12 dS m^{-1} and 6 dS m^{-1} compared to 3 dS m^{-1}, whereas *SLAH3* was upregulated in roots at 12 dS m^{-1} compared to the control (Figure 7d). In leaves, *CLCg* was upregulated at 6 dS m^{-1} as compared to 3 dS m^{-1} (Figure 7c).

4. Discussion

4.1. Effects of Na and Cl Accumulation on Shoot Biomass

According to the salinity tolerance of several species under hydroponics, species that tolerated an EC_w = 5.5 dS m^{-1} were considered tolerant [25]. In addition, young passion fruit plants that tolerated irrigation-water salinity of EC_w = 4.43 dS m^{-1} were considered moderately tolerant to salinity [26]. As our yellow passion fruit plants had significant increases in the accumulation of both Na and Cl in leaves, but had no significant decrease in shoot biomass or visual salt toxicity symptoms up to EC_w = 6.0 dS m^{-1}, we believe that our plants can be classified as moderately tolerant to salinity. However, the accumulation of either Na or Cl (or both) was toxic for the vines and reduced their leaf biomass production at the EC_w = 12.0 dS m^{-1}. After two years of exposure to the highest irrigation water salinity of 12.0 dS m^{-1}, although the plants had enough macronutrients to sustain their growth, four of the sixteen plants under this treatment died during the last months of the third year. Additionally, all surviving plants had their shoot biomass severely reduced and leaf marginal burning typical of chloride toxicity (Figure 4c), also consistent with chloride toxicity in strawberry leaves [27]. However, it is impossible to exclude the possibility that leaf damage could also be caused by sodium unless a separate study were conducted, with sodium and chloride salts used separately. As the yellow passion fruit plants evolved in the Brazilian Amazon, where soils are very low in sodium and chloride and excessive salts are leached by the abundant rains, we believe that plants have no effective mechanism to avoid either Na$^+$ or Cl$^-$ absorption, as it was reported and discussed recently for habanero peppers irrigated with saline waters [28]. In fruit crops, such as avocado rootstocks, salt toxicity has been reported at much lower salinity levels. When Hass avocado plants field-grown in Riverside were grafted onto 13 rootstocks and irrigated with water salinity of 1.5 dS m^{-1} for 23 months, a significant variation in mortality rate was observed for different rootstocks with Hass plants that were grafted onto more salt-tolerant rootstocks, accumulating less leaf chloride [29]. Results prompted the authors to conclude that plant mortality was highly correlated to chloride (but not sodium) accumulation in avocado leaves [29]. As our passion fruit vines presented similar chloride toxicity symptoms to those seen in avocado leaves (necrosis of leaf tips extending to leaf edges, Figure 4c), maybe NaCl-tolerant passion fruit rootstocks could increase the yellow passion fruit salt tolerance, as it was found to do for avocados in Southern California.

4.2. Effects of Na and Cl Accumulation, Drying Method, and Leaf Age on Antioxidant Capacity

The antioxidant capacity of leaves, measured by both ORAC and TP methods, did not change in response to salinity, and ranged from 600–800 µmoles TE g^{-1} (ORAC) and 15–18 mg GAE g^{-1} (TP) of leaf dry matter (Figure S1). However, antioxidant capacity (ORAC and TP) decreased consistently and significantly from the youngest (1200 µmole TE g^{-1} DM) to the oldest (600 µmoles TE g^{-1} DM) leaves (Figure 6). The leaf with the lowest antioxidant capacity was the leaf from the previous year, at the base of the vine generated that year. These results indicate that young leaves are physiologically more active and generate more antioxidants, possibly as protection from sunlight stress, while the lower leaves may be more protected from sunlight and less physiologically active, and produce smaller concentrations of antioxidants. Regardless of the variation in antioxidant capacities based on the age of the leaf, passion fruit leaves have a higher hydrophilic antioxidant capacity than alfalfa, the ORAC of which ranged from 244–287 µmoles TE g^{-1} DM and TP from 5–5.6 mg GAE g^{-1} DM at 300 days after sowing and cultivation under moderately to highly saline irrigation [22]. As the yellow passion fruit leaves had 3- to 6-fold higher ORAC and 4- to 5-fold higher TP than alfalfa, its leaves should be evaluated as a rich source of minerals, NaCl, and antioxidants that would be beneficial for small ruminants and that could be provided after each annual pruning when stems and leaves are discarded. Although *Passiflora* sp. leaves contain cyanogenic glycosides, we could find no reliable toxicity studies on leaf extracts in animals, nor any report of toxicity by animals that would preclude its feeding to ruminants.

When the antioxidant capacity (both ORAC and TP) of leaves dried through freeze-drying and oven drying at 50 °C was assessed, there were no significant differences (Figure 5). This indicates that oven drying leaves at 50 °C for 48 h can effectively substitute for freeze-drying when liquid nitrogen and a freeze-drier are unavailable. These results with passion fruit confirmed our previous results with the antimalarial plant *Artemisia annua*, where leaves dried by oven drying at 50 °C for 48 h had the same antioxidant capacity (ORAC) as leaves that were frozen in liquid nitrogen and freeze-dried [14]. However, shade-drying and sun-drying, mainly the latter, destroyed most of the antioxidant capacity of *A. annua* leaves.

4.3. Effect of Na and Cl Accumulation on Gene Expression

Expression analyses showed that the genes involved in Na and Cl transport studied in this investigation had higher expression in roots compared to leaves (Figure 7), suggesting that roots play a more critical role than leaves under salinity stress. Similar observations were made in alfalfa, where most genes involved in Na and Cl transport were expressed at higher levels in roots compared to leaves [30].

Under salinity stress, a higher Na$^+$ concentration is sensed by SOS3, which physically interacts with SOS2 [31]. SOS3-SOS2 complex then activates a Na$^+$/H$^+$ antiporter, SOS1, by phosphorylating it, causing extensive Na$^+$ exclusion from the cytoplasm [32]. *SOS1* and *SOS3*, known to be involved in Na$^+$ exclusion, were upregulated in roots under salinity, indicating that the Na$^+$ exclusion is vital during salinity stress in passion fruit (Figure 7b). At the same time, the lower expression of *SOS3* in leaves under salinity than the control may suggest an additional role of SOS3 in the cellular metabolism of passion fruit (Figure 7a).

ALMT12 is found in root stelar cells, and is predicted to be involved in the loading of Cl$^-$ from root to xylem [33]. *ALMT12* was significantly downregulated in roots under salinity compared to the control in our study (Figure 7d). Downregulation of this gene will restrict Cl$^-$ to roots and will protect leaves from higher Cl concentrations.

CLCg is localized to the tonoplast of mesophyll cells, and is involved in sequestering Cl$^-$ to vacuoles to protect the cytoplasm [34]. *CLCg* was upregulated at 6 dS m^{-1} compared to the control, suggesting its role in tissue tolerance under medium salinity. The high salinity of 12 dS m^{-1} does not show induction of *CLCg* compared to the control in leaves (Figure 7c).

Some of the genes, such as *AKT1*, *NHX1*, *NHX2*, *SOS2*, *ALMT9*, *CCC*, and *CLCc*, which do not show differential expression under control and saline treatments may be differentially expressed in different genotypes. Although not tested in this investigation, genotypic-specific differences may also be critical during salinity stress in passion fruit.

5. Conclusions

This work is first to report on the effect of saline irrigation on the biomass, mineral status, antioxidant capacity, and genetic response of *Passiflora edulis* f. cv flavicarpa, as well as the first to compare low-temperature oven drying with freeze-drying and lyophilization prior to leaf antioxidant analyses. Our experiment produced the main findings: Saline waters up to 6.8 dS m^{-1} (equivalent to a soil-paste salinity (EC$_e$) of 3.1 dS m^{-1}) were perfectly tolerated by young passion fruit vines in their first year. During the second and third years, salinities from 3.0 dS m^{-1} (equivalent to an EC$_e$ = 1.36 dS m^{-1}) to 6.0 dS m^{-1} (equivalent to an EC$_e$ = 2.7 dS m^{-1}) produced no visual toxicity symptoms, leaf biomass reduction, major changes in leaf mineral status, or decreased leaf antioxidant capacity under the hot summer temperatures of southern California. However, plants irrigated with water with a salinity of 12.0 dS m^{-1} (equivalent to an EC$_e$ = 5.45 dS m^{-1}) resulted in significantly reduced shoot biomass and some plant death in the third year, confirming previous reports that the salinity threshold for the yellow passion fruit in the field had an EC$_e$ of approximately 4.0 dS m^{-1}. However, our experience with salinity tolerance in other crops suggests that salinity trials must involve other cultivars, which may increase salt tolerance in a crop of Amazonian origin such as *Passiflora edulis*. The high antioxidant capacity found in young leaves makes them appealing to the nutraceutical market. We also showed that freezing-drying in liquid nitrogen and lyophilization at −50 °C can be substituted by drying leaves in a forced-air oven set at 50 °C for 48 h, which is a more effective, rapid, and economical way to preserve leaf antioxidant capacity. However, oven-dried leaves should be further analyzed for individual flavonoid glucosides by HPLC-UV in order to guarantee that key flavonoids are preserved by low-temperature oven drying. The similar trends in TP and ORAC analyses indicate that the passion fruit flavonoids have high antioxidant activity. Finally, the gene expression analyses of passion fruit leaves of plants submitted to irrigation waters of low to high salinity confirmed a more critical role of roots compared to leaves under salinity stress. The efflux of Na$^+$ from roots to the soil, loading of Cl$^-$ from root to xylem, and sequestration of Cl$^-$ into vacuoles of mesophyll cells are vital components of salinity tolerance in passion fruit. In order to increase our understanding of *Passiflora edulis* f. flavicarpa responses to salinity, future studies should involve different cultivars, of both edible and wild species, and separate saline waters dominant in either sodium or chloride salts to determine which salt is most toxic to the crop.

Supplementary Materials: The following supporting information can be downloaded at: https://www.mdpi.com/article/10.3390/agriculture12111856/s1, Figure S1: Leaf hydrophilic oxygen radical absorbance capacity (ORAC) and total phenolics (TP) concentration of passion fruit leaves of plants grown under three salinities and of leaves facing east or west. (a) Leaf ORAC at different salinities. (b) Leaf ORAC for samples taken from leaves facing east or west. (c) Leaf TP at different salinities. (d) Leaf TP for samples taken from leaves facing east or west. Error bars represent standard errors of the means. Table S1: The list of primers used for the expression analysis.

Author Contributions: Conceptualization, J.F.S.F. and D.S.; methodology, J.F.S.F., D.S., X.L., and S.R.P.S.; formal analysis, X.L. and C.N.; investigation, J.F.S.F. and D.S.; data curation, J.F.S.F., D.S., X.L., and C.N.; writing—original draft preparation, J.F.S.F., X.L., and D.S.; writing—review and editing, J.F.S.F. and D.S.; visualization, J.F.S.F. and D.S.; supervision, J.F.S.F.; project administration, J.F.S.F. and D.S. All authors have read and agreed to the published version of the manuscript.

Funding: This was funded by the USDA Project Number 2036-13210-011-00D (Enhancing Specialty Crop Tolerance to Saline Irrigation Waters).

Institutional Review Board Statement: Not applicable.

Informed Consent Statement: Not applicable.

Data Availability Statement: Not applicable.

Acknowledgments: To Noah Gangoso, for his valuable contribution in taking the lead in monitoring irrigation-water salinity and pH and for overseeing the irrigation of plants since its inception and final data collection. Thanks are also due to Jason Verd and Alysia Sorya for general care and pruning, and to Dennise Jenkins for her valuable help installing the irrigation system for seedling greenhouse production and for overseeing the installation of the trellis system outdoors. Thanks are also due to Jaime Barros for his help during the shoot biomass data collection during the COVID-19 pandemic.

Conflicts of Interest: The authors declare no conflict of interest.

References

1. Santos, J.T.C.; Petry, F.C.; Tobaruela, E.C.; Mercadante, A.Z.; Gloria, M.B.A.; Costa, A.M.; Lajolo, F.M.; Hassimotto, N.M.A. Brazilian native passion fruit (*Passiflora tenuifila* Killip) is a rich source of proanthocyanidins, carote-noids, and dietary fiber. *Food Res. Int.* **2021**, *147*, 110521. [CrossRef] [PubMed]
2. Bernacci, L.C.; Soares-Scott, M.D.; Junqueira, N.T.V.; Passos, I.R.S.; Meletti, L.M.M. *Passiflora edulis* Sims: The correct taxonomic way to cite the yellow passion fruit (and of others colors). *Rev. Bras. Frutic.* **2008**, *30*, 566–576. [CrossRef]
3. Sá, C.P.; Neto, R.C.A.; Negreiros, J.R.S.; Nogueira, S.R. *Coeficientes Técnicos, Custos de Produção e Indicadores Economicos Para o Cultivo do Maracujá BRS Gigante Amarelo, No Acre*; Embrapa: Rio Branco, Acre, Brazil, 2015.
4. Altendorf, S. *Minor Tropical Fruits: Mainstreaming a Niche Market*; Food and Agricultural Organization: Rome, Italy, 2018; pp. 67–75. Available online: http://www.fao.org/fileadmin/templates/est/COMM_MARKETS_MONITORING/Tropical_Fruits/Documents/Minor_Tropical_Fruits_FoodOutlook_1_2018.pdf (accessed on 10 September 2022).
5. Li, H.; Zhou, P.; Yang, Q.; Shen, Y.; Deng, J.; Li, L.; Zhao, D. Comparative studies on anxiolytic activities and fla-vonoid compositions of *Passiflora edulis* 'edulis' and *Passiflora edulis* 'flavicarpa'. *J. Ethnopharmacol.* **2011**, *133*, 1085–1090. [CrossRef] [PubMed]
6. Newall, C.A.; Anderson, L.A.; Phillipson, J.D. *Herbal Medicines. A Guide for Health-Care Professionals*; The Pharmaceutical Press: London, UK, 1996; p. 296.
7. Fisher, A.A.; Purcell, P.; Couteur, D.G.L. Toxicity of *Passiflora incarnata* L. *J. Toxicol. Clin. Toxicol.* **2000**, *38*, 63–66. [CrossRef]
8. Machado, C.F.; Jesus, F.N.; Ledo, C.A.S. Divergencia genetica de acessos de maracuja utilizando descritores quantitativos e qualitativos. *Rev. Bras. Frutic.* **2015**, *37*, 442–449. [CrossRef]
9. Dias, T.J.; Cavalcante, L.F.; Pereira, W.E.; Freire, J.L.D.O.; Souto, A.G.D.L. Irrigação com água salina em solo com biofertilizante bovino no crescimento do maracujazeiro amarelo. *Semin. Cienc. Agrar.* **2013**, *34*, 1639–1652. [CrossRef]
10. Freire, J.L.O.; Cavalcante, L.F.; Rebequi, A.M.; Dias, T.J.; Brehm, M.A.S.; Santos, J.B. Quality of yellow passion fruit juice with cultivation using different organic sources and saline water. *IDESIA* **2014**, *32*, 79–87. [CrossRef]
11. Cavalcante, L.F.; Cavalcante, Í.H.L.; Júnior, F.R.; Beckmann-Cavalcante, M.Z.; Santos, G.P. Leaf-macronutrient status and fruit yield of biofertilized yellow passion fruit plants. *J. Plant Nutr.* **2012**, *35*, 176–191. [CrossRef]
12. Lima, L.K.S.; Jesus, O.N.; Soares, T.L.; Santos, I.S.; Oliveira, E.J.; Filho, M.A.C. Growth, physiological, ana-tomical and nutritional responses of two phenotypically distinct passion fruit species (*Passiflora* L.) and their hybrid under saline conditions. *Sci. Hort.* **2020**, *263*, 109037. [CrossRef]
13. Reis, L.C.R.; Facco, E.M.P.; Salvador, M.; Flôres, S.H.; Rios, A.O. Antioxidant potential and physicochemical characterization of yellow, purple and orange passion fruit. *J. Food Sci. Technol.* **2018**, *55*, 2679–2691. [CrossRef]
14. Ferreira, J.F.S.; Luthria, D.L. Drying affects artemisinin, dihydroartemisinic acid, artemisinic acid, and the anti-oxidant capacity of *Artemisia annua* L. leaves. *J. Agric. Food Chem.* **2010**, *58*, 1691–1698. [CrossRef] [PubMed]
15. Cerqueira-Silva, C.B.M.; Jesus, O.N.; Santos, E.S.L.; Corrêa, R.X.; Souza, A.P. Genetic breeding and diversity of the genus Passiflora: Progress and perspectives in molecular and genetic studies. *Int. J. Mol. Sci.* **2014**, *15*, 14122–14152. [CrossRef] [PubMed]
16. Cerqueira-Silva, C.B.M.; Faleiro, F.G.; Jesus, O.N.; Santos, E.S.L.; Souza, A.P. The Genetic Diversity, Conservation, and Use of Passion Fruit (*Passiflora* spp.). In *Genetic Diversity and Erosion in Plants: Case Histories*; Ahuja, M.R., Jain, S.M., Eds.; Springer International Publishing: Cham, Switzerland, 2016; pp. 215–231.
17. Hoagland, D.R.; Arnon, D.I. The water-culture method for growing plants without soil. *Circ. Calif. Agric. Exp. Stn.* **1950**, *347*, 32.
18. Suarez, D.L.; Simunek, J. UNSATCHEM: Unsaturated water and solute transport model with equilibrium and kinetic chemistry. *Soil Sci. Soc. Am. J.* **1997**, *61*, 1633–1646. [CrossRef]
19. Prior, R.L.; Wu, X.; Schaich, K. Standardized methods for the determination of antioxidant capacity and phenolics in foods and dietary supplements. *J. Agric. Food Chem.* **2005**, *53*, 4290–4302. [CrossRef] [PubMed]
20. Prior, R.L.; Hoang, H.; Gu, L.; Wu, X.; Bacchiocca, M.; Howard, L.; Hampsch-Woodill, M.; Huang, D.; Ou, B.; Jacob, R. Assays for hydrophilic and lipophilic antioxidant capacity [oxygen radical absorbance capacity (ORAC)] of plasma and other biological and food samples. *J. Agric. Food Chem.* **2003**, *51*, 3273–3279. [CrossRef]
21. Ferreira, J.F.S.; Sandhu, D.; Liu, X.; Halvorson, J.J. Spinach (*Spinacea oleracea* L.) response to salinity: Nutritional value, physiologi-cal parameters, antioxidant capacity, and gene expression. *Agriculture* **2018**, *8*, 163. [CrossRef]
22. Ferreira, J.F.S.; Cornacchione, M.V.; Liu, X.; Suarez, D.L. Nutrient composition, forage parameters, and antioxi-dant capacity of alfalfa (*Medicago sativa*, L.) in response to saline irrigation water. *Agriculture* **2015**, *5*, 577–597. [CrossRef]
23. Munhoz, C.F.; Santos, A.A.; Arenhart, R.A.; Santini, L.; Monteiro-Vitorello, C.B.; Vieira, M.L.C. Analysis of plant gene expression during passion fruit–*Xanthomonas axonopodis* interaction implicates lipoxygenase 2 in host de-fence. *Ann. Appl. Biol.* **2015**, *167*, 135–155. [CrossRef]
24. Freitas, M.S.M.; Monnerat, P.H.; Vieira, I.J.C. Mineral deficiency in *Passiflora alata* Curtis: Vitexin bioproduction. *J. Plant Nutr.* **2008**, *31*, 1844–1854. [CrossRef]
25. Hurtado-Salazar, A.; Silva, D.F.P.; Ceballos-Aguirre, N.; Ocampo-Pérez, J.; Bruckner, C.H. Promissory *Passiflora* L. species (Passifloraceae) for tolerance to water-salt stress. *Rev. Colomb. Cienc. Hortic.* **2020**, *14*, 44–49. [CrossRef]
26. Soares, F.A.L.; Gheyi, H.R.; Viana, S.B.A.; Uyeda, C.A.; Fernandes, P.D. Water salinity and initial development of yellow passion fruit. *Sci. Agric.* **2002**, *59*, 491–497. [CrossRef]

27. Ferreira, J.F.S.; Liu, X.; Suarez, D.L. Fruit yield and survival of five commercial strawberry cultivars under field cultivation and salinity stress. *Sci. Hort.* **2019**, *243*, 401–410. [CrossRef]
28. Suarez, D.L.; Celis, N.; Ferreira, J.F.S.; Reynolds, T.; Sandhu, D. Linking genetic determinants with salinity toler-ance and ion relationships in eggplant, tomato and pepper. *Sci. Rep.* **2021**, *11*, 16298. [CrossRef] [PubMed]
29. Celis, N.; Suarez, D.L.; Wu, L.; Li, R.; Arpaia, M.L.; Mauk, P. Salt tolerance and growth of 13 avocado rootstocks related best to chloride uptake. *HortScience* **2018**, *53*, 1737. [CrossRef]
30. Sandhu, D.; Cornacchione, M.V.; Ferreira, J.F.; Suarez, D.L. Variable salinity responses of 12 alfalfa genotypes and comparative expression analyses of salt-response genes. *Sci. Rep.* **2017**, *7*, 42958. [CrossRef]
31. Hasegawa, P.M.; Bressan, R.A.; Zhu, J.-K.; Bohnert, H.J. Plant cellular and molecular responses to high salinity. *Ann. Rev. Plant Physiol. Plant Mol. Biol.* **2000**, *51*, 463–499. [CrossRef]
32. Qiu, Q.-S.; Guo, Y.; Dietrich, M.A.; Schumaker, K.S.; Zhu, J.-K. Regulation of SOS1, a plasma membrane Na+/H+ exchanger in *Arabidopsis thaliana*, by SOS2 and SOS3. *Proc. Natl. Acad. Sci. USA* **2002**, *99*, 8436–8441. [CrossRef]
33. Li, B.; Tester, M.; Gilliham, M. Chloride on the Move. *Trends Plant Sci.* **2017**, *22*, 236–248. [CrossRef]
34. Nguyen, C.T.; Agorio, A.; Jossier, M.; Depré, S.; Thomine, S.; Filleur, S. Characterization of the chloride channel-like, AtCLCg, involved in chloride tolerance in *Arabidopsis thaliana*. *Plant Cell Physiol.* **2016**, *57*, 764–775. [CrossRef]

Article

Salt Tolerance Indicators in 'Tahiti' Acid Lime Grafted on 13 Rootstocks

Gabriel O. Martins [1], Stefane S. Santos [2], Edclecio R. Esteves [2], Raimundo R. de Melo Neto [2], Raimundo R. Gomes Filho [1], Alberto S. de Melo [3], Pedro D. Fernandes [4], Hans R. Gheyi [4], Walter S. Soares Filho [5] and Marcos E. B. Brito [2,*]

[1] Post Graduate Program in Water Resources, Universidade Federal de Sergipe (UFS), Campus of São Cristóvão, São Cristóvão 49100-000, SE, Brazil
[2] Center of Agrarian Sciences of Sertão (CCAS), Universidade Federal de Sergipe (UFS), Campus of Sertão, Nossa Senhora da Glória 49680-000, SE, Brazil
[3] Center of Biological and Health Sciences, Universidade Estadual da Paraíba (UEPB), Campus I, Campina Grande 58429-500, PB, Brazil
[4] Center of Tecnologia and Natural Resources (CTRN), Universidade Federal de Campina Grande (UFCG), Campina Grande 58429-900, PB, Brazil
[5] National Research Center of Cassava and Fruit Crops (CNPMF), Empresa Brasileira de Pesquisa Agropecuaria, Cruz das Almas 44380-000, BA, Brazil
* Correspondence: marcoseric@academico.ufs.br

Abstract: The citrus yield is limited by soil and/or water salinity, but appropriate rootstocks can ensure the sustainability of the production system. Therefore, the objective of the present research was to evaluate the salt content in the soil and the production and physiological aspects of the 'Tahiti' acid lime combined with thirteen rootstocks, irrigated with saline water in the first two production years to identify indicators of salt tolerance. The rootstocks evaluated were: 'Santa Cruz Rangpur' lime, 'Indio', 'Riverside' and 'San Diego' citrandarins, 'Sunki Tropical' mandarin, and eight hybrids, obtained from the Citrus Breeding Program of Embrapa Cassava and Fruits. The waters used had three saline levels: 0.14, 2.40, and 4.80 dS m^{-1}, in a randomized block adopting a split-plot design, with rootstocks in the plots and saline waters in the subplots, with four replicates. From August 2019 to February 2021, fruit harvests and agronomic traits were measured. At the end of each production year, the soil characteristics, leaf gas exchange, and chlorophyll a fluorescence analysis were performed. It was concluded that: (1) the effects of water salinity on citrus are of osmotic nature, reducing gas exchange, (2) the salinity did not significantly damage the photosynthetic apparatus until the second year of production, and (3) using more stable, salt-tolerant rootstocks makes it possible to cultivate 'Tahiti' acid lime under irrigation with waters of 2.4 dS m^{-1} electrical conductivity.

Keywords: chlorophyll a fluorescence; *Citrus* spp.; fruit production; *Poncirus* hybrids; salt balance

1. Introduction

Brazilian citrus is composed of several species, especially those belonging to the genus *Citrus*, which is formed, among other species, by sweet oranges (C. × *sinensis* (L.) Osbeck) [1] and Tahiti acid lime (C. × *latifolia* (Yu. Tanaka) Tanaka). Regarding 'Tahiti' acid lime, known as 'Tahiti' lemon, it stands out for its increase in terms of both harvested area (58,438 ha) and production, with a total of 1,585,215 tons of fruits, highlighting the states of São Paulo, Pará, Minas Gerais, Bahia, Rio de Janeiro, Ceará, Paraná, Espírito Santo, Rio Grande do Sul, and Sergipe, the last-mentioned at the tenth position on the national scene [2].

'Tahiti' acid lime production is very important in Brazil, not only from an economic point of view but also socially, with the generation of employment and income, because it allows families to remain in the field, with work and dignity [3]. Despite its socio-economic importance, its cultivation in the Northeast region faces some adversities, which lead to a low yield, about 11.9 t ha^{-1} [2], considering its potential of 40 t ha^{-1} [4].

The yield of citrus plants in northeastern Brazil may be optimized with the use of irrigation [5]. However, the waters available in this region of the country, mainly groundwaters, have high salt contents [6], which can cause problems since citrus plants are considered sensitive to salinity [7,8], which reduces their production capacity as their threshold salinity is around 1.4 dS m^{-1} in saturation extract of soil and 1.1 dS m^{-1} in irrigation water [6].

However, this response may vary depending on the scion/rootstock combination used and the management of the production system [9,10], which denotes the importance of identifying rootstocks that can confer tolerance to the scion variety, to obtain economically viable yields, even under saline conditions.

Thus, the screening of rootstocks may make it possible to use saline waters and soils, as they enable the sustainability of citrus cultivation in areas subject to this abiotic stress, common in northeastern Brazil, which is a large citrus-producing center. It is important to choose a rootstock that can give the scion desirable agronomic characteristics, such as early fruit production, low height, resistance to pests and diseases, and tolerance to abiotic stresses [11]. Few studies have been conducted on scion/rootstock combinations with salt tolerance for Brazilian citrus production [12].

Regarding tolerance to salt stress, researchers have studied the effects of salinity on nutritional imbalance and ionic interactions in plant tissue [8–10,13]. In glycophytes, this tolerance/adaptation to salt stress can be observed even under small accumulation of sodium (Na$^+$) and chloride (Cl$^-$) in the aerial parts or in the plant as a whole, a process that is related to the ability to exclude ions, especially in the root system [13], which makes the rootstock an essential component in the formation of the citrus plant.

Thus, the objective of this work was to study the salt accumulation in the soil, in addition to the production and physiological aspects of combinations of citrus rootstocks with 'Tahiti' acid lime irrigated with saline water during the first two years of cultivation, aiming to identify the indicators of salt tolerance.

2. Materials and Methods

The experiment was carried out at the Experimental Farm of the Embrapa semi-arid region, located in the municipality of Nossa Senhora da Glória, Sergipe, Brazil (10°12′18″ S, 37°19′39″ W, and 294 m altitude). Using the spreadsheet of water balance integration with climate classification proposed by Sousa and Brito [14], it is possible to observe the 'As' climate classification, relative to tropical climate. The rainy season is between April and August, with a concentration in May, June, and July. The region has low relative air humidity and wide thermal variation between day (between 28 and 35 °C) and night (between 18 and 21 °C).

The accumulated precipitation in the 26 months of the study was 1387.4 mm, being 554.3 mm in 2019, 750.9 mm in 2020, and 82.2 mm in January and February 2021, values that are within the average for the region, but below the water requirements of most citrus species, within the range from 900 to 1500 mm annually [15].

The experimental design was randomized blocks, with 13 scion/rootstock combinations under 3 levels of saline water, using the split-plot scheme with 4 replicates, as follows:

(a) Plot: 13 scion/rootstock combinations (genotypes), with 'Tahiti' acid lime grafted onto 13 rootstocks (Table 1), all from the Citrus Breeding Program of Embrapa Cassava and Fruits—CBP.

(b) Subplot: Three types of water (salinities), with electrical conductivities (ECw) of 0.14, 2.4, and 4.8 dS m^{-1}, with the first corresponding to water from the São Francisco River and the other two obtained by diluting tube well water. The chemical characteristics are presented in Table 2, with the water from the São Francisco River, until reaching the desired EC levels, with values measured using a portable microprocessor conductivity meter with automatic temperature adjustment at 25 °C.

Table 1. Rootstocks (genotypes) studied under water salinity when grafted on 'Tahiti' acid lime (*Citrus* × *latifolia* (Yu. Tanaka) Tanaka).

	Rootstock	Origin
1.	'Rangpur Santa Cruz' lime	*Citrus* × *limonia* Osbeck
2.	'Indio' citrandarin	*C. sunki* (Hayata) hort. ex Tanaka × *Poncirus trifoliata* (L.) Raf.
3.	'Riverside' citrandarin	*C. sunki* × *P. trifoliata*
4.	'San Diego' citrandarin	*C. sunki* × *P. trifoliata*
5.	'Sunki Tropical' mandarin	*C. sunki*
6.	TSKC × TRBK—007	*C. sunki* × *P. trifoliata*
7.	TSKFL × TRBK—030	*C. sunki* × *P. trifoliata*
8.	TSKC × CTTR—012	*C. sunki* × [*C.* × *sinensis* (L.) Osbeck × *P. trifoliata*]
9.	TSKFL × CTTR—013	*C. sunki* × (*C.* × *sinensis* × *P. trifoliata*)
10.	HTR—069 [1]	*C.* × *sinensis* × (*C.* × *sinensis* × *P. trifoliata*)
11.	TSKC × (LCR × TR)—040 [2]	*C. sunki* × (*Citrus* × *limonia* × *P. trifoliata*)
12.	TSKC × (LCR × TR)—059 [3]	*C. sunki* × (*Citrus* × *limonia* × *P. trifoliata*)
13.	TSKC × CTARG—019	*C. sunki* × (*C.* × *sinensis* × *P. trifoliata*)

TSKC = common 'Sunki' mandarin; TRBK = *P. trifoliata* 'Benecke'; TSKFL = 'Sunki of Florida' mandarin; CTTR = 'Troyer' citrange; HTR - 069 = trifoliate hybrid, called citrangor due crossing of 'Pera' sweet orange with 'Yuma' citrange; LCR = 'Rangpur' lime; TR = P. trifoliata; CTARG = 'Argentina' citrange. [1] Citrangor in the registration process as a rootstock variety, by Embrapa, in the National Cultivar Register (RNC) of the Brazilian Agriculture, Livestock and Supply Ministry (MAPA), called 'BRS Santana'. [2] Citrimoniandarin registered as a rootstock variety, by Embrapa, in the RNC-MAPA under the name 'BRS Tabuleiro'. [3] Citrimoniandarin registered as a rootstock variety, by Embrapa, in the RNC-MAPA under the name 'BRS Bravo'.

Table 2. Chemical characteristics of the water from the tube well used in the preparation of water of 2.4 and 4.8 dS m^{-1}.

EC	pH	Ca^{2+}	Mg^2	Na$^+$	K$^+$	CO$_3^{2-}$	HCO$_3^-$	SO$_4^{2-}$	Cl$^-$
dS m^{-1}					mmol$_c$ dm^{-3}				
30.80	7.20	30.80	78.88	148.22	2.35	0.00	7.36	3.37	289.0

EC = electrical conductivity at 25 °C.

The experimental unit consisted of one plant per pot, and the application of waters with different salinity levels began 30 days after transplanting (DAT) of the seedlings in pots adapted as lysimeters, which continued throughout the evaluation period along with the soil water balance.

The plants of nucellar origin, identified based on leaf morphological characteristics, with good formation and representative of each rootstock, were grafted on the acid lime clone 'Tahiti CNPMF-01'. The seedlings were produced by the Tamafe® Nursery, following the certified seedlings process. These were produced in plastic bags with a capacity of 2 L, filled with the commercial substrate (Basa-plant®). The seedlings remained under this condition until they were suitable to be transplanted, and the period before transplanting was approximately 300 days.

The seedlings were taken to the experimental farm of the Embrapa semi-arid region, where they were transplanted into pots adopted as lysimeters with a capacity of 60 L. The lysimeters were filled with sieved (10 mesh) Ultisol from the nearby area with the addition of organic manure.

Until 30 DAT, the plants received water of low electrical conductivity (ECw), coming from the local supply system; after this period, water of different types according to treatments was applied. Irrigations were performed with a drip irrigation system every two days.

Irrigation management was carried out based on the water balance method to replenish daily mean water consumption by the plants, plus a leaching fraction (LF) of 0.10, to avoid excessive accumulation of salts in the root zone, using Equation (1) to calculate the volume. The drained water was collected by a hose connected to the base of each lysimeter.

$$Vi = \frac{(Va - Vd)}{1 - LF} \quad (1)$$

where: Vi = volume to be applied in the irrigation event (mL), Va = volume applied in the previous irrigation event (mL), Vd = volume of water drained (mL), and LF = leaching fraction (10% = 0.10).

The nutritional management of plants was based on soil analysis and followed the recommendations presented in [15]. In the first year of production, in fertilization (N, P_2O_5, and K_2O), at weekly intervals, fertigation was performed applying 4.27 g plant^{-1} of urea, 7.69 g plant^{-1} of purified monoammonium phosphate (MAP), and 1.28 g plant^{-1} of KCl.

In the second year of production, after a new soil analysis, weekly fertilization was adjusted to 9.40 g plant^{-1} of urea, 5.12 g plant^{-1} of purified MAP, and 3.84 g plant^{-1} of KCl. In both years of production, the plants received all the micronutrients necessary for their development, also via irrigation water, as recommended in [15].

Other tillage practices related to the control of weeds and pests were adopted, particularly manual removal of invasive plants that appeared in the pots and application of pesticides when pests occurred [15]. Additionally, cleaning operations in the irrigation system, preparation of the irrigation water, formative pruning of the plants, mowing in the inter-rows of the pots, as well as cleaning and periodic maintenance of lysimeters were carried out when necessary.

At the end of each production year, soil samples (0–0.20 m) were collected in each experimental plot to obtain the saturation extract, to determine EC, pH, and the contents of Ca^{2+}, K^+, Na^+, and Mg^{2+}, to calculate the sodium adsorption ratio (SAR) and estimate the exchangeable sodium percentage (ESP), using the relationship between ESP and SAR reported by Richards [16].

During the reproductive stage, starting around 300 DAT, as the fruits reached the harvest stage [17] they were collected for subsequent counting and weighing to determine the number of fruits per plant (NFPL), the weight of fruits per plant (WFPL) (g plant^{-1}), using a scale with a resolution of 0.01 g, and to calculate the average fruit weight (AFW) (g fruit^{1}).

Physiological analyses of the plants were also performed at 270 and 720 DAT in the morning. Chlorophyll a fluorescence analysis was determined in the first mature leaf from the stem apex, in good phytosanitary condition and fully expanded, using an Opti Science OS5p pulse-modulated fluorometer, based on the OJIP protocol to determine the quantum efficiency of the photosystem II (PSII) (Fv/Fm). Thus, the fluorescence induction variables were determined: initial fluorescence (F0) and maximum fluorescence (Fm).

These data were then used to calculate the variable fluorescence (Fv, where Fv = Fm − F0) and maximum quantum efficiency of photosystem II, using the Fv/Fm ratio [18]. The analyses were performed after adaptation of the leaves to the dark for 40 min, based on a previous test, using a clip of the device to ensure that all the primary acceptors were oxidized, i.e., with the reaction centers opened.

In the same period of fluorescence determination, gas exchange analysis was performed using an infrared gas analyzer (IRGA) (LCpro+), with constant light of 1200 μmol of photons m^{-2} s^{-1}, on the third leaf of the plant counted from the apex, obtaining the following photosynthetic variables: CO_2 assimilation rate (A) (μmol$_{CO2}$ m^{-2} s^{-1}), transpiration (E) (mol H_2O m^{-2} s^{-1}), stomatal conductance (gs) (mol H_2O m^{-2} s^{-1}), and internal CO_2 concentration (Ci) (μmol$_{CO2}$ mol^{-1}). These data were used to quantify the intrinsic water use efficiency (WUEi) by the A/E ratio ((μmol m^{-2} s^{-1}) (mol H$_2$O m^{-2} s^{-1})$^{-1}$) and the intrinsic carboxylation efficiency (Φc (CEi)) by the A/Ci ratio [19].

The data obtained were subjected to ANOVA by the F test ($p \leq 0.05$); following this, the production data were analyzed using boxplots from package ggplot2 of the RStudio software. For the significant effect of the water salinity, the Tukey test ($p \leq 0.05$) was employed, and for the genotype factor, the cluster test was carried out (Scott–Knott, $p \leq 0.05$).

The obtained production data were correlated with the soil attributes, using the corrplot package in RStudio®.

3. Results

3.1. Chemical Analysis of Soil

In the first year, there were variations in the electrical conductivity of the saturation extract (ECse), 1.5 to 10.8 dS m^{-1}, for pH 5.6 to 6.0, and in SARse between 2.16 and 35.04 (mmol L^{-1})$^{0.5}$, characterizing the soil as non-saline non-sodic (without salinity problems) under irrigation with water of 0.14 dS m^{-1}, and those that received waters of 2.4 and 4.8 dS m^{-1} as saline-sodic soil, with the ESP values $\geq$ 15% and ECse $\geq$ 4 dS m^{-1} (Table 3) [20].

Table 3. Mean values of soil chemical attributes (0–0.20 m depth) at each salinity level. Samples were taken from the plots in the first and second years of production.

Water Salinity	Year	pH	ECse	Ca^{2+}	K$^+$	Na$^+$	Mg^{2+}	SARse	ESP
dS m^{-1}			dS m^{-1}		mmol$_c$ L^{-1}			(mmol L^{-1})$^{0.5}$	
0.14	1	5.6	1.5	1.1	7.4	2.4	1.1	2.2	1.70
2.40	1	5.9	5.5	0.8	13.2	36.2	1.6	34.5	34.5
4.80	1	6.0	10.8	9.6	15.6	81.5	2.7	35.0	35.0
0.14	2	5.3	2.0	2.2	13.3	1.2	2.5	0.5	0.3
2.40	2	4.9	6.6	6.1	11.8	21.7	24.1	5.1	5.8
4.80	2	4.9	9.5	12.9	9.4	32.2	38.9	6.3	7.5

pH = hydrogen potential; ECse = electrical conductivity of saturation extract; SARse = sodium adsorption ratio of saturation extract; ESP = exchangeable sodium percentage, estimated based on the relationship between SARse and ESP according to Richards [16].

In the second production year, the ECse, pH, and SARse ranged from 2.0 to 9.5 dS m^{-1}, 5.3 and 4.9, and 0.5 and 6.3 (mmol L^{-1})$^{0.5}$, respectively (Table 3). Another difference that draws attention is the reduction of K$^+$ contents in the soil in the second year (13.3 to 9.4 mmol$_c$ dm^{-3}).

It was also observed that the Na$^+$/K$^+$ ratio in the saturation extract increased with the increase in the salinity of the applied water, being, in the first year, from 0.32 in soil irrigated with an EC of 0.14 dS m^{-1}, to 5.22 in soil under irrigation using waters with an EC of 4.8 dS m^{-1}. In the same line, in the second year, the increase of the relation Na$^+$/K$^+$ was from 0.09 to 3.42 in soil under irrigation water of 0.14 and 4.8 dS m^{-1}, respectively.

3.2. Analysis of Production

The increase in salinity significantly affected the yield of 'Tahiti' acid lime grafted onto the rootstocks in the first year of production (Table 4), with effects of the interaction between rootstocks and water salinity levels on the number of fruits per plant (NFPL), weight of fruits per plant (WFPL), and average fruit weight (AFW) ($p \leq 0.05$). When considering the single factors, significant effects of rootstocks were observed for AFW ($p \leq 0.01$), and salinity affected ($p \leq 0.01$) all variables analyzed.

Table 4. Summary of analysis of variance of number of fruits per plant (NFPL), weight of fruits per plant (WFPL) (g plant^{-1}), and average fruit weight (AFW) (g fruit^{-1}) of combinations of 'Tahiti' acid lime (Citrus × latifolia (Yu. Tanaka) Tanaka) with 13 rootstocks 270 days after the onset of saline water irrigation.

Variation Factors		Mean Squares		
		NFPL	WFPL	AFW
Block	3	58.99 ns	231,500.83 ns	323.33 **
Genotype (Gen)	12	119.92 ns	298,621.30 ns	47.91 **
Error 1	36	85.84	191,893.86	40.37
Salinity (Sal)	2	10,281.82 **	27,051,122.46 **	251.53 **
Gen × Sal	24	175.66 *	284,288.68 *	72.519 *
Error 2	78	72.50	168,967.88	29.59
CV 1 (%)		31.71	32.96	13.97
CV 2 (%)		30.52	30.93	11.96
Mean		29.215	1329.011	45.471

ns = not significant; * and ** = significant at 0.05 and 0.01 probability levels, respectively; CV = coefficient of variation; DF = degrees of freedom; Genotype = combination of scion/rootstock.

The increased water salinity reduces the number of fruits per plant, but differently among the rootstocks (Figure 1). Under 0.14 dS m^{-1} (Figure 1A), there were no differences between genotypes, according to the means from the cluster test, although genotypes 6 and 7, corresponding to hybrids between the common 'Sunki' mandarin (TSKC) and *P. trifoliata* Benecke (TRBK)—007 (TSKC × TRBK—007) and between the 'Sunki of Florida' mandarin (TSKFL) and the selection of trifoliate orange (*P. trifoliata*) (TSKFL × TRBK—030), respectively, in addition to the citrandarins 'San Diego' (genotype 4) and 'Indio' (genotype 2), had lower variability in the number of fruits (Figure 1).

Water with 2.4 dS m^{-1} (Figure 1B) caused a reduction in the number of fruits per plant, with the distinction of two groups of genotypes, with the lowest means observed in genotypes 8, 9, 11, and 13, corresponding to the hybrids TSKC × 'Troyer' citrange (CTTR)—012, TSKFL × CTTR—013, TSKC × ('Rangpur' lime (LCR) × *P. trifoliata* (TR))—040, and citrimoniandarin and TSKC × 'Argentina' citrange (CTARG)—019, respectively, which denotes higher sensitivity, already at this salinity level. On the other hand, citrandarins TSKC × TRBK—007, 'San Diego' and TSKFL × TRBK—030, and the citrimoniandarin TSKC × (LCR × TR)—059, besides being in the group of genotypes with the highest number of fruits, showed greater stability in the first year.

Irrigation with waters of 4.8 dS m^{-1} (Figure 1C) did not show distinction among genotypes according to the Scott–Knott test; however, in genotypes such as 'Rangpur Santa Cruz' lime, TSKFL × TRBK—030, TSKC × (LCR × TR)—059, and TSKC × CTARG—019, maximum reductions were observed in the number of fruits per plant compared to the values obtained when the plants were irrigated with waters of 0.14 dS m^{-1}. Furthermore, under the condition of higher water salinity, the mean values observed mostly showed less variation, which can be observed by the size of the boxplot.

Weight of fruits per plant (Figure 2A–C) was reduced by salinity in all scion/rootstock combinations; however, distinct groups were formed only when applying water with lower salinity, with the highest values of fruit weight in plants grafted with 'Rangpur Santa Cruz' lime, 'Indio' and 'Riverside' citrandarins, 'Sunki Tropical' mandarin, and with hybrids TSKFL × TRBK—030, HTR—069, and TSKC × (LCR × TR)—059, according to the Scott–Knott test ($p \leq 0.05$).

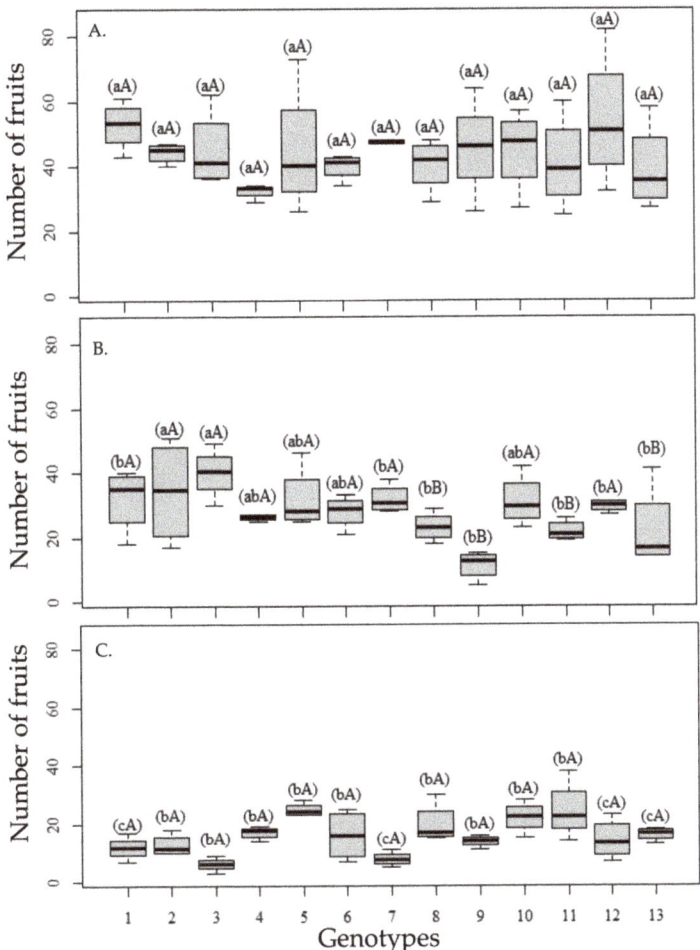

Figure 1. Boxplot relative to the mean number of fruits per plant of 13 citrus rootstocks in combination with 'Tahiti' acid lime (*Citrus* × *latifolia* (Yu. Tanaka) Tanaka) irrigated with water of 0.14 (**A**), 2.40 (**B**), and 4.80 dS m^{-1} (**C**) in the first year. For identification of genotypes, see Table 1. Boxplots with the same lowercase letter do not differ statistically, according to the Tukey test between salinity levels ($p \leq 0.05$), and those with the same uppercase letter belong to the same genotype group, according to the Scott–Knott test ($p \leq 0.05$).

When analyzing the effect of salinity on the mean fruit production per rootstock plant, it is possible to verify greater relative reductions in plants grafted with 'Santa Cruz Rangpur' lime, 'Indio' citrandarin, and the hybrid TSKFL × TRBK—030, with a reduction in fruit weight greater than 60% when the plants were irrigated with water of 4.8 dS m^{-1} (Figure 2C) compared to the results obtained in these combinations irrigated with water of 0.14 dS m^{-1} (Figure 2A).

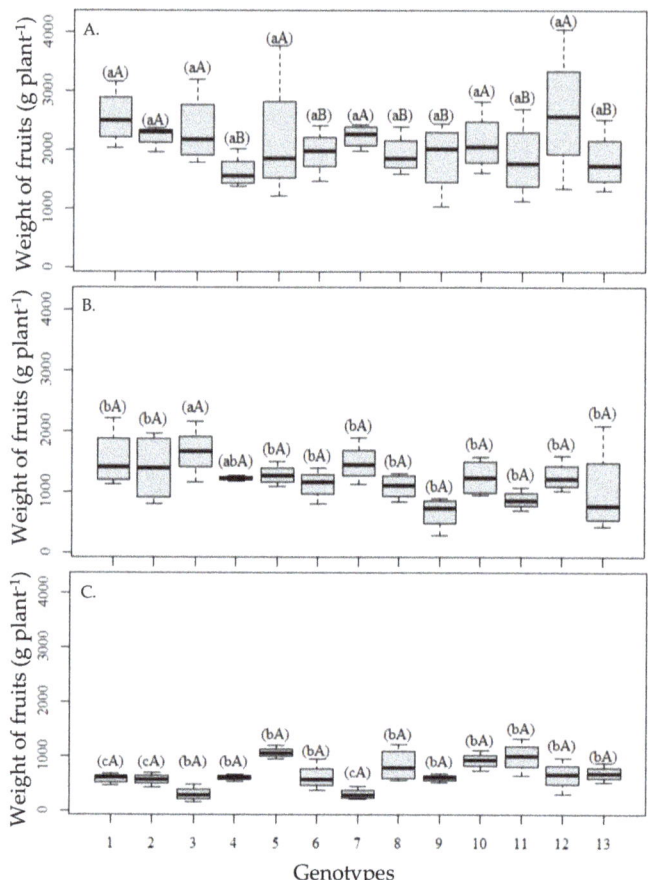

Figure 2. Boxplot relative to the average weight of fruits per plant of 13 citrus rootstocks in combination with 'Tahiti' acid lime (*Citrus* × *latifolia* (Yu. Tanaka) Tanaka) under irrigation with waters of 0.14 (**A**), 2.40 (**B**) and 4.80 dS m^{-1} (**C**), in the first year. For identification of genotypes, see Table 1. Boxplots with the same lowercase letter do not differ statistically, according to the Tukey test between salinity levels ($p \leq 0.05$), and those with the same uppercase letter belong to the same genotype group, according to the Scott–Knott test ($p \leq 0.05$).

The 'Sunki Tropical' mandarin, HTR—069 citrangor, and TSKC × (LCR × TR)—059 citrimoniandarin, on the other hand, gave the 'Tahiti' acid lime greater stability in variable weight of fruits (Figure 2), even with the increase in the salinity level in the first year of cultivation, i.e., there was a smaller reduction in the mean production of fruits with the increase in water salinity.

The salinity caused a loss of production in the number and weight of fruits per plant, but in general, the plants tried to maintain the mean weight of fruit in the first year of cultivation (Figure 3A–C), even with the increase in water salinity, with a significant reduction in the mean weight of 'Tahiti' fruit when grafted on 'Riverside' and 'San Diego' citrandarins.

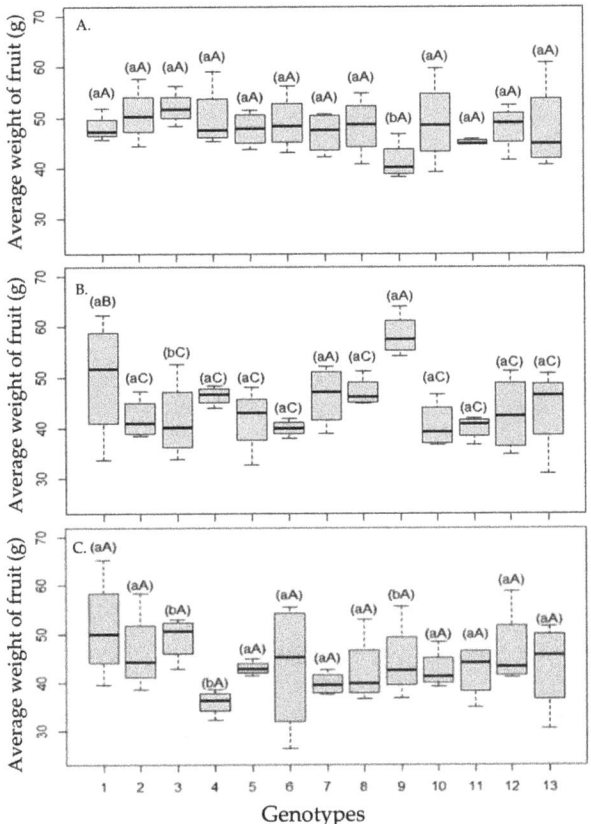

Figure 3. Boxplot relative to the average weight of fruit of 13 citrus rootstocks in combination with 'Tahiti' acid lime (Citrus × latifolia (Yu. Tanaka) Tanaka) under irrigation with waters of 0.14 (**A**), 2.40 (**B**), and 4.80 dS m^{-1} (**C**), in the first year. For identification of genotypes, see Table 1. Boxplots with the same lowercase letter do not differ statistically, according to the Tukey test between salinity levels ($p \leq 0.05$), and those with the same uppercase letter belong to the same genotype group, according to the Scott–Knott test ($p \leq 0.05$).

As for the distinction between genotypes used as rootstocks in salinity levels, differentiation is highlighted when irrigated with water of 2.4 dS m^{-1}, with the formation of three groups of genotypes, highlighting the 'Santa Cruz Rangpur' lime and TSKFL × CTTR—013 as materials that conferred the highest average weight of fruit to the plants, thus maintaining the quality.

When analyzing the production of the second year (Table 5), there were effects of the interaction between rootstocks and water salinity levels on the number of fruits per plant (NFPL), fruit production (WFPL), and average fruit weight (AFW) ($p \leq 0.05$). When considering the factors independently, significant effects were not observed only for the genotypes used as rootstocks in the average fruit weight (AFW) ($p \leq 0.01$), whereas salinity ($p \leq 0.01$) caused effects on all production variables studied.

Table 5. Summary of analysis of variance of number of fruits per plant (NFPL), weight of fruits per plant (WFPL) (g plant^{-1}), and average fruit weight (AFW) (g fruit^{-1}) of combinations of 'Tahiti' acid lime (*Citrus × latifolia* (Yu. Tanaka) Tanaka) with 13 rootstocks under water salinity, at 720 days after the onset of saline water irrigation.

Variation Factors	DF	Mean Squares		
		NFPL	WFPL	AFW (g)
Block	3	456.65 *	1,003,605.85 ns	229.155975 *
Genotype (Gen)	12	950.14 **	2,213,721.24 **	104.240321 ns
Error 1	36	110.48	284,424.46	68.466528
Salinity (Sal)	2	19,320.75 **	54,851,922.32 **	861.357813 **
Gen × Sal	24	528.51 **	1,524,565.20 **	109.237338 **
Error 2	78	127.10	323,892.75	39.697871
CV 1 (%)		27.60	31.20	18.81
CV 2 (%)		29.61.78	33.29	14.32
Mean		38.077	1709.559	44.0009814

ns = not significant; * and ** = significant at 0.05 and 0.01 probability levels, respectively; CV = coefficient of variation; DF = degrees of freedom; Gen = combination of scion/rootstock.

As occurred in the first year of production, the number of fruits per plant was reduced by the increase in ECw (Figure 4), and the means grouping test showed a higher number of fruits when the rootstocks were 'Sunki Tropical' mandarin and the hybrid TSKC × CTTR—012, under lower salinity (0.14 dS m^{-1}) (Figure 4A). It can also be verified that there was no significant difference between the other rootstocks studied, and the hybrid TSKC × CTARG led to a lower average number of fruits per plant, while 'Rangpur Santa Cruz' lime and the citrandarin TSKC × TRBK—007 were related to the lower variability in the production of the second year.

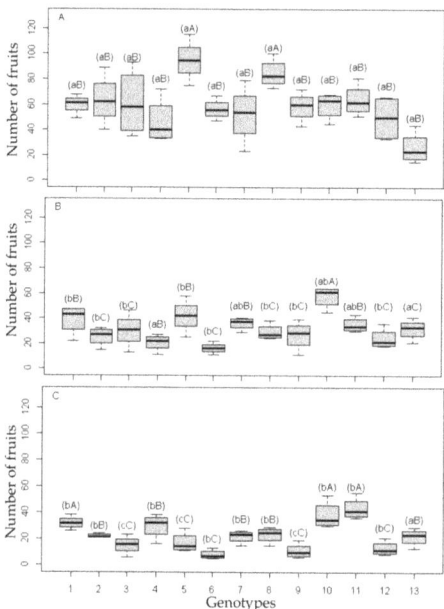

Figure 4. Boxplot relative to the average number of fruits per plant of 13 citrus rootstocks in combination with 'Tahiti' acid lime (*Citrus × latifolia* (Yu. Tanaka) Tanaka) under irrigation with waters of 0.14 (**A**), 2.40 (**B**), and 4.80 dS m^{-1} (**C**), in the secund year. For identification of genotypes, see Table 1. Boxplots with the same lowercase letter do not differ statistically, according to the Tukey test between salinity levels ($p \leq 0.05$), and those with the same uppercase letter belong to the same genotype group, according to the Scott–Knott test ($p \leq 0.05$).

Application of waters with 2.4 dS m^{-1} (Figure 4B) caused, in addition to the overall reduction in the number of fruits per plant, the formation of three groups of genotypes, with the highest means observed when the rootstock was the citrangor HTR—069. The TSKC × TRBK—007 and TSKFL × TRBK—030 citrandarins, TSKC × CTTR—012 citrangedarin, and TSKC × (LCR × TR)—040 citrimoniandarin led to lower variability in production in the second year of cultivation.

Water with salinity of 4.8 dS m^{-1} caused reductions in the number of fruits in all genotypes (Figure 4C). However, as in the first year of production, lower reductions were observed when the rootstocks were HTR—069 and TSKC × (LCR × TR)—040, which indicates that these rootstocks are better indicated for 'Tahiti' acid lime under salinity.

The weight of fruits per plant (Figure 5) was also reduced by the increase in water salinity in all scion/rootstock combinations, and it was possible to identify three distinct groups of combinations at the three salinity levels.

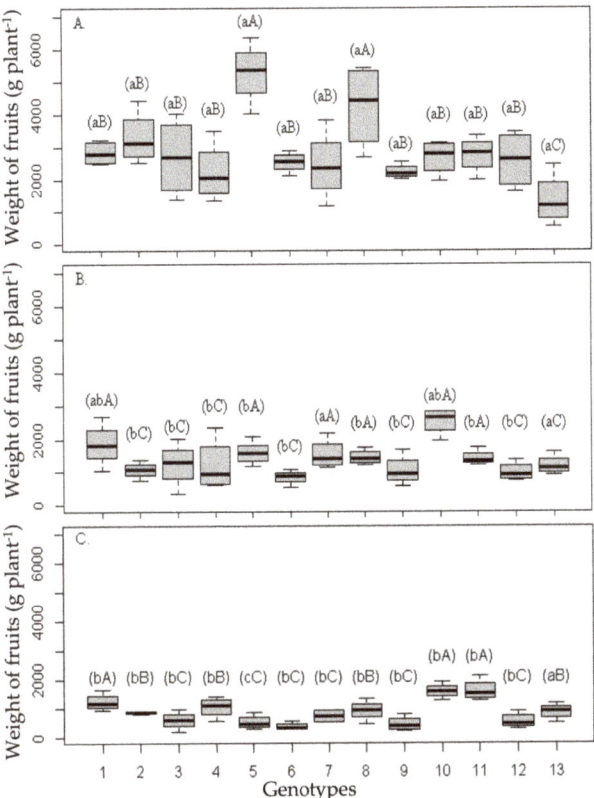

Figure 5. Boxplot relative to the average weight of fruits per plant of 13 citrus rootstocks in combination with 'Tahiti' acid lime (*Citrus* × *latifolia* (Yu. Tanaka) Tanaka) under irrigation with waters of 0.14 (**A**), 2.40 (**B**), and 4.80 dS m^{-1} (**C**), in the second year. For identification of genotypes, see Table 1. Boxplots with the same lowercase letter do not differ statistically, according to the Tukey test between salinity levels ($p \leq 0.05$), and those with the same uppercase letter belong to the same genotype group, according to the Scott–Knott test ($p \leq 0.05$).

Regarding irrigation with water of 0.14 dS m^{-1} (Figure 5A), the highest means of the weight of fruits per plant were observed for the rootstocks 'Sunki Tropical' mandarin and TSKC × CTTR—012 citrangedarin.

When the water of 2.4 dS m^{-1} was used in irrigation (Figure 5B), higher means of weight of fruits per plant were observed for 'Santa Cruz Rangpur' lime, 'Sunki Tropical' mandarin, TSKFL × TRBK—030, TSKC × CTTR—012, HTR—069, and TSKC × (LCR × TR)—040.

Conversely, when applying water of 4.8 dS m^{-1} (Figure 5C), the rootstocks 'Rangpur Santa Cruz' lime, HTR—069, and TSKC × (LCR × TR)—040 remained in the group of higher means according to the Scott–Knott test ($p \leq 0.05$), especially the trifoliate hybrid HTR—069, which showed lower variability, proving to have a good level of salinity tolerance.

The mean weight of fruits in the second year of production was significantly reduced by the salinity of the water in some genotypes, different from what occurred in the first year (Figure 6A–C), with the greatest reductions in the weight of the fruits being verified. However, the 'Tahiti' acid lime grafted on 'San Diego' and TSKFL × TRBK—030 citrandarins were also highlighted, which were sensitive to salinity in the first year, too, in addition to plants grafted on 'Sunki Tropical' mandarin, TSKC × CTTR—012, and TSKC × (LCR × TR)—059, with an estimated mean reduction of 50 g fruit^{-1} when irrigated with water of 0.14 dS m^{-1} (Figure 6A) and a mean fruit mass between 35 and 40 g when they were irrigated with water of 4.8 dS m^{-1} (Figure 6C).

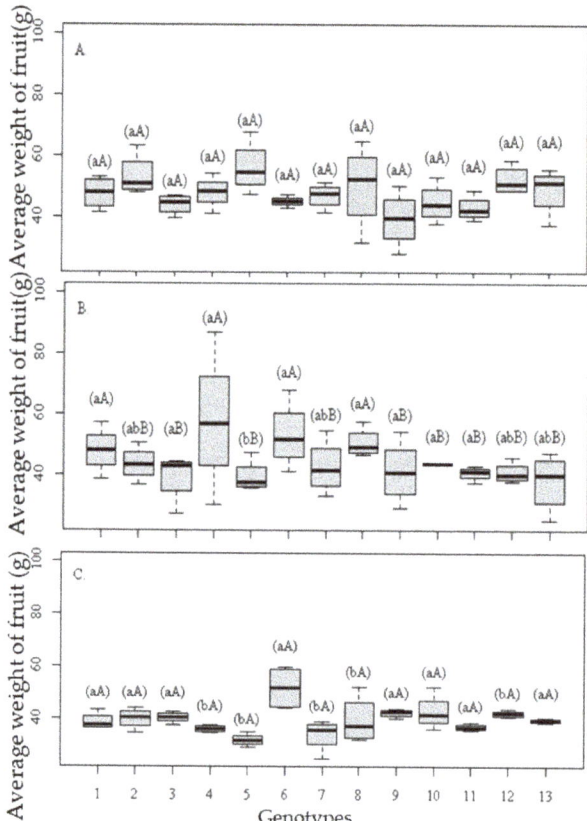

Figure 6. Boxplot relative to the average weight of fruit of 13 citrus rootstocks in combination with 'Tahiti' acid lime (*Citrus × latifolia* (Yu. Tanaka) Tanaka) under irrigation with waters of 0.14 (**A**), 2.40 (**B**), and 4.80 dS m^{-1} (**C**), in the second year. For identification of genotypes, see Table 1. Boxplots with the same lowercase letter do not differ statistically, according to the Tukey test between salinity levels ($p \leq 0.05$), and those with the same uppercase letter belong to the same genotype group, according to the Scott–Knott test ($p \leq 0.05$).

The matrices of the analytical performance of the production data with the soil analyses (Figure 7A,B) show the correlations of the first and second years of cultivation, where it is possible to observe negative numbers (without highlighting) related to the negative correlation and positive numbers (highlighted in bold) related to the positive correlation. It was also observed that the higher value means the prediction of the correlation between the variables; for example, the increase in ECse had a positive and predictive correlation with Ca^{2+}, Mg^{2+}, and Na^+ contents in both years, while on the other hand, ECse had a negative and predictive correlation with production variables. However, highlighting the behavior of K^+ in the soil during the first and the second years, correlations were predictive with ECse, but positive in the first year and negative in the second year.

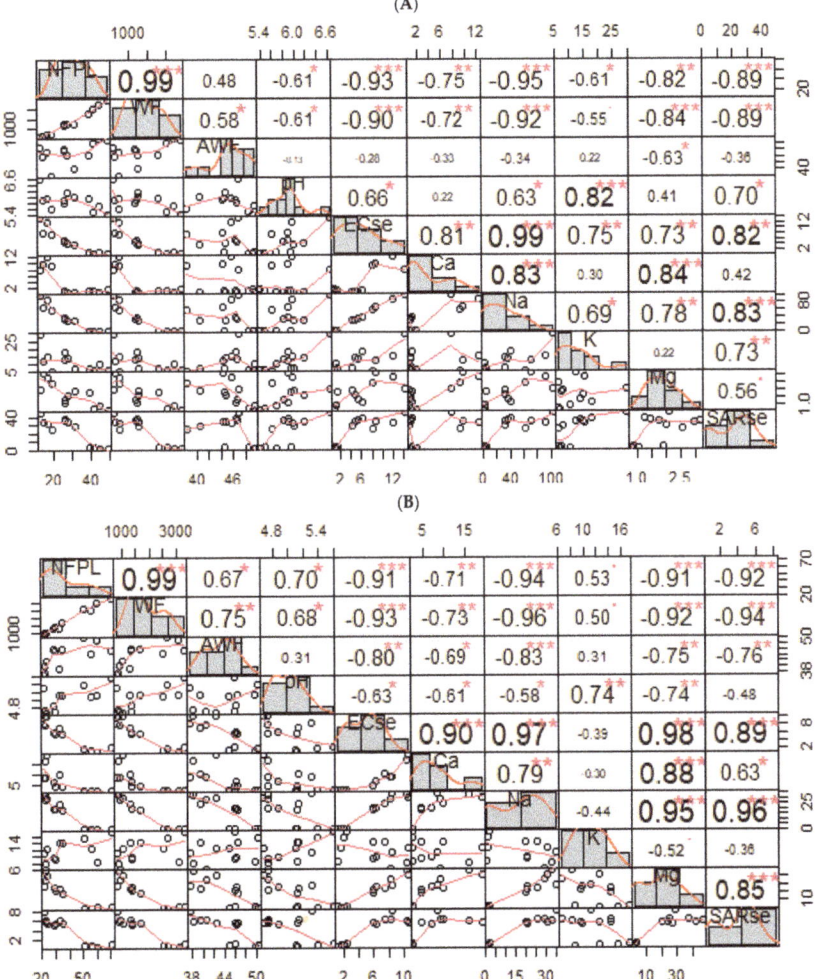

Figure 7. Analytical performance of Pearson correlation matrix of production variables (number of fruits per plant (NFPL), weight of fruits (WF), and average weight of fruit (AWF)) of genotypes studied under different water salinity levels, with the chemical characteristics of the soil saturation extract (pH, Ca^{2+}, Mg^{2+}, Na^+, K^+, ECes, and SARse) in the first (A) and second years (B). ·, *, ** and *** = significant at 0.1, 0.05, 0.01 and 0.001 probability level, respectively.

In the first year, a positive correlation was observed between K^+ (0.75) and ECse. However, the opposite was observed in the second year, with a negative correlation (-0.39). In the two years of cultivation, the interaction between water salinity and the soil variables pH, ECse, Ca^{2+}, Mg^{2+}, and SARse was positive and predictive, with a very strong correlation, close to 1, as water with higher electrical conductivity was employed in irrigation.

3.3. Chlorophyll a Fluorescence Analysis

The interaction between rootstocks and water salinity levels did not affect chlorophyll *a* fluorescence analysis after adaptation to the dark, and no significant differences were observed among rootstocks or between salinity levels in initial fluorescence (F_0), maximum fluorescence (Fm), variable fluorescence (Fv), and quantum efficiency of photosystem II (Fv/Fm), evaluated at 270 days after the beginning of stress (Table 6).

Table 6. Summary of analysis of variance of chlorophyll *a* fluorescence analysis on dark stage: initial fluorescence (F0), maximum fluorescence (Fm), variable fluorescence (Fv), and quantum efficiency of photosystem II (Fv/Fm), of combinations of 'Tahiti' acid lime (*Citrus* × *latifolia* (Yu. Tanaka) Tanaka) with 13 rootstocks under water salinity, at 270 days after the onset of saline water irrigation.

Variation Factors	DF	Mean Squares			
		F_0	Fm	Fv	Fv/Fm
Block	3	164,284.95 **	1,855,164.769 **	938,945.117 **	0.0048 ns
Genotypes (Gen)	12	3035.368 ns	64,857.381 ns	54,188.286 ns	0.0007 ns
Error 1	36	4167.117	74,885.949	79,660.237	0.0018
Salinity (Sal)	2	4458.083 ns	139,959.480 ns	172,505.237 ns	0.0043 ns
Gen × Sal	24	3631.208 ns	49,066.730 ns	46,836.771 ns	0.0011 ns
Error 2	78	3942.771	62,927.730	58,535.619	0.0013
CV 1 (%)		14.14	12.99	17.11	5.55
CV 2 (%)		13.76	11.91	14.67	4.71
Mean		456.4679	2105.9230	1649.4551	0.7812

ns = not significant; ** = significant at 0.01 probability levels, respectively; CV = coefficient of variation; DF = degrees of freedom; Gen = combination of scion/rootstock.

The fluorescence variables in the second year of cultivation, initial fluorescence (F_0), maximum fluorescence (Fm), and variable fluorescence (Fv), showed significant effects caused by the genotype x salinity interaction ($p \leq 0.05$) (Table 7), unlike the first year of cultivation.

Table 7. Summary of analysis of variance of chlorophyll *a* fluorescence analysis after dark stage, initial fluorescence (F0), maximum fluorescence (Fm), variable fluorescence (Fv), and quantum efficiency of photosystem II (Fv/Fm), of combinations of 'Tahiti' acid lime (*Citrus* × *latifolia* (Yu. Tanaka) Tanaka) with 13 rootstocks under water salinity, at 720 days after the onset of saline water irrigation.

Variation Factors	GL	Mean Squares			
		F_0	Fm	Fv	Fv/Fm
Block	3	1245.21 ns	328,638.98 **	355,262.19 **	0.017931 **
Genotype (Gen)	12	3657.48 ns	67,213.51 ns	59,530.95 ns	0.003377 ns
Error 1	36	3116.35	52,108.23	51,488.36	0.004039
Salinity (Sal)	2	2864.95 ns	33,451.08 ns	16,915.90 ns	0.000137 ns
Gen × Sal	24	5687.24 *	111,264.15 *	86,869.26 *	0.003540 ns
Error 2	78	3087.59	59,136.33	51,321.87	0.002324
CV 1 (%)		14.56	14.61	19.26	8.49
CV 2 (%)		14.49	15.57	19.22	6.44
Mean		383.53	1561.91	1178.38	0.7481

ns = not significant; * and ** = significant at 0.05 and 0.01 probability levels, respectively; CV = coefficient of variation; DF = degrees of freedom; Gen = combination of scion/rootstock.

The effect of salinity on initial fluorescence (F_0), maximum fluorescence (Fm), and variable fluorescence (Fv) varied among genotypes. When initial fluorescence was analyzed at the conductivity of 2.4 dS m^{-1} (Figure 8B), two groups of genotypes (scion/rootstock combinations) were formed, with the highest means observed in those in which the rootstocks were 'San Diego' citrandarin, 'Sunki Tropical' mandarin, TSKC × TRBK—030 citrandarin, and TSKFL × CTTR—013 citrangedarin.

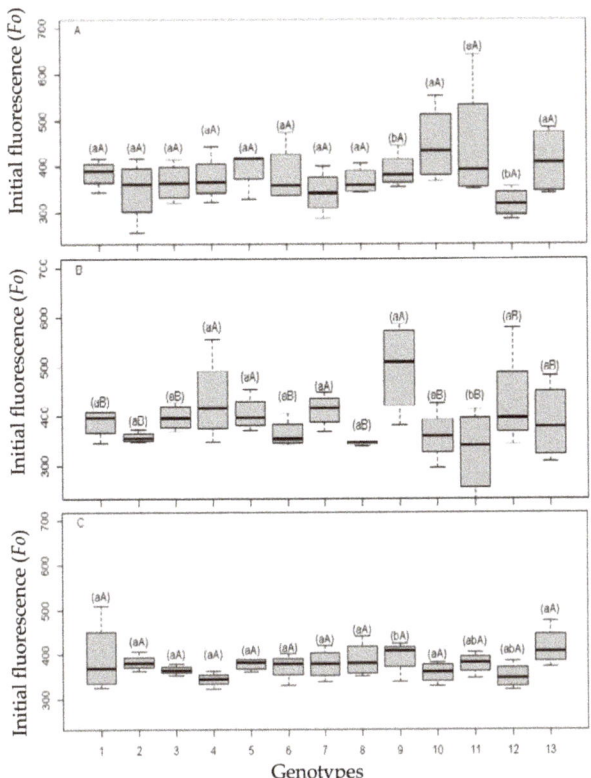

Figure 8. Boxplot relative to the average initial fluorescence (F0) of 13 citrus rootstocks in combination with 'Tahiti' acid lime (*Citrus × latifolia* (Yu. Tanaka) Tanaka) under irrigation with waters of 0.14 (**A**), 2.40 (**B**), and 4.80 dS m^{-1} (**C**), in the second year. For identification of genotypes, see Table 1. Boxplots with the same lowercase letter do not differ statistically, according to the Tukey test between salinity levels ($p \leq 0.05$), and those with the same uppercase letter belong to the same genotype group, according to the Scott–Knott test ($p \leq 0.05$).

It was also possible to notice a distinction between salinity levels in plants grafted onto TSKFL × CTTR—013 citrangedarin and TSKC × (LCR × TR)—040 and TSKC × (LCR × TR)—059 citrimoniandarins. The highest values were observed when applying water of 2.4 dS m^{-1} in the combinations of the hybrids TSKFL × CTTR—013 and TSKC × (LCR × TR)—059, and when applying water of 0.14 dS m^{-1} in plants which had the hybrid TSKC × (LCR × TR)—040 as a rootstock (Figure 8A).

The maximum fluorescence of plants (Fm) (Figure 9) was significantly reduced only in 'Tahiti' plants grafted onto TSKC × (LCR × TR)—040 citrimoniandarin and TSKC × CTARG—019 citrangedarin, especially when they were irrigated with waters of 4.8 dS m^{-1}, highlighting this variable as an indicator of ionic stress.

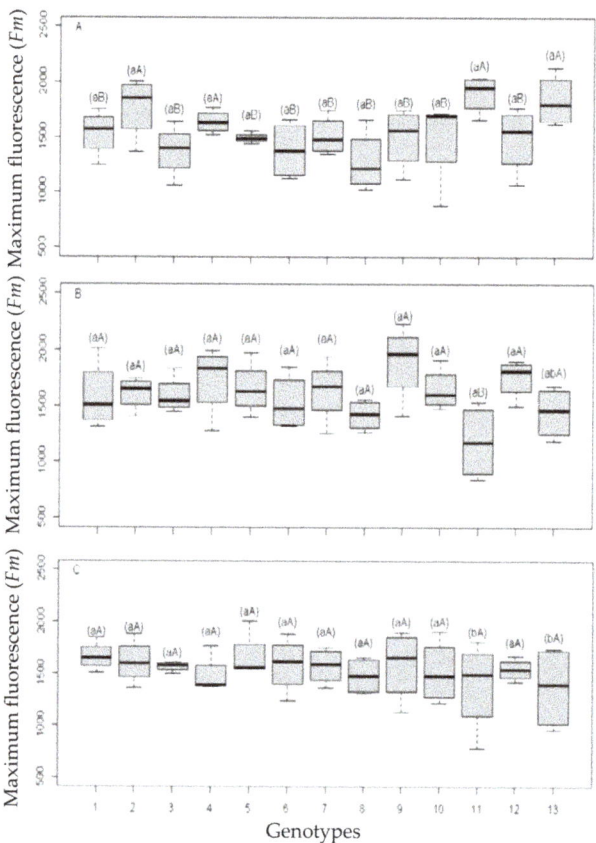

Figure 9. Boxplot relative to the average maximum fluorescence (Fm) of 13 citrus rootstocks in combination with 'Tahiti' acid lime (*Citrus* × *latifolia* (Yu. Tanaka) Tanaka) under irrigation with waters of 0.14 (**A**), 2.40 (**B**), and 4.80 dS m^{-1} (**C**), in the second year. For identification of genotypes, see Table 1. Boxplots with the same lowercase letter do not differ statistically, according to the Tukey test between salinity levels ($p \leq 0.05$), and those with the same uppercase letter belong to the same genotype group, according to the Scott–Knott test ($p \leq 0.05$).

Variable fluorescence (Fv) was different among the genotypes only at the lowest level of water salinity, highlighting two groups of genotypes (Figure 10). However, the increase in water salinity reduced ($p \leq 0.05$) the variable fluorescence in 'Tahiti' plants grafted onto the hybrids TSKC × (LCR× TR)—040 and TSKC × CTARG—019, which could be related to the reduction in the maximum fluorescence values observed in these genotypes and to the increase in the minimum fluorescence values recorded in TSKC × (LCR × TR)—040 citrimoniandarin.

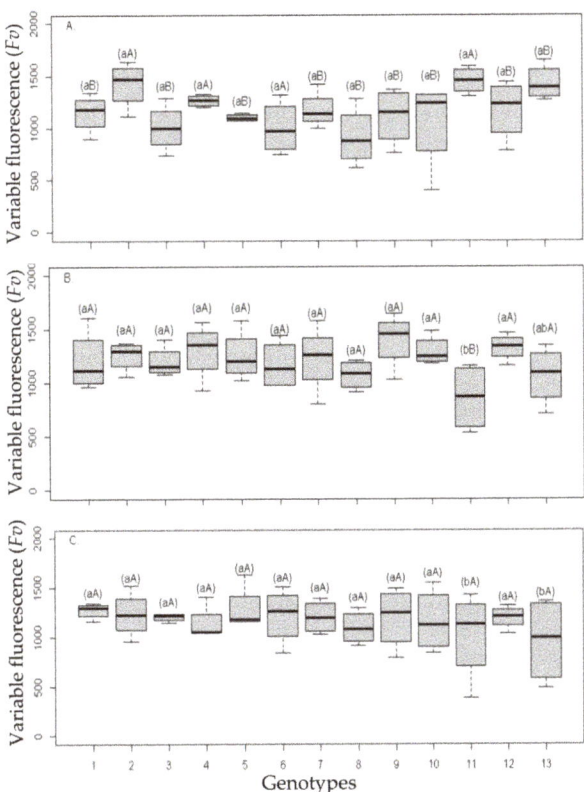

Figure 10. Boxplot relative to the average variable fluorescence (Fv) of 13 citrus rootstocks in combination with 'Tahiti' acid lime (*Citrus* × *latifolia* (Yu. Tanaka) Tanaka) under irrigation with waters of 0.14 (**A**), 2.40 (**B**), and 4.80 dS m^{-1} (**C**), in the second year. For identification of genotypes, see Table 1. Boxplots with the same lowercase letter do not differ statistically, according to the Tukey test between salinity levels ($p \leq 0.05$), and those with the same uppercase letter belong to the same genotype group, according to the Scott–Knott test ($p \leq 0.05$).

3.4. Gas Exchange

There was a significant effect ($p \leq 0.05$) of the interaction between rootstocks and water salinity levels (Table 8) on the CO_2 assimilation rate (*A*), stomatal conductance (*gs*), transpiration (*E*), and intrinsic carboxylation efficiency (*CEi*) (Table 8). Similarly, there was also a significant effect ($p \leq 0.01$) of the salinity factor on the internal CO_2 concentration (*Ci*), with no significant effect of any factor on the intrinsic water use efficiency (*WUEi*).

For the genotype factor, significant effects were found only in the variables *A*, *E*, *gs*, and *CEi* ($p \leq 0.01$), highlighting that water salinity affects citrus rootstocks differently in most of the gas exchange variables studied.

The CO_2 assimilation rate (*A*) varied among the 13 scion/rootstock combinations according to the water salinity level (Figure 11), leading to the formation of 3 groups of genotypes when the plants were irrigated with waters of 0.14 dS m^{-1} (Figure 11A) and 2 groups of genotypes when the plants were irrigated with waters of 2.40 dS m^{-1} (Figure 11B) and 4.80 dS m^{-1} (Figure 11C) according to the Scott–Knott test ($p \leq 0.05$). However, it was found that salinity only reduced the net photosynthesis of 'Tahiti' plants grafted onto 'Santa Cruz Rangpur' lime, 'Riverside' citrandarin, TSKC × CTTR—012 and TSKC × CTARG—019 citrangedarins, and TSKC × (LCR × TR)—040 citrimoniandarin.

Table 8. Summary of analysis of variance of gas exchange—assimilation rate (A) (μmol m^{-2} s^{-1}), transpiration (E) (mmol$_{H2O}$ m^{-2} s^{-1}), stomatal conductance (gs) (mol$_{H2O}$ m^{-2} s^{-1}), CO_2 concentration (Ci) (μmol mol^{-1}), intrinsic water use efficiency ($WUEi$) ((μmol$_{CO2}$ m^{-2} s^{-1}) (mmol$_{H2O}$ m^{-2} s^{-1})$^{-1}$), and intrinsic carboxylation efficiency (CEi) (μmol$_{CO2}$ m^{-2} s^{-1}) of combinations of 'Tahiti' acid lime (*Citrus* × *latifolia* (Yu. Tanaka) Tanaka) with 13 rootstocks under water salinity at 270 days after the onset of saline water irrigation.

Variation Factors	DF	Mean Squares					
		A	E	gs	Ci	$WUEi$	CEi
Block	3	15.5781 **	0.1015 ns	0.0057 **	6624.621 **	11.456 ns	0.00006 ns
Genotype (Gen)	12	13.4054 **	0.4518 **	0.0039 **	1116.853 ns	6.1785 ns	0.0002 **
Error 1	36	1.6734	0.1565	0.0006	1102.876	6.4194	0.00004
Salinity (Sal)	2	13.0319 **	0.5770 **	0.0030 **	2166.160 *	1.3231 ns	0.0002 **
Gen × Sal	24	3.4415 **	0.1569 *	0.0006 **	398.861 ns	4.9946 ns	0.00007 **
Error 2	78	1.3732	0.0884	0.0003	510.670	6.0904	0.00003
CV 1 (%)		21.86	30.53	38.42	15.32	51.97	24.99
CV 2 (%)		19.80	22.96	27.09	10.42	50.62	21.98
Mean		5.9188	1.2955	0.0641	216.8397	4.8753	0.0275

ns = not significant; * and ** = significant at 0.05 and 0.01 probability levels, respectively; CV = coefficient of variation; DF = degrees of freedom; Gen = combination of scion/rootstock.

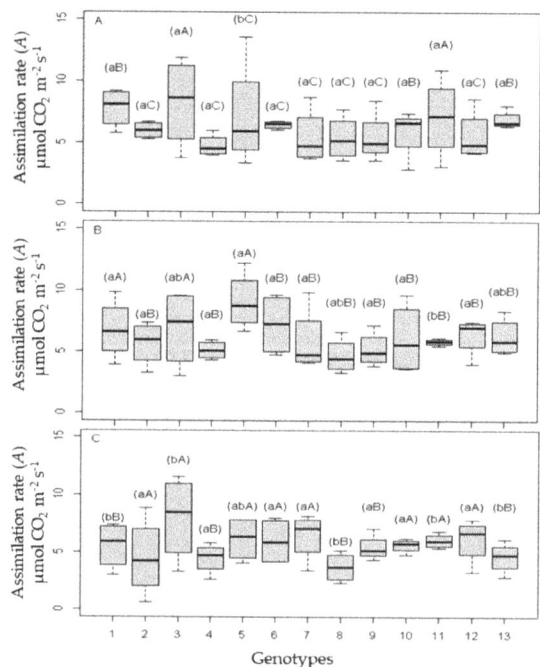

Figure 11. Boxplot relative to the assimilation rate (A) of 13 citrus rootstocks in combination with 'Tahiti' acid lime (*Citrus* × *latifolia* (Yu. Tanaka) Tanaka) under irrigation with waters of 0.14 (**A**), 2.40 (**B**), and 4.80 dS m^{-1} (**C**), in the first year. For identification of genotypes, see Table 1. Boxplots with the same lowercase letter do not differ statistically, according to the Tukey test between salinity levels ($p \leq 0.05$), and those with the same uppercase letter belong to the same genotype group, according to the Scott–Knott test ($p \leq 0.05$).

When irrigation was performed with waters of 4.8 dS m^{-1}, it was found that 'Indio' citrandarin, 'Sunki Tropical' mandarin, and the hybrids TSKC × TRBK—007, TSKFL × TRBK—030,

HTR—069, and TSKC × (LCR × TR)—059 led to a higher value of assimilation rate in the scion variety, being grouped among the genotypes with the highest means.

On the other hand, the largest reductions in assimilation rate (A) were noted for Tahiti grafted on 'Rangpur Santa Cruz' lime, 'Riverside' citrandarin, TSKC × CTTR—012 citrangedarin, TSKC × (CSF × TR)—040 citrimoniandarin, and TSKC × CTARG—019 citrangedarin.

Figure 12 contains the boxplots related to the means of transpiration (E) of the 13 scion/rootstock combinations under irrigation with waters of 0.14 dS m^{-1} (Figure 12A), 2.40 dS m^{-1} (Figure 12B), and 4.80 dS m^{-1} (Figure 12C), and it is possible to verify the rootstocks 'Santa Cruz Rangpur' lime and the 'Riverside' citrandarin between the genotypes with the highest mean values at all salinity levels. However, the increase in salinity, in general, caused a reduction in plant transpiration, but this reduction was more significant when water with an ECw of 4.8 dS m^{-1} was applied to the hybrids TSKC × CTTR—012, TSKC × (LCR × TR)—040, and TSKC × CTARG—019, and the rootstock 'Santa Cruz Rangpur' lime.

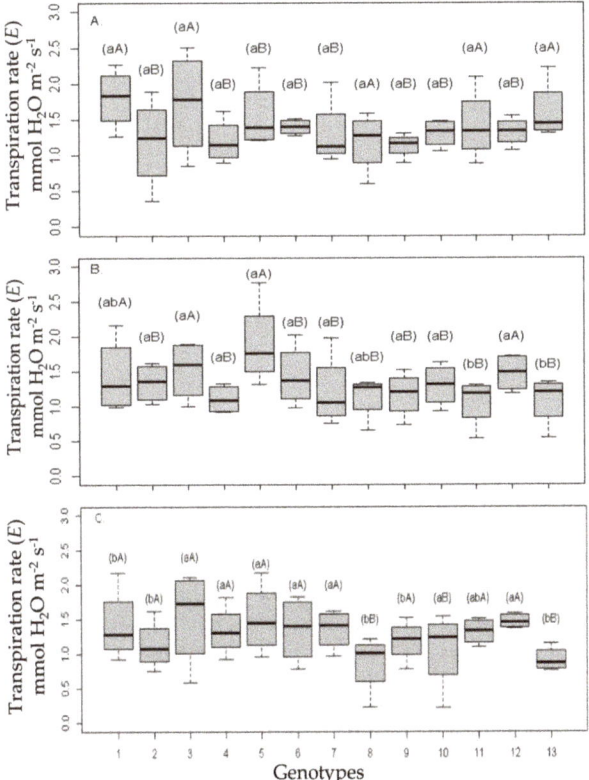

Figure 12. Boxplot relative to the average transpiration rate (E) of 13 citrus rootstocks in combination with 'Tahiti' acid lime (*Citrus* × *latifolia* (Yu. Tanaka) Tanaka) under irrigation with waters of 0.14 (**A**), 2.40 (**B**), and 4.80 dS m^{-1} (**C**), in the first year. For identification of genotypes, see Table 1. Boxplots with the same lowercase letter do not differ statistically, according to the Tukey test between salinity levels ($p \leq 0.05$), and those with the same uppercase letter belong to the same genotype group, according to the Scott–Knott test ($p \leq 0.05$).

Stomatal conductance (gs) (Figure 13) was reduced by salinity differently among the rootstocks, and higher values of gs mean that the stomata of the plant are more open, which allows the influx of CO_2 and, consequently, substrate for photosynthesis. In this context, although 'Santa Cruz Rangpur' lime and 'Riverside' citrandarin were in the group of the highest means when irrigated with waters of 0.14 dS m^{-1} (Figure 13A), when irrigated with water of 4.8 dS m^{-1} (Figure 13C), they showed reductions of the order of 54.5% and 30.8% in gs values, respectively.

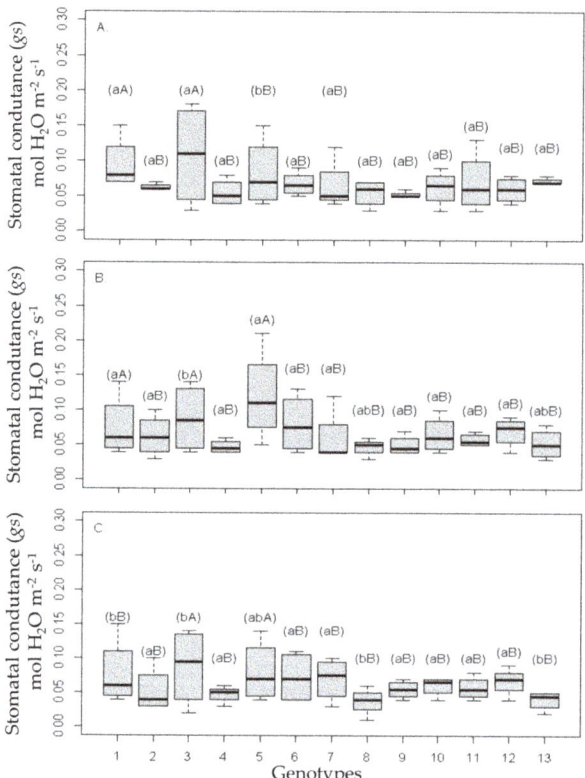

Figure 13. Boxplot relative to the average stomatal conductance (gs) of 13 citrus rootstocks in combination with 'Tahiti' acid lime (*Citrus* × *latifolia* (Yu. Tanaka) Tanaka) under irrigation with waters of 0.14 (**A**), 2.40 (**B**), and 4.80 dS m^{-1} (**C**), in the first year. For identification of genotypes, see Table 1. Boxplots with the same lowercase letter do not differ statistically, according to the Tukey test between salinity levels ($p \leq 0.05$), and those with the same uppercase letter belong to the same genotype group, according to the Scott–Knott test ($p \leq 0.05$).

The evaluation of gas exchange (Table 9) in the second year of cultivation revealed a significant effect of the interaction between rootstocks and salinity levels only for stomatal conductance (gs) ($p \leq 0.01$). For the genotype factor, there was no significant effect in any of the gas exchange variables. Salinity as an isolated factor affected the CO_2 assimilation rate (*A*) and transpiration (*E*), indicating, besides the ionic effect already observed in fluorescence reactions, osmotic effects on the plants, a situation that confirms the stress condition.

Table 9. Summary of analysis of variance of gas exchange—assimilation rate (A) (μmol m^{-2} s^{-1}), transpiration (E) (mmol$_{H2O}$ m^{-2} s^{-1}), stomatal conductance (gs) (mol$_{H2O}$ m^{-2} s^{-1}), CO_2 concentration (Ci) (μmol mol^{-1}), intrinsic water use efficiency ($WUEi$) ((μmol$_{CO2}$ m^{-2} s^{-1}) (mmol$_{H2O}$ m^{-2} s^{-1})$^{-1}$), and intrinsic carboxylation efficiency (CEi) (μmol$_{CO2}$ m^{-2} s^{-1}) of combinations of 'Tahiti' acid lime (*Citrus* × *latifolia* (Yu. Tanaka) Tanaka) with 13 rootstocks under water salinity at 720 days after the onset of saline water irrigation.

Variation Factors	DF	Mean Squares					
		A	E	gs	Ci	WUEi	CEi
Block	3	0.0935 ns	0.0009 ns	1.6405 **	14.9985 *	1.0832 **	0.0015 ns
Genotype (Gen)	12	0.0632 ns	0.0006 ns	0.3186 ns	7.8753 ns	0.0833 ns	0.0012 ns
Error 1	36	0.0423	0.0037	0.3597	4.7952	0.1037	0.0014
Salinity (Sal)	2	0.1704 **	0.0011 *	0.8505 ns	0.0513 ns	0.0813 ns	0.0003 ns
Gen × Sal	24	0.0494 ns	0.0004 ns	0.6301 **	4.3351 ns	0.1114 ns	0.0012 ns
Error 2	78	0.0329	0.0003	0.2848	3.5694	0.0988	0.0009
CV 1 (%)		12.07	1.86	19.66	17.26	13.90	3.62
CV 2 (%)		10.65	1.61	17.50	14.89	13.56	2.87
Mean		1.7043	1.0364	3.0500	12.6887	2.3168	1.0327

ns = not significant; * and ** = significant at 0.05 and 0.01 probability levels, respectively; CV = coefficient of variation; DF = degrees of freedom; Gen = combination of scion/rootstock.

The CO_2 assimilation rates (A) were not statistically different among the 13 scion/rootstock combinations at any of the salinity levels of irrigation water (Figure 14A–C), i.e., no groups of genotypes with higher means were formed. However, the effect of salinity was different among the genotypes, with a reduction in net photosynthesis in 'Tahiti' plants grafted onto the hybrid TSKC × CTARG—019, indicative of its sensitivity since decreases were observed in gas exchange and fluorescence variables.

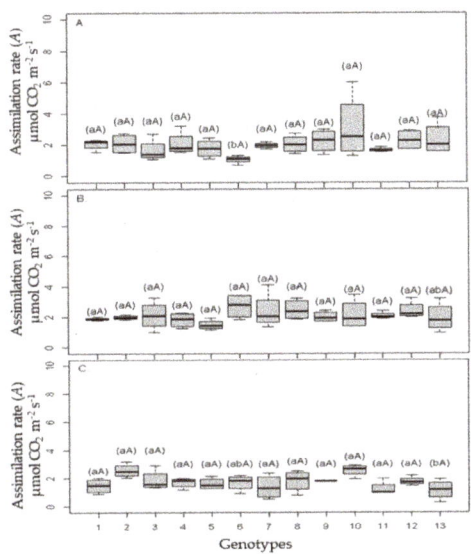

Figure 14. Boxplot relative to the assimilation rate (A) of 13 citrus rootstocks in combination with 'Tahiti' acid lime (*Citrus* × *latifolia* (Yu. Tanaka) Tanaka) under irrigation with waters of 0.14 (A), 2.40 (B), and 4.80 dS m^{-1} (C), in the second year. For identification of genotypes, see Table 1. Boxplots with the same lowercase letter do not differ statistically, according to the Tukey test between salinity levels ($p \leq 0.05$), and those with the same uppercase letter belong to the same genotype group, according to the Scott–Knott test ($p \leq 0.05$).

There was no difference ($p > 0.05$) between rootstocks in terms of transpiration rate (E) and CO_2 assimilation rate of 'Tahiti' acid lime (Figure 15A–C). Even with the application of saline water with electrical conductivity levels of 2.4 and 4.8 dS m^{-1}, the recorded values of E were between 1.0 and 3.0 mmol H_2O m^{-2} s^{-1}, except for the values observed in 'Indio' citrandarin, which showed greater variation under an ECw of 4.8 dS m^{-1}. The reduction in transpiration values should be highlighted, especially in plants grafted onto TSKC × CTARG—019 citrangedarin.

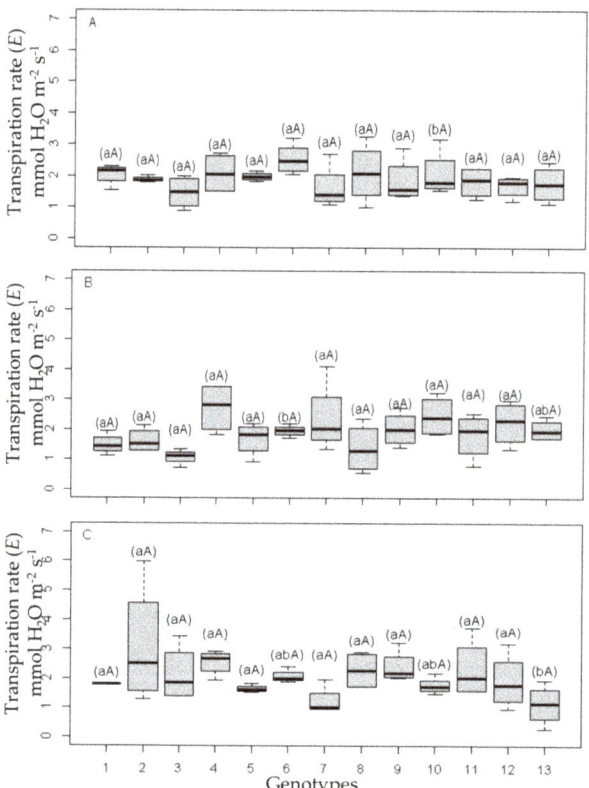

Figure 15. Boxplot relative to the average transpiration rate (E) of 13 citrus rootstocks in combination with 'Tahiti' acid lime (Citrus × latifolia (Yu. Tanaka) Tanaka) under irrigation with waters of 0.14 (**A**), 2.40 (**B**), and 4.80 dS m^{-1} (**C**), in the second year. For identification of genotypes, see Table 1. Boxplots with the same lowercase letter do not differ statistically, according to the Tukey test between salinity levels ($p \leq 0.05$), and those with the same uppercase letter belong to the same genotype group, according to the Scott–Knott test ($p \leq 0.05$).

Stomatal conductance (gs) did not vary between the rootstocks at each water salinity level applied (Figure 16), but it was observed that the effect of salinity was different among genotypes, especially because there was a significant reduction in the values of stomatal conductance in the hybrid TSKC × CTARG—019 under an ECw of 4.8 dS m^{-1}.

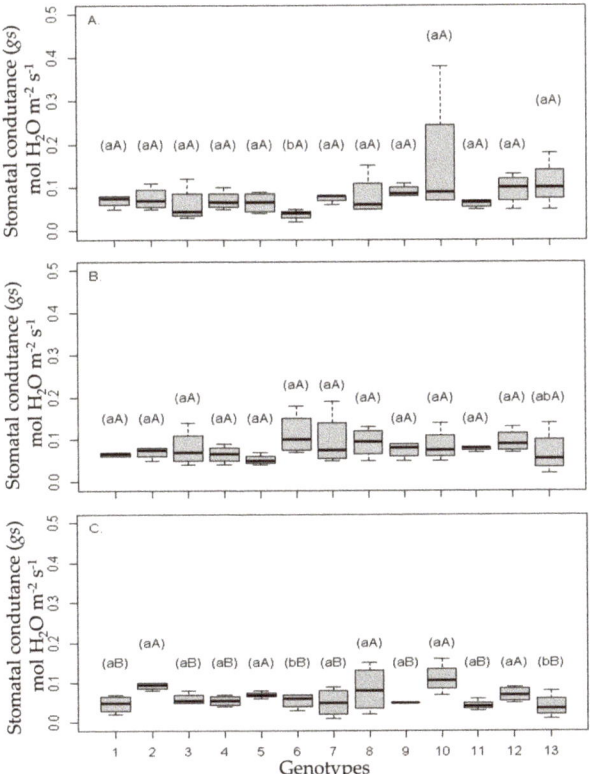

Figure 16. Boxplot relative to the average stomatal conductance (gs) of 13 citrus rootstocks in combination with 'Tahiti' acid lime (*Citrus* × *latifolia* (Yu. Tanaka) Tanaka) under irrigation with waters of 0.14 (**A**), 2.40 (**B**), and 4.80 dS m^{-1} (**C**), in the second year. For identification of genotypes, see Table 1. Boxplots with the same lowercase letter do not differ statistically, according to the Tukey test between salinity levels ($p \leq 0.05$), and those with the same uppercase letter belong to the same genotype group, according to the Scott–Knott test ($p \leq 0.05$).

4. Discussion

4.1. Soil Analysis

Soil analysis in the first year showed an increase in Na$^+$ contents, which can alter soil physical characteristics due to clay dispersion and reduce porosity, hydraulic conductivity, and infiltration rate, as well as causing soil destructuring, thereby affecting plant development [19,20].

However, the reduction in the infiltration rate was not observed in pots used as lysimeters, despite the records of higher sodium contents to the detriment of potassium contents (Table 3).

It should be emphasized that the potassium contents observed in the soil, in all treatments, can be classified as very high [15]. However, the increase in sodium content can cause a nutritional imbalance, since these ions have similar ionic radii, competing for similar exchange sites, making those that are in a greater quantity more available. On the other hand, for citrus plants, only in some genotypes, especially the most sensitive ones, was it possible to observe symptoms characteristic of potassium deficiency and sodium accumulation in the cell vacuole.

The exchangeable sodium percentage and salt content in the soil recorded during the second year allowed classifying the soil as saline, but non-sodic, which can be related to a more significant rainy season, with precipitation in the region slightly above average (751 mm), which even reduced the frequency and duration of saline water irrigation events in addition to the increase in nutrient requirements in the second year of citrus plants (Table 3).

The analysis of soil chemical characteristics in the second year was affected by the reduction in K^+ contents, which indicates a greater absorption of this nutrient by plants. In the experiment, the plants well-supplemented with this chemical element showed no symptoms of mineral deficiency.

In the same line, potassium (K^+) is an essential element for cell osmotic adjustment, as well as participating in stomatal opening and closure, which influences the water status of the plants and confers greater tolerance to salt stress [21]. Additionally, it is important to observe the decreased pH value of about 1 unit in the second year, characteristics that influence the availability of micronutrients.

4.2. Production Analysis

The increase in salt concentration in the soil causes negative effects on the plant, especially the reduction in yield, attributed to several physiological and biochemical processes [22], also observed in the present study in the first year of cultivation (Table 5), but with different behaviors of the rootstocks under salinity. These findings may be related to the distinct responses shown by citrus under salt stress, as salt tolerance varies with species, scion/rootstock combination, and plant development stage [8], under the influence of specific genes and environmental factors [23,24].

The hypothesis presented was verified when water salinity was increased to 4.8 dS m^{-1}, which caused reductions in the number of fruits, during the first year, in all genotypes (Figure 1C). However, in genotypes 5, 6, 8, 10, and 11, corresponding to 'Sunki Tropical' mandarin, TSKC × TRBK—007 citrandarin, TSKC × CTTR—012 citrangedarin, HTR—069 citrangor, and TSKC × (LCR × TR)—040 citrimoniandarin, lower reductions were observed, which characterized the lower sensitivity of 'Tahiti' acid lime when grafted onto these rootstocks. It is worth highlighting that TSKC × TRBK—007 citrandarin showed greater stability in production when irrigated with water of 4.80 dS m^{-1}.

When studying the formation of citrus rootstocks (before grafting) under water salinity, some authors [8,25] state that water of up to 2.0 dS m^{-1} can be used, similar to this study, in which water of electrical conductivity 2.4 dS m^{-1} was used.

In addition, during the first year of production, the number of fruits per plant was not different among the genotypes when irrigation was performed with waters of 2.4 and 4.8 dS m^{-1}. However, in 'Tahiti' acid lime plants grafted onto 'Riverside' citrandarin, 'Sunki Tropical' mandarin, and the trifoliate hybrid HTR—069 citrangor, there was greater stability of production between the salinity levels (Figure 2). The results obtained highlight these materials as promising since in the agricultural production system it is of great importance that production is satisfactory in quantity and for a longer time. However, it is also emphasized that such distinctions in the number of fruits per plant may be related to the early production of some genotypes, which shows the importance of having more years of evaluation.

The evaluation of the second production year made it possible to confirm the information obtained in the first year when the distinction of genotypes irrigated with waters of higher salinity levels was verified. The observations can be confirmed particularly for HTR—069, a hybrid of 'Pera' sweet orange with 'Yuma' citrange, as a rootstock that confers greater tolerance to salinity, corroborating the results obtained with this hybrid in the seedling formation stage [10]. On the other hand, highlighting this rootstock does not disqualify the potential of the others, since citrus plants show stabilization of production between six and seven years [15]. Considering that this study looked at the second year of production and third year of cultivation, the scion/rootstock combinations under analysis should be studied for a longer period.

Regarding the production and number of fruits per plant, the effects of salinity were different among the rootstocks, which points to the need to highlight the variation in the juvenility period of citrus, and it can be verified that the different production may be related to the precocious or later onset of fruiting of one or another scion/rootstock combination. Therefore, it is not possible to accurately predict fruit yield in the future, a situation that requires the monitoring of five or more seasons.

However, when evaluating the effect of salinity levels, it was possible to observe differences between the scion/rootstock combinations regarding the tolerance to this abiotic stress, since less variation in production between salinity levels, for a certain combination, means the obtaining of viable yields, even under stress conditions, which was found when the rootstocks were the hybrids HTR—069 and TSKC × (LCR × TR)—040. This result indicates tolerance to the salinity of these varieties and their potential for cultivation under irrigation in the Brazilian semi-arid region.

The correlations between production variables and salinity levels observed in the soil during the first and second years of production generated information that explains the increase in the need for some nutrients, especially potassium, by plants so that they could maintain physiological and biochemical processes and survival, since the ability of citrus plants to develop in saline soils is generally more associated with exclusion than with compartmentalization of ions in their leaves [26]; that is, more efficient plants avoid the accumulation of toxic ions, reducing the absorption of ions.

Another inference that can be made is that, both in the first and the second year of cultivation, salinity negatively affected the production variables NFPL, WFPL, and AFW (Figure 7A,B); that is, as salinity increased, production decreased, corroborating what was shown in the boxplots presented in Figures 1–6, as well as in studies conducted by several authors [8,10,25,27].

The process of formation of saline soils involves the increase in the concentration of soluble salts in the soil solution and results in the accumulation of Na^+, Ca^{2+}, Mg^{2+}, and K^+, so the constant use of saline water must be well-managed to avoid or delay salinization phenomena. In this context, it was verified that the water used in irrigation had high contents of cations, especially Ca^{2+}, Mg^{2+}, and Na^+, which caused an increase in the contents of these elements in the soil and in the relationships between them. The increase in sodium, compared to the others, should be highlighted, since this is the cation in the greatest proportion, as observed in the two years through correlation analyses.

In general, the main salts found were the chlorides and sulfates of Na^+, Ca^{2+}, and Mg^{2+}, however carbonates and nitrates were also found [28,29]. In this work, the water used for irrigation had some elements mentioned, which were reflected in the contents found in the soil cultivated with citrus in the two years, although in a more significant way in the second year.

4.3. Chlorophyll a Fluorescence

The effects of salinity on plants may be osmotic or ionic, with the former corresponding to the decrease in osmotic potential and the latter related to nutritional imbalance, due to the high concentration of specific ions, which results in a series of harmful effects on plant physiology, including chlorophyll fluorescence and gas exchange, which are severely affected [18,27].

The effects of salinity on fruit production generated physiological disorders in the first year of cultivation; however, the effects were osmotic since, in the fluorescence analyses performed in the first year, the damage to the photosynthetic apparatus was not significant, indicating that the ionic effect was not deleterious in the scion/rootstock combinations to the point of causing damage to the photosynthetic apparatus, as observed by Zang et al. [30].

In stressful environments, a decrease in the potential quantum efficiency of photosystem II can occur and is detected by the reduction in the Fv/Fm ratio [31]. Although the plants were irrigated with waters of 4.8 dS m^{-1}, which caused an accumulation of salts in the soil of the order of 14.7 dS m^{-1} in the first year and 9.5 dS m^{-1} in the second year, values well above the salinity threshold of the citrus crop (1.4 dS m^{-1}), there was no significant effect on the quantum efficiency of photosystem II to the point of causing deleterious effects.

In the process of photosynthesis, the absorbed light can be transferred to the photosystems or, if there is excess energy, it is dissipated in the form of heat or fluorescence [31,32]. Considering that there were no changes in fluorescence with the increase in salt concentration in the irrigation water applied to plants and that values of quantum efficiency of photosystem II (Fv/Fm) were close to 0.78 in the first year (Table 7), it is possible to assume that the photosynthetic apparatus of the scion/rootstock combinations was not damaged in the present study, with possible acclimatization, or that this response was a consequence of the short period of stress, or due to the more regular distribution of rainfall, which caused a decrease in salt concentration since irrigation was performed based on water balance and a leaching fraction of 10% was used, which enabled a reduction in the accumulation of salts in the soil in the second year.

However, according to Barbosa et al. [25] and Fernandes et al. [33], in studies on the growth and physiology of hybrids and some varieties recommended as rootstock for citrus, when plants are irrigated with saline waters, the response may be diverse due to the genetic load of the material, as well as to the ability to adapt to stress conditions, which may have occurred in the present study and thus preserved the PSII. Therefore, it is necessary to conduct studies for longer periods under controlled environments, where ambient conditions can be maintained so as not to cause other stresses or other types of interference.

The photochemical quantum efficiency of PSII (FV/Fm) reflects the efficiency related to the absorption of light energy by the PSII antenna complex and its conversion into chemical energy. Therefore, as there was no significant effect of salinity on this variable, especially because the mean value was higher than 0.72, a limit that indicates damage to the photosynthetic apparatus [19], there was no damage caused by salt stress [34,35].

The intensification of stress, in the second year of plants under irrigation with saline waters, made it possible to observe more significant physiological disorders, as reported by Zang et al. [30], who concluded that the deleterious effects of irrigation water salinity on scion/rootstock combinations may be more intense with longer exposure to stress. This fact was verified in the present study since the salinity in the saturation extract was maintained in the second year (Table 3), and it was possible to notice exhaustion of plants that were under prolonged stress, although there was a period of rainfall.

The most noticeable effect mentioned here is related to the different behaviors of the genotypes regarding chlorophyll fluorescence when irrigated with saline waters, although there were no effects of the factors on the quantum efficiency of photosystem II (Fv/Fm) (Table 8), which can be related to acclimatization, although the plants were under stress.

The hypothesis is that salinity triggered reactions, which were noticed through the increase in the values of initial, maximum, and variable fluorescence, although this does not mean higher or lower sensitivity to salinity, as there may be physiological adjustments that allow the plant to maintain energy formation for metabolic processes, even with lower efficiency.

For the sake of understanding, the initial fluorescence is measured after the tissue is exposed to darkness for a certain period, which causes the oxidation of electrons, optimizing the absorption of light energy, i.e., this is the moment when there is the lowest loss by fluorescence, so the values are lower. However, if the photosystem is under stress, the values tend to increase, becoming an indicator of the stress condition [18,34–36], as observed in the present study.

On the other hand, the opposite occurs in the maximum fluorescence: the plant under stress conditions cannot reach the energy boost limit, and rootstocks such as TSKC × (LCR × TR)—040 and TSKC × CTARG—019 showed a condition of exhaustion of the photosynthetic apparatus, which may result from an attempt to make greater use of the energy received.

The changes observed in minimum fluorescence and maximum fluorescence resulted in an effect on the variable fluorescence, which is obtained by their subtraction (Figures 8–10). However, the values of quantum efficiency of photosystem II were not affected by the interaction between the factors, which leads to the hypothesis that the photosynthetic apparatus did not suffer significant damage, i.e., despite being under stress conditions, the plants were able to maintain photosynthetic metabolism. Despite that, compared to the first year of cultivation, it was possible to observe a small reduction in the quantum efficiency of photosystem II, which changed from 0.78 in the first year to 0.75 in the second year (Table 8), a value within the normal range [37]. In similar studies with control plants, these authors reported the range of variation for the quantum efficiency of photosystem II between 0.72 and 0.81 as ideal for citrus plants.

4.4. Gas Exchange

The effect of salinity on gas exchange was significant in the first year of cultivation, although the values of net photosynthesis were within the range described for citrus plants, from 4 to 10 µmol CO_2 m^{-2} s^{-1} [15]. The values found can be considered normal, even with irrigation with water of higher salinity (4.8 dS m^{-1}), except for TSKC × CTTR—012 citrangedarin, whose value was below 4 µmol CO_2 m^{-2} s^{-1}.

The gas exchange variables were more sensitive to the effect of salinity and allowed the selection of rootstocks with greater physiological adaptation capacity, especially through the evaluation of the relative reduction in stomatal conductance and transpiration of plants, as observed in 'Sunki Tropical' mandarin, which was classified among the best genotypes when irrigated with water of 4.8 dS m^{-1}, and the hybrids HTR—069 citrangor and TSKC × (LCR × TR)—040 citrimoniandarin, which did not suffer reductions when subjected to this salinity level, denoting the lower sensitivity of these individuals [8,33], the opposite of what occurred with TSKC × CTARG—019 citrangedarin, the most sensitive genotype to salinity.

On the other hand, it was possible to observe that most genotypes kept similar means, with no significant differences until the use of an ECw of 2.4 dS m^{-1}, and the values were reduced when the salinity level was increased to 4.8 dS m^{-1}, which points to the condition of stability or acclimatization since plants were in their second year of production and under irrigation with saline water. Furthermore, it should be noted that the mean values of net photosynthesis, at all salinity levels, were below 4.0 µmol CO_2 m^{-2} s^{-1}, which can be interpreted as a condition of plant exhaustion, in the face of multiple stresses, notably due to the cultivation in the pot, because even if adequate nutritional management was applied, the restrictive conditions of the pots may have generated other stresses, such as those related to temperature [23,24], leading to a reduction in physiological potential.

The reduction caused by salinity in the CO_2 assimilation rate is reported in the literature as an osmotic effect [19] since the increase in the salt concentration reduces the water potential in the soil and reduces water absorption, causing stomata closure to ensure the maintenance of the plant's turgor, thereby reducing the CO_2 influx into the substomatal chamber.

The higher sensitivity to salinity observed in the hybrid TSKC × CTARG—019, based on physiological data, was accompanied by a lower fruit production by this rootstock. Although the application of water with an ECw of 4.8 dS m^{-1} did not cause a significant loss of yield compared to plants irrigated with water of 0.14 dS m^{-1} (Figure 4), there was a significant reduction in the production of plants grafted onto this citrangedarin.

This fact, however, only points to the importance of physiological variables in the understanding and indication of stress situations, which can be a very useful tool for decision-making, since the reduction caused by salinity was also observed in citrus plants under water salinity, with deleterious effects [37].

Considering that stomatal conductance reflects the process of CO_2 entry and water vapor exit for the performance of net photosynthesis, the results were coherent since the citrangedarin TSKC × CTARG—019 was the rootstock that also conferred the scion variety the lowest rates of net photosynthesis and stomatal conductance when it was irrigated with waters of higher salinity. In line with the results obtained in the present study, Sousa et al. [38] highlight that stomatal conductance is responsible for controlling water exit and CO_2 entry, which in turn is one of the main substrates for photosynthesis. Thus, as the mean of stomatal conductance is modified by the salinity level of irrigation water, the stomata begin to be hampered in their natural activities, compromising CO_2 assimilation.

5. Conclusions

Water salinity reduced the production of citrus plants, especially the number of fruits per plant (NFPL), the average fruit weight (AFW), and consequently the total weight of fruits (WFPL), and gradual reductions were observed in the two years of cultivation.

'Riverside' citrandarin, 'Sunki Tropical' mandarin, and HTR—069 citrangor showed greater production stability, even with the increase in irrigation water salinity.

Genotypes 'Tahiti' grafted on 'Sunki Tropical' mandarin, TSKC × (LCR × TR)—040 citrimoniandarin, and HTR—069 citrangor were more tolerant to an ECw of 4.8 dS m^{-1}.

The effect of salinity on citrus plants was osmotic, reducing their net photosynthesis, transpiration, and stomatal conductance.

The photosynthetic apparatus of the scion/rootstock combinations was not affected by salinity since the quantum efficiency of photosystem II (Fv/Fm) was higher than 0.78 in the first year of cultivation and equal to 0.75 during the second year of cultivation.

Citrangedarins TSKC × CTTR—012, TSKFL × CTTR—013, and TSKC × CTARG—019 citrangors showed higher sensitivity to salt stress, with the lowest yields when irrigated with saline water of 2.4 dS m^{-1} and with the largest reductions in quantum efficiency of photosystem II (PSII) and gas exchange.

Irrigation water of up to an electrical conductivity of 2.4 dS m^{-1} can be used in the cultivation of citrus plants without significantly compromising their physiology, provided that salt-tolerant rootstocks are used and a leaching fraction of 0.10 is applied.

Author Contributions: Writing—review and editing, G.O.M., M.E.B.B., H.R.G., A.S.d.M. and W.S.S.F.; visualization, R.R.G.F., A.S.d.M., P.D.F., H.R.G., W.S.S.F. and M.E.B.B.; supervision, A.S.d.M., P.D.F. and H.R.G.; Data curation, G.O.M., S.S.S., E.R.E., R.R.d.M.N. and M.E.B.B.; project administration, M.E.B.B.; funding acquisition, M.E.B.B. All authors have read and agreed to the published version of the manuscript.

Funding: National Council for Scientific and Technological Development: 406460/2018-3.

Institutional Review Board Statement: Not applicable.

Data Availability Statement: Not applicable.

Conflicts of Interest: The authors declare no conflict of interest.

References

1. Maçorano, R.P. Impacto das Mudanças Climáticas na Dinâmica da Citricultura no Estado de São Paulo. Master's Thesis, State University of Campinas, São Paulo, Brazil, 2017.
2. IBGE—Instituto Brasileiro de Geografia e Estatística. Levantamento Sistemático da Produção Agrícola: Agosto. 2020. Available online: https://sidra.ibge.gov.br (accessed on 10 December 2021).
3. Diana, E.H.L.; Isidoro, P.F.; Ikefuti, C.V. Agribusiness and the productivity of Tahiti lemon: A study in the municipality of Marinópolis, in Northwest Paulista. Braz. J. Bus. 2021, 3, 3208–3219. [CrossRef]
4. da Silva, S.R.; Stuchi, E.S.; Girardi, E.A.; Cantuarias-Avilés, T.; Bassan, M.M. Desempenho da tangerineira 'Span Americana' em diferentes porta-enxertos. Rev. Bras. Frutic. 2013, 35, 1052–1058. [CrossRef]
5. Santos, M.R.; Donato, S.L.R.; Coelho, E.F.; Arantes, A.M.; Coelho Filho, M.A. Irrigação lateralmente alternada em lima ácida 'Tahiti' na região norte de Minas Gerais. Irriga 2016, 1, 71–88. [CrossRef]
6. Medeiros, J.F.D.; Lisboa, R.D.A.; Oliveira, M.D.; Silva Júnior, M.J.D.; Alves, L.P. Caracterização das águas subterrâneas usadas para irrigação na área produtora de melão da Chapada do Apodi. Rev. Bras. Eng. Agríc. Ambient. 2003, 7, 469–472. [CrossRef]

7. Maas, E.V. Salinity and citriculture. *Tree Physiol.* **1993**, *12*, 195–216. [CrossRef] [PubMed]
8. Brito, M.E.B.; Fernandes, P.D.; Gheyi, H.R.; de Melo, A.S.; dos Santos Soares Filho, W.; dos Santos, R.T. Sensibilidade à salinidade de híbridos trifoliados e outros porta-enxertos de citros. *Rev. Caatinga* **2014**, *27*, 17–27. Available online: https://periodicos.ufersa.edu.br/caatinga/article/view/2610/pdf_82 (accessed on 4 September 2022).
9. Brito, M.E.B.; Soares, L.A.A.; Soares Filho, W.S.; Fernandes, P.D.; Silva, E.C.B.; Sá, F.V.S.; Silva, L.A. Emergence and morpho-physiology of Sunki mandarin and other citrus genotypes seedlings under saline stress. *Span. J. Agric. Res.* **2018**, *16*, e0801. [CrossRef]
10. Brito, M.E.B.; Fernandes, P.D.; Gheyi, H.R.; Soares, L.A.A.; Soares Filho, W.S.; Suassuna., J.F. Screening of citrus scion-rootstock combinations for tolerance to water salinity during seedling formation. *Acta Sci.-Agron.* **2021**, *43*, e48163. [CrossRef]
11. Bastos, D.C.; Ferreira, E.A.; Passos, O.S.; Sá, J.F.; Ataíde, E.M.; Calgaro, M. Cultivares copa e porta-enxertos para a citricultura brasileira. *Inf. Agropecu.* **2014**, *35*, 36–45. Available online: https://www.alice.cnptia.embrapa.br/alice/bitstream/doc/1007492/1/Debora214.pdf (accessed on 15 July 2021).
12. Soares, L.A.A.; Brito, M.E.B.; Fernandes, P.D.; de Lima, G.S.; Soares Filho, W.S.; Oliveira Filho, E.S.D. Crescimento de combinações copa-porta-enxerto de citros sob estresse hídrico em casa de vegetação. *Rev. Bras. Eng. Agríc. Ambient.* **2015**, *19*, 211–217. [CrossRef]
13. Aquino, A.J.S.; de Lacerda, C.F.; Bezerra, M.A.; Gomes Filho, E.; Costa, R.N.T. Crescimento, partição de matéria seca e retenção de Na^+ e Cl^- em dois genótipos de sorgo irrigados com águas salinas. *Rev. Bras. Ciênc. Solo* **2007**, *31*, 961–971. [CrossRef]
14. Sousa, J.S.C.; Brito, M.E.B. Programa computacional BHCN&CCTK: Balanço hídrico climatológico normal e classificação climática de Thornthwaite e Köppen. *Braz. J. Dev.* **2022**, *8*, 35877–35898. [CrossRef]
15. Mattos Junior, D.; de Negri, J.D.; Pio, R.S.; Pompeu Junior, J. *Citros*; Instituto Agronômico e Fundag: Campinas, Brazil, 2005.
16. Richard, L.A. *Diagnosis and Improvement of Saline and Alkalis Soils*; Agric. Handbook 60; United States Department of Agriculture: Washington, DC, USA, 1954.
17. Gayet, J.P.; Salvo Filho, A. Colheita e beneficiamento. In *Lima Ácida Tahit*; Mattos Junior, D., Negri, J.D., Figueiredo, J.O., Eds.; Instituto Agronômico: Campinas, Brazil, 2003; pp. 147–162.
18. Genty, B.; Briantais, J.; Baker, N.R. The relationship between the quantum yield of photosynthetic electron transport and quenching of chlorophyll fluorescence. *Biochim. Biophys. Acta (BBA)—Gen. Subj.* **1989**, *990*, 87–92. [CrossRef]
19. Silva, L.D.A.; Brito, M.E.B.; Sá, F.V.D.S.; Moreira, R.C.L.; Soares Filho, W.D.S.; Fernandes, P.D. Mecanismos fisiológicos em híbridos de citros sob estresse salino em cultivo hidropônico. *Rev. Bras. Eng. Agríc. Ambient.* **2014**, *18*, 1–7. [CrossRef]
20. Ayers, R.S.; Westcot, D.W. *A Qualidade da Água na Agricultura*; Universidade Federal da Paraíba: Campina Grande, Brazil, 1999.
21. Homem, B.G.C.; de Almeida Neto, O.B.; Condé, M.S.; Silva, M.D.; Ferreira, I.M. Efeito do uso prolongado de água residuária da suinocultura sobre as propriedades químicas e físicas de um Latossolo Vermelho-Amarelo. *Rev. Cient.* **2014**, *42*, 299–309. [CrossRef]
22. Taleisnik, E.; Grunberg, K. Ion balance in tomato cultivars differing in salt tolerance. I. Sodium and potassium accumulation and fluxes under moderate salinity. *Physiol. Plant.* **1994**, *92*, 528–534. [CrossRef]
23. Munns, R.; Tester, M. Mechanisms of salinity tolerance. *Annu. Rev. Plant Biol.* **2008**, *59*, 651–681. [CrossRef] [PubMed]
24. Syvertsen, J.P.; Garcia-Sanchez, F. Multiple abiotic stresses occurring with salinity stress in citrus. *Environ. Exp. Bot.* **2014**, *103*, 128–137. [CrossRef]
25. Barbosa, R.C.A.; Brito, M.E.B.; Sá, F.V.S.; Soares Filho, W.S.; Fernandes, P.D.; Silva, L.A. Gas exchange of citrus rootstocks in response to intensity and duration of saline stress. *Semin. Ciênc. Agrár.* **2017**, *38*, 725–738. [CrossRef]
26. Storey, R.; Walker, R.R. Citrus and salinity. *Sci. Hortic.* **1999**, *8*, 39–81. [CrossRef]
27. Almeida, J.F. Ecofisiologia de Limeira Ácida 'Tahiti' Condicionada a Porta—Enxertos de Citros e Salinidade da Água. Master's Thesis, Universidade Federal de Campina, Grande Pombal, Brazil, 2019.
28. Resende, R.S.; Amorim, J.R.A.; Cruz, M.A.S.; Meneses, T.N. Distribuição espacial e lixiviação natural de sais em solos do Perímetro Irrigado Califórnia, em Sergipe. *Rev. Bras. Eng. Agríc. Ambient.* **2014**, *18*, S46–S52. [CrossRef]
29. Pedrotti, A.; Chagas, R.M.; Ramos, V.C.; do NascimentoPrata, A.P.; Lucas, A.A.T.; dos Santos, P.B. Causas e consequências do processo de salinização dos solos. *Rev. Eletrôn. Gestão Educ. Tecnol. Ambient.* **2015**, *19*, 1308–1324. [CrossRef]
30. Zhang, H.S.; Zai, X.M.; Wu, X.H.; Qin, P.; Zhang, W.M. An ecological technology of coastal saline soil amelioration. *Ecol. Eng.* **2014**, *67*, 80–88. [CrossRef]
31. Krause, G.H.; Weis, E. Chlorophyll fluorescence and photosynthesis: The basics. *Annu. Rev. Plant Physiol. Plant Mol. Biol.* **1991**, *42*, 313–349. [CrossRef]
32. Young, A.L.; Frank, H.A. Energy transfer reactions involving carotenoids: Quenching of chlorophyll fluorescence. *J. Photochem. Photobiol. B Biol.* **1996**, *36*, 3–15. [CrossRef]
33. Fernandes, P.D.; Brito, M.E.B.; Gheyi, H.R.; Soares Filho, W.D.S.; de Melo, A.S.; Carneiro, P.T. Crescimento de híbridos e variedades porta-enxerto de citros sob salinidade. *Acta Sci. Agron.* **2011**, *33*, 259–267. [CrossRef]
34. Baker, N.R. Chlorophyll fluorescence: A probe of photosynthesis in vivo. *Annu. Rev. Plant Biol.* **2008**, *59*, 89–113. [CrossRef]
35. Lucena, C.C.; Siqueira, D.L.; Martinez, H.E.P.; Cecon, P.R. Salt stress change chlorophyll fluorescence in mango. *Rev. Bras. Frutic.* **2012**, *34*, 1245–1255. [CrossRef]
36. da Silva Sá, F.V.; Brito, M.E.B.; Pereira, I.B.; Antônio Neto, P.; de Andrade Silva, L.; da Costa, F.B. Balanço de sais e crescimento inicial de mudas de pinheira (*Annona squamosa* L.) sob substratos irrigados com água salina. *Irriga* **2015**, *20*, 544–556. [CrossRef]

37. López Climent, M.F.; Arbona, V.; Pérez Clemente, R.M.; Gómez Cadenas, A. Relationship between salt tolerance and photosynthetic machinery performance in citrus. *Environ. Exp. Bot.* **2008**, *62*, 176–184. [CrossRef]
38. Sousa, J.R.M.; Gheyi, H.R.; Brito, M.E.B.; Xavier, D.A.; Furtado, G.F. Impact of saline conditions and nitrogen fertilization on citrus production and gas exchanges. *Rev. Caatinga* **2016**, *29*, 415–424. [CrossRef]

Article

Calcium Lignosulfonate Can Mitigate the Impact of Salt Stress on Growth, Physiological, and Yield Characteristics of Two Barley Cultivars (*Hordeum vulgare* L.)

Hayam I. A. Elsawy [1], Khadiga Alharbi [2,*], Amany M. M. Mohamed [1], Akihiro Ueda [3], Muneera AlKahtani [2], Latifa AlHusnain [2], Kotb A. Attia [4,5], Khaled Abdelaal [6] and Alaa M. E. A. Shahein [1]

[1] Seed Technology Department, Field Crops Research Institute, Agriculture Research Center (ARC), Giza 12619, Egypt
[2] Department of Biology, College of Science, Princess Nourah bint Abdulrahman University, P.O. Box 84428, Riyadh 11671, Saudi Arabia
[3] Graduate School of Integrated Sciences for Life, Hiroshima University, Higashi-Hiroshima 739-8528, Japan
[4] Center of Excellence in Biotechnology Research, King Saud University, P.O. Box 2455, Riyadh 11451, Saudi Arabia
[5] Rice Biotechnology Lab, Rice Department, Field Crops Research Institute, ARC, Sakha 33717, Egypt
[6] Excellence Center (EPCRS), Plant Pathology and Biotechnology Laboratory, Faculty of Agriculture, Kafrelsheikh University, Kafr Elsheikh 33516, Egypt
* Correspondence: kralharbi@pnu.edu.sa

Abstract: The current study was conducted in a pot experiment with sand bed soil for two winter seasons (2019/20, 2020/21) to illuminate the impact of calcium lignosulfonate (Ca-LIGN) (100 mg/L) in alleviating various levels of NaCl (0, 100, 200, and 300 mM) on two barley cultivars, Giza132 and Giza133. Giza133 outgrew Giza132 under salinity stress by accumulating less Na^+ content and retaining more K^+ content. Surprisingly, Ca-LIGN was shown to be involved in both cultivars' capacity to efflux Na^+ in return for greater K^+ influx under 100 and 200 mM NaCl, resulting in an increased dry weight of shoots and roots as well as leaf area compared with the untreated salinity levels. Physiological parameters were measured as relative water content (RWC), electrolyte leakage rate (ELR), peroxidase activity (POD) in leaf and root and grain yield, and grain protein content were evaluated. Adding Ca-LIGN ameliorated both cultivars' growth in all the recorded characteristics. Under salinity stress, Ca-LIGN induced a higher RWC in both cultivars compared to those without Ca-LIGN. Although the ELR increased significantly in Giza132 leaves under the different NaCl concentrations compared to in Giza133 leaves, applying Ca-LIGN for both cultivars reduced the deterioration in their leaf and root by significantly lowering the ELR. As a result, applying Ca-LIGN to the salinity-affected plants (Giza133 and Giza132) under (100 and 200 mM NaCl), respectively, inhibited POD activity by about (10-fold, 6-fold, and 3-fold, 5-fold). The impact of Ca-LIGN on grain yield was more effective in Giza133 than in Giza132, with (61.46, 35.04, 29.21% and 46.02, 24.16, 21.96%) at various salinity levels. Moreover, while both cultivars recorded similar protein content under normal conditions, adding Ca-LIGN increased protein accumulation by raising salinity concentration until it reached 3% and 2% increases in both cultivars, Giza133 and Giza132, respectively, under 300 mM NaCl. It can be concluded that applying Ca-LIGN on barley can help to alleviate the ionic stress by excluding the harmful ions, resulting in higher grain yield and protein content.

Keywords: barley; Ca lignosulfonate; Na^+ extrusion; K^+ influx; salinity tolerance

1. Introduction

Salinity stress is one of the most harmful abiotic stress factors limiting crop growth, development, and yield [1–4]. Accumulation of sodium (Na^+) excessively causes ion toxicity and ion imbalance by restricting the competitive absorption of some mineral nutrients such as potassium (K^+) [5].

Salinity affects practically every part of plant morphology, physiology, and biochemistry, resulting in considerable agricultural production loss. A greater salt content in the soil limits plant roots' ability to absorb water and critical nutrients. The greater ion concentration (Na^+) in the root produces osmotic stress, reduces water potential, and disrupts nutritional balance. A larger Na^+ concentration outside the plant cell also has a detrimental influence on intracellular K^+ influx, which is necessary for plant development [6].

Under salinity conditions, there are many osmolytes and osmoprotectants, such as proline and jasmonic acid, which are already used to improve crop growth in calendula and faba bean plants [2,7]. Application of these osmoprotectants led to improved leaf number, chlorophyll content, relative water content, and yield characteristics [2,7].

Under salt stress, salt-tolerant plant species and genotypes must maintain decreased Na^+ accumulation in shoots. Shoot Na^+ exclusion or vascular compartmentalization are both responsible for lower shoot accumulative Na^+ in plants [8]. The ability of plants to exclude the damaging Na^+ ions is one of the greatest strategies to resist salt stress, as Na+ sequestration in plant tissues is thought to be the major cause of plant deterioration under salinity stress conditions. Low leaf Na^+ concentration and K^+/Na^+ discrimination is important for salinity tolerance in barley under mild salt stress. Both leaf and root Na^+ contents are negatively linked with plant dry matter during long-term salt stress [9–11].

Nowadays, a decrease in fresh water sources poses a threat to modern agriculture, which faces a number of issues as the world's population continues to expand while the available land that is suitable for food production continues to decline [12]. Hence, one of the issues encountered in barley farming in newly recovered regions, especially in Egypt, is salinity [13]. The characteristics most affected by salinity stress were number of leaves and stem height [14], chlorophyll content, RWC%, and yield [14–17]. So, identifying saline limits requires research on salt tolerance throughout the vegetative growth phase of plants [18] and also at harvest time. In the newly reclaimed lands, barley is regarded as a desirable crop to sow due to its ability to grow under stress, and some studies have been carried out to increase its grain yield under various stress conditions [19–23] because it is one of the most indispensable crops in covering the gap in access to food due to insufficient cereal production [24].

According to Kanbar and El-Drussi, [25] rising salt levels reduced barley seed germination considerably. Even though one of the examined cultivars benefited from salinity stress at 25 mM in terms of shoot–root length and weight, all cultivars were salt-sensitive at 50, 75, and 100 mM in terms of all features studied. Moreover, Allel et al. [26] and El-Wakeel et al. [27] indicated that a salinity stress of 200 mM NaCl had a detrimental effect on the parameters studied in barley. However, some wild barley cultivars can be grown under more than 500 mM NaCl; glycophytic species such as *Hordeum vulgare* can tolerate more than 200 mM NaCl [28]. Therefore, treating barley plants with substances that enhance its growth under high levels of salinity is now a crucial method.

Lignosulfonates (low-cost LIGNs) are byproducts from the paper industry; these complex polymers are formed by the solubilization of lignin under alkaline conditions, and numerous chelated forms, such as Fe-, Ca-, or K-chelated lignosulfonates, have been identified depending on the nature of the base involved during this chemical process (Fe-LIGN, Ca-LIGN, K-LIGN). Investigations in the laboratory, glasshouse, and field have also shown their stimulating effects on both vegetative and reproductive growth, as well as on fructification; their favorable influence on the root system development was particularly apparent [29,30]. Calcium lignosulfonate (Ca-LIGN), as a plant biostimulant, positively affects plant growth and fruit-fixed expression, plant pigments, shoot and root development, nutritional efficiency, crop production, rhizospheric and soil microorganisms, general soil health, and plant–environment interactions. Biostimulants are obtained from natural sources and can assist in decreasing the consumption of chemical imports while also reducing the negative environmental effects of hazardous substances [31]. Biostimulants have recently been utilized to help stressed plants and enhance their growth and production [32], and these biostimulants are not nutrients in and of themselves; rather, they aid in the absorption of nutrients or advantageously promote growth or stress resistance [33].

The application of Ca lignosulfonate compounds has the potential to promote agricultural yield [34]; in addition, lignosulfonate has been demonstrated to promote plant development in maize [35] and rice [35] because their hydrophilic parts have negative charges, and Ca lignosulfonate surfactant has a short-chain structure that makes reactions with salt simpler, has good water solubility and promotes root and leaf development, and boosts chlorophyll content and crop yield [36].

Approximately 30 crop plant species currently produce 90% of plant-based human nourishment, and the majority of the essential glycophytic crops have 50–80% lower average yields when grown under salinity conditions [37]. Because of the changing climate, problems related to soil salinity are expected to develop in many areas, and to deal with such effects, a variety of adaptations and mitigation methods are necessary. Therefore, lignosulfonate is used to ameliorate plant growth, while there has not been any research conducted on the use of calcium lignosulfonate as a mitigating substance to alleviate the salinity stress. Hence, the objective of this study was to elucidate the indispensable role of adding Ca lignosulfonate to two barley cultivars, Giza132 and Giza133, in alleviating the detrimental effects of different salinity stress levels.

2. Materials and Methods

2.1. Plant Materials and Growth Conditions

The current experiment was conducted in the greenhouse of the Seed Technology Research Department at Sakha Agriculture Research Station, ARC, Kafr Elsheikh, Egypt, under the environmental conditions shown in (Table 1). Two barley cultivars (Giza133 and Giza132) were obtained from the Agricultural Research Center, Egypt, based on their behavior under salinity stress conditions; the Giza133 cultivar is more tolerant than the Giza132 cultivar.

Table 1. The environmental conditions of the experiment during both seasons 2019/2020 and 2020/2021.

Year	Months	Temperature (°C)		Relative Humidity (RH%)		Wind Velocity (km/24 h)
		Min.	Max.	7:30 a.m.	13:30 p.m.	
2019	November	25.1	27.4	82.8	48.3	36.6
	December	13.4	21.4	86.9	58.9	38.5
2020	January	11.8	18.4	86.7	62.7	30.0
	February	12.7	20.4	84.6	56.5	51.0
	March	15.6	22.6	81.1	53.9	80.1
	April	18.9	26.0	80.0	45.1	98.8
	May	23.8	31.0	68.9	38.4	14.4
	November	17.5	25.0	80.7	56.8	46.9
	December	13.7	22.9	87.7	55.7	44.9
2021	January	13.5	21.0	86.7	59.5	39.2
	February	12.5	21.5	87.5	55.9	58.3
	March	16.5	23.8	83.6	55.2	81.5
	April	19.5	27.6	78.6	42.0	95.9
	May	25.1	32.1	65.3	33.8	12.4

Max.: maximum, Min.: minimum (these numbers are calculated as the grand mean of 30 days of data).

Over two productive seasons (2019/2020 and 2020/2021), seeds were sown and grown in acid-washed quartz sand (saturation percent 20 percent). A total of 6 kg of sand was placed inside the 25 cm-diameter, 7 L plastic pots, which also included a bottom vent for free drainage. Full-strength nutrient solution (75 g/L ZnSO$_4$·H$_2$O, 45 g/L MnSO$_4$·H$_2$O, 100 g/L FeSO$_4$·5H$_2$O, 55 g/L CuSO$_4$·7H$_2$O, 25 g/L MgSO$_4$·5H$_2$O, 5 g/L, potassium fulvate,

5 g/L citric acid, and 20 g/L potassium alginate) in addition to the recommended NPK fertilization was used as a source of nitrogen and phosphorous for the plants. After germination, 10 days later, seedlings were thinned to 6 plants per pot and watered with either nutritional solution with 0 mM NaCl (control) or nutrient solution with 100 mM, 200 mM, and 300 mM NaCl. The pots of the plants that were irrigated with salty solutions were later divided into two groups, and one of them was provided with calcium lignosulfonate (Saigo Ligno Sal compound) (100 mg/L) (Ca-LIGN + salt treatments) due to the symptoms of salinity stress impact on the plants after 20 days of those three salinity treatments. Without using calcium lignosulfonate throughout the vegetative stage, the other group continued to be watered as needed with various salt solution treatments (100, 200, and 300 mM NaCl), and each treatment was replicated 4 times. The environmental conditions of the experiment are recorded in Table 1.

2.2. Preparation of Calcium Lignosulfonate (Ca-LIGN) Solution and Applying the Salt Treatments

The calcium lignosulfonate substance used in the current study was purchased from (Saigo chemical company, Kafr Elsheikh, Egypt) as an active ingredient in a commercial compound (Saigo Ligno Sal). The chemical compositions in this compound are (10% CaO, 11% N, 10% potassium fulvate, 10% calcium lignosulfonate, 30% organic acids, 2% zeolite, and 1% MgO). The calcium lignosulfonate (Ca-LIGN) was prepared by adding (100 mg/L) to salt solutions for each salinity concentration (100 mM NaCl, 200 mM NaCl, and 300 mM NaCl). In addition, salinity solution treatments (NaCl treatments) were applied gradually (25, 50, 75, and finally 100 mM) for the first salinity level (25, 50, 75, 100, and finally 200 mM) for the second salinity level, and (25, 50, 75, 100, 200, and finally 300 mM) for the third salinity level over 2-day intervals to avoid causing osmotic shock to the plants.

2.3. Growth Parameters

After 30 days of applying Ca lignosulfonate, the plants were taken and divided into 2 parts (roots and shoots (sheaths and leaves)), and two sets of samples were prepared. The dry weights (DWs) were recorded for each plant part from the first set of samples after oven drying at 70 °C for 3 days, and the other set was kept at −80 °C for peroxidase enzyme activity analysis. Leaf area (LA) was measured using a leaf area meter (Model: LI-3000A Portable leaf area meter, LI-COR Biosciences, Lincoln, NE, USA), and plant height was measured. To determine the yield characteristics, a different set of plants was allowed to develop till harvest.

2.4. Plant Pigment Content Determination

Chlorophyll a, chlorophyll b, and total chlorophyll as well as carotenoid contents were extracted from 0.5 g of fresh leaves samples in 10 mL of 100% N,N-dimethyl formamide, and then the concentration was determined with a spectrophotometer using the methods of Porra et al. [26] and Porra [38].

2.5. Relative Water Content (RWC)

Leaf samples were placed in vials and weighed (FW); after the sample had been weighed, the piece of leaf was transferred to a 1.5 mL Eppendorff tube containing deionized water. Then, tubes were placed for 4 h in a fridge to allow the tissue to take up water. The sample was instantly transferred and reweighed to measure the fully turgid fresh weight (TW). The samples oven-dried at 80 °C for 72 h and then reweighed to determine the dry weight (DW). The relative water content was calculated according to Gonzalez and Gonzalez-Vilar [39] as follows:

$$RWC (\%) = [(FW - DW)/(TW - DW)] \times 100 \quad (1)$$

where FW (fresh weight), TW (turgid weight), and DW (dry weight).

2.6. Electrolyte Leakage Rate (ELR)

The electrolyte leakage rate (ELR) was measured in the third leaf from the top of each plant (0.5 g), and 0.5 g roots were well-washed using the method of Murray et al. [40] with some modifications. The leaf and root samples were soaked in distilled water and shaken for 12 h. Electrical conductivity EC (EC1) was measured with a portable EC meter (9V-1AmP, Thermo Electron Corporation, USA). The samples were later autoclaved and cooled to determine total EC (EC2). The ELR was calculated as follows: ELR (%) = (EC1/EC2) × 100.

2.7. Determination of Peroxidase Activity (POD)

The extraction was carried out using 0.5 g fresh samples according to the method of Takagi et al. [41]. Fresh (leaf and root) samples were ground in liquid nitrogen with adding phosphate buffer (pH 7.0). The homogenate was centrifuged, and then sPOD activity was determined in the supernatant. The 1 mL of reaction mixture contained 15 mM guaiacol, 10 mM H_2O_2, 73 mM phosphate buffer, and 2% enzyme extract. The absorbance was observed at 470 nm for 1 min, and the activity of sPOD was calculated according to Chance and Maehly, [42], and one-unit sPOD activity was defined as µmol tetraguaiacol/min.

2.8. Na^+ and K^+ Concentrations

The selected leaves and roots samples were oven-dried and fine-grounded. The fine powder was digested with a mixture of HNO_3 and $HClO_4$ (5:1, v/v), and the Na^+ and K^+ concentrations in the extracts were measured using an atomic absorption spectrophotometer (Shimadzu, Kyoto, Japan) according to Rahman et al. [43].

2.9. Post-Harvest Measurements

At harvest time, plant height (cm) was measured. The remained spikes were harvested and threshed to obtain grains. Length of spike (cm), grain number/spike, 1000-grain weight (gm), and grain yield/pot (gm) were measured.

2.10. Crude Protein Content

The known weight of the fine-ground seeds (ca. 0.1 g) was digested using a micro Kjeldahl apparatus by (98% H_2SO_4) and (30% H_2O_2). The crude protein was calculated by multiplying the total nitrogen by 5.85 [44].

2.11. Statistical Analysis

The collected data of the two seasons were subjected to combined analysis of variance (ANOVA) for the completely randomized block design of each experiment (n = 4), as mentioned by Gomez and Gomez [45] using (MSTAT-C 1990) computer software, and the means of genotypes were compared using Fisher's protected LSD at a 0.05 probability level. The physiological data are shown as means, and the means were separated using the Duncan's Multiple Range Test at p = 0.05 [46].

3. Results

After two months of various treatments, it was clear that they had different aesthetic impacts on the cultivars Giza133 and Giza132. The combined analysis revealed that there was no significant difference between the two seasons in root length, Na^+ concentrations, or peroxide enzyme activity in either tissue (leaf or root), but there was a significant difference between the two seasons in shoot length, shoot, and root dry weights (Table 2).

In addition, there was no significant effect of these seasons on the relative water content (RWC) and electrolyte leakage rate (ELR) of either tissue (leaf and root) or total leaf area (TLA) (Table 3). Nevertheless, all the post-harvest investigated parameters (Table 4) and the remaining physiological and growth characteristics were significantly impacted by both seasons (Tables 2 and 3). According to the influence of the cultivars, both cultivars, Giza133 and Giza132, performed quite differently in all the studied characteristics that are shown in Tables 2 and 4, with the exception of shoot and root lengths (Table 2), TLA (Table 2), and the number of grains/spike (Table 4). Furthermore, the various salinity treatments, whether they involved the addition of Ca lignosulfonate or not, had a significant impact on all the examined features (Tables 2–4).

Table 2. Means of shoot and root lengths, shoot and root dry weights, Na^+ and K^+ concentrations, and peroxide enzyme activity in both leaf and root tissues of Giza133 and Giza132 barley cultivars under different treatments (control, NaCl concentrations, and NaCl concentrations + Ca-LIGNO) during two seasons 2019/2020 and 2020/2021.

Characteristics Treatments	Shoot Length (cm)	Root Length (cm)	Shoot Dry Weight (gm)	Root Dry Weight (gm)	Na^+ Concentration (mg gm^{-1} DW)		K^+ Concentration (mg gm^{-1} DW)		POD Enzyme Activity in (µmol gm^{-1} min^{-1})	
					Leaf	Root	Leaf	Root	Leaf	Root
Seasons										
First (2019/2020)	40.68	42.09	3.86	3.13	67.18	43.65	21.21	16.04	8.95	7.85
Second (2020/2021)	42.90	43.86	4.67	4.95	68.23	44.65	22.31	17.00	9.70	8.51
F. Test	*	N.S	*	*	N.S	N.S	N.S	N.S	**	**
Cultivars										
Giza133	41.67	42.30	3.24	3.72	54.24	33.82	23.42	18.27	12.33	7.20
Giza132	41.90	43.65	2.29	2.36	80.13	53.49	19.01	13.80	6.32	9.15
F. Test	N.S	N.S	**	**	**	**	**	**	**	**
Treatments										
Control	41.19	48.06	2.56	2.94	0.70	0.25	35.96	23.36	4.10	4.01
100 mM NaCl + Ca-LIGNO	47.25	47.38	3.81	4.77	36.91	13.71	29.75	20.90	2.84	6.23
200 mM NaCl + Ca-LIGNO	44.28	42.10	3.76	3.72	43.94	35.23	27.37	18.67	3.82	7.31
300 mM NaCl + Ca-LIGNO	44.09	44.44	2.75	2.45	58.71	43.70	21.37	14.99	9.41	9.93
100 mM NaCl	40.60	41.39	2.34	2.75	62.64	35.75	14.10	14.16	11.37	6.68
200 mM NaCl	38.09	37.21	2.25	2.44	79.02	61.51	11.13	11.70	13.82	6.83
300 mM NaCl	37.00	40.25	1.90	2.23	188.36	115.42	8.83	8.50	19.93	16.23
F. Test	**	**	**	**	**	**	**	**	**	**
L.S.D	2.782	6.033	0.2334	0.4356	12.02	17.55	6.343	2.927	3.367	3.778

L.S. D$_{0.05}$, least significant differences at 0.05 probability level and *, ** and N.S state the significant, highly significant and not significant respectively.

Table 3. Means of relative water content (RWC), leaf and root electrolyte leakage rate (ELR), leaf area (LA), and total leaf area (TLA), and different pigments concentrations (chlorophyll a, chlorophyll b, total chlorophyll, and total carotenoids) in the leaves of Giza133 and Giza132 barley cultivars under different treatments (control, NaCl concentrations, and NaCl concentrations + Ca-LIGNO) during two seasons, 2019/2020 and 2020/2021.

Characteristics Treatments	Relative Water Content (RWC) (%)	Electrolyte Leakage Rate (ELR) (%)		Leaf Area (LA) (cm^2 Plant^{-1})	Total Leaf Area (TLA) (cm^2 Plant^{-1})	Chlorophyll a (Chll a) (µg mg^{-1} FW)	Chlorophyll b (Chll b) (µg mg^{-1} FW)	Total Chlorophyll Content (µg mg^{-1} FW)	Total Carotenoids Content (µg mg^{-1} FW)
		Leaf	Root						
Seasons									
First (2019/2020)	92.11	78.49	78.25	30.08	270.75	1.20	1.22	1.42	1.05
Second (2020/2021)	91.11	77.46	76.43	32.22	264.07	2.41	2.42	2.62	2.26
F. Test	N.S	N.S	N.S	*	N.S	**	**	**	**
Cultivars									
Giza133	92.86	61.28	85.61	32.16	290.61	2.29	2.33	2.53	2.14
Giza132	90.36	94.66	70.70	30.14	250.21	0.32	0.31	0.51	0.18
F. Test	*	**	**	**	N.S	**	**	**	**
Treatment									
Control	98.17	46.34	80.29	30.76	267.59	1.53	1.51	1.71	1.39
100 mM NaCl + Ca-LIGNO	97.65	27.02	57.91	40.94	326.08	2.30	2.35	2.56	2.16
200 mM NaCl + Ca-LIGNO	96.86	48.63	76.22	35.92	224.47	1.99	2.01	2.20	1.84
300 mM NaCl + Ca-LIGNO	92.48	72.79	73.09	31.35	379.01	1.50	1.52	1.71	1.35
100 mM NaCl	87.21	100.99	75.54	26.77	200.03	0.76	0.77	0.98	0.61

Table 3. Cont.

Characteristics Treatments	Relative Water Content (RWC) (%)	Electrolyte Leakage Rate (ELR) (%)		Leaf Area (LA) (cm² Plant⁻¹)	Total Leaf Area (TLA) (cm² Plant⁻¹)	Chlorophyll a (Chll a) (µg mg⁻¹ FW)	Chlorophyll b (Chll b) (µg mg⁻¹ FW)	Total Chlorophyll Content (µg mg⁻¹ FW)	Total Carotenoids Content (µg mg⁻¹ FW)
		Leaf	Root						
200 mM NaCl	88.58	113.16	77.92	28.50	219.85	0.55	0.59	0.79	0.41
300 mM NaCl	80.33	136.88	103.90	23.80	275.85	0.49	0.50	0.69	0.36
F. Test	**	**	**	**	**	**	**	**	**
L.S.D	4.171	11.45	10.50	2.867	93.89	0.0223	0.0223	0.0224	0.0315

L.S.D$_{0.05}$, least significant differences at 0.05 probability level and *, ** and N.S state the significant, highly significant and not significant respectively.

Table 4. Means of plant height, spike length, number of grains spike⁻¹, grain yield pot⁻¹, 1000-grain weight, and grain protein content of Giza133 and Giza132 barley cultivars under different treatments (control, NaCl concentrations, and NaCl concentrations + Ca-LIGNO) during, two seasons 2019/2020 and 2020/2021.

Characteristics Treatments	Plant Height (cm)	Spike Length (cm)	No. Grains Spike⁻¹	Grain Yield Pot⁻¹ (gm)	1000-Grain Weight (gm)	Grain Protein Content (%)
			Seasons			
First (2019/2020)	48.39	6.70	7.73	15.53	29.97	10.78
Second (2020/2021)	52.16	8.68	9.28	18.08	31.94	10.87
F. Test	*	*	*	*	*	*
			Cultivars			
Giza133	51.82	8.19	9.81	18.35	33.13	11.70
Giza132	48.73	9.18	10.19	13.26	24.77	10.59
F. Test	**	**	N.S	**	**	*
			Treatments			
Control	56.42	9.11	10.05	15.41	37.3	9.38
100 mM NaCl + Ca-LIGNO	59.96	9.91	10.28	29.15	44.3	11.20
200 mM NaCl + Ca-LIGNO	46.57	9.13	10.69	17.64	29.7	11.90
300 mM NaCl + Ca-LIGNO	47.26	7.57	8.83	13.30	22.1	13.06
100 mM NaCl	51.92	8.49	10.09	12.96	24.3	9.54
200 mM NaCl	48.72	8.66	10.35	12.19	21.5	10.08
300 mM NaCl	41.07	7.93	9.73	9.98	23.6	10.67
F. Test	**	**	**	**	**	**
L.S.D	2.924	0.3623	0.5351	2.365	4.144	0.224

L.S.D$_{0.05}$, least significant differences at 0.05 probability level and *, ** and N.S state the significant, highly significant and not significant respectively.

3.1. Effect of the Treatments on Plant Growth and Biomass

In both genotypes, there was a rise in NaCl concentration accompanied by a reduction in growth (Figures 1–4). By adding Ca lignosulfonate (Ca-LIGN) to salty solutions, the second treatment, which consisted of applying Ca-LIGN to 100 mM NaCl followed by adding it to 200 mM NaCl, produced the greatest shoot and root lengths (Figure 2), leaf area (LA), and total leaf area (TLA) (Figure 2) and increased the production of shoot and root dry weights (DWs) (Figure 1), outperforming the control treatment. When comparing the three salinity levels with the same levels of Ca-LIGN, both cultivars recorded a significant increase in all the growth features, especially in shoot DW, by (52.8%, 49.3%, 33.5% and 31.8%, 16.2%, and 27.0%) in Giza133 and Giza132 with the three salinity levels 100, 200, and 300 mM NaCl, respectively (Figure 1A). The same trend was recorded in root DW in both cultivars; however, a slight increase in DW was shown with the highest salinity level, 300 mM, using the Ca-LIGN substance (Figure 1B). The shoot height decreased slightly in both cultivars under the first salinity level, 100 mM NaCl, compared to control conditions (Figure 2A); however, in the other salinity levels, 200 mM NaCl and 300 mM NaCl, Giza133 decreased more than Giza132 than when under control conditions. The results in Figure 3 show that applying the Ca-LIGN substance induced an upsurge in both shoot and root lengths in the Giza133 cultivar by (22.7%,17.2%) and (22.3%,15.17%) under both salinity levels (100 mM and 200 mM NaCl), respectively. Giza132 cultivar did not have significantly altered shoot and root lengths when Ca-LIGN was added (Figure 2). We examined the leaf area for both cultivars under the different studied treatments, noting the Ca-LIGN substance prompted an elongation in the leaf tissues, which was observed in both cultivars compared with control and salinity stress without adding the Ca-LIGN substance (Figure 3A). The Giza133 cultivar recorded an increase of about 50% in LA when Ca-LIGN was added to 100 mM NaCl compared to the control and also 100 mM NaCl when that substance was not added; additionally, the LA increased by 23.5% and 34.2% when the Ca-LIGN substance was added to 200 mM NaCl and 300 mM NaCl, respectively, compared to these salt levels when it was not added (Figure 3A). Adding Ca-LIGN to 100 mM NaCl in Giza132 increased the LA significantly by 17.7% and 21.9% compared to the control conditions and 100 mM NaCl without that substance, respectively; however, adding Ca-LIGN did not significantly affect its LA under the highest salinity level, 300 mM NaCl (Figure 3A).

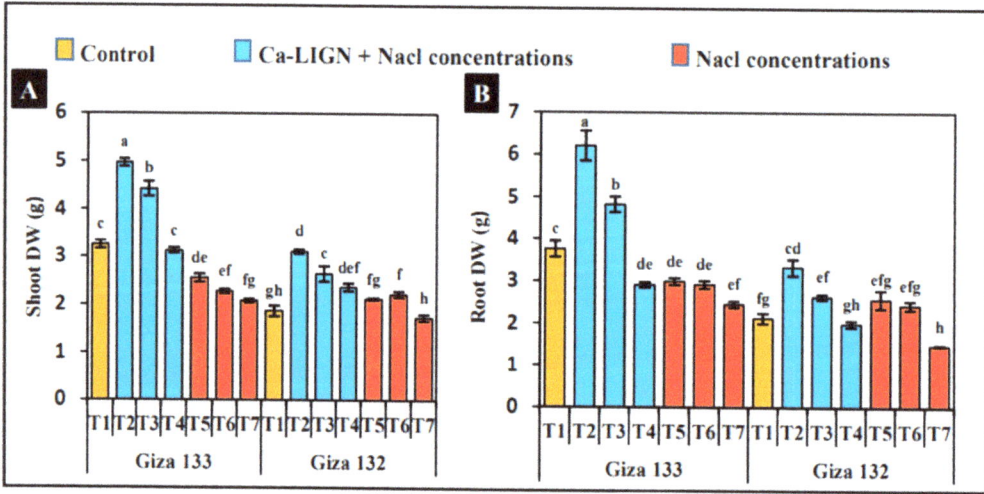

Figure 1. Effect of different treatments: (T1) control, (T2) 100 mM NaCl + Ca-LIGNO, (T3) 200 mM NaCl + Ca-LIGNO, (T4) 300 mM NaCl + Ca-LIGNO, (T5) 100 mM NaCl, (T6) 200 mM NaCl, and (T7) 300 mM NaCl on the average (**A**) shoot dry weight (DW) and (**B**) root dry weight (DW) of the two barley cultivars, Giza133 and Giza132, during two seasons, 2019/2020 and 2020/2021. Data represent the mean ± SE (n = 4). The same letter indicates no significant difference ($p \leq 0.05$).

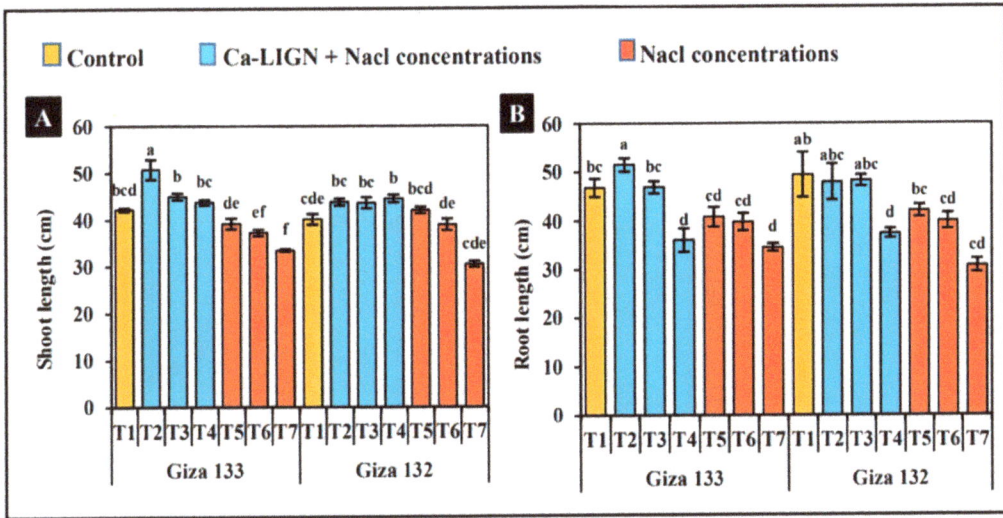

Figure 2. Effect of different treatments: (T1) control, (T2) 100 mM NaCl + Ca-LIGNO, (T3) 200 mM NaCl + Ca-LIGNO, (T4) 300 mM NaCl + Ca-LIGNO, (T5) 100 mM NaCl, (T6) 200 mM NaCl, and (T7) 300 mM NaCl on the average (**A**) shoot length and (**B**) root length of the two barley cultivars, Giza133 and Giza132, during two seasons, 2019/2020 and 2020/2021. Data represent the mean ± SE (n = 4). The same letter indicates no significant difference ($p \leq 0.05$).

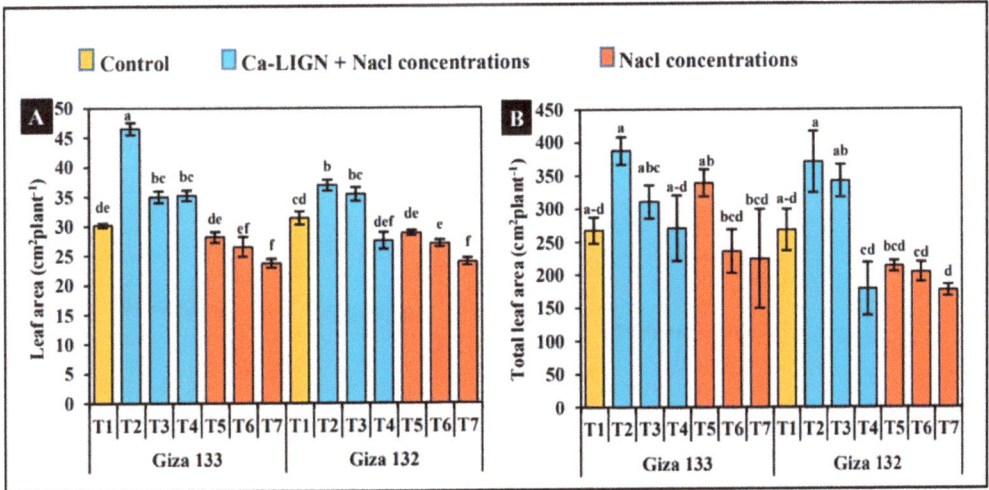

Figure 3. Effect of different treatments: (T1) control, (T2) 100 mM NaCl + Ca-LIGNO, (T3) 200 mM NaCl + Ca-LIGNO, (T4) 300 mM NaCl + Ca-LIGNO, (T5) 100 mM NaCl, (T6) 200 mM NaCl, and (T7) 300 mM NaCl on the average (**A**) leaf area (LA) and (**B**) total leaf area (TLA) of the two barley cultivars, Giza133 and Giza132, during two seasons, 2019/2020 and 2020/2021. Data represent the mean ± SE (n = 4). The same letter indicates no significant difference ($p \leq 0.05$).

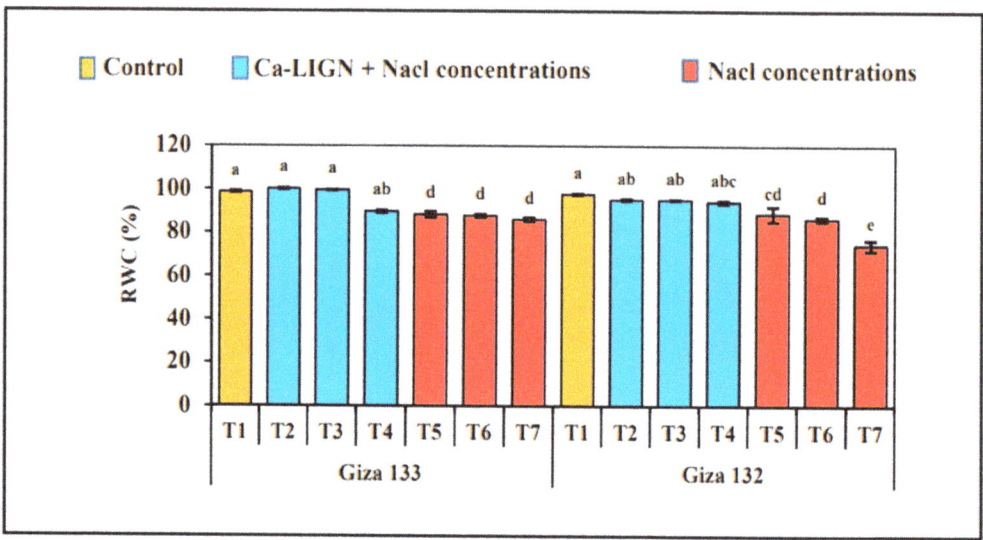

Figure 4. Effect of different treatments: (T1) control, (T2) 100 mM NaCl + Ca-LIGNO, (T3) 200 mM NaCl + Ca-LIGNO, (T4) 300 mM NaCl + Ca-LIGNO, (T5) 100 mM NaCl, (T6) 200 mM NaCl, and (T7) 300 mM NaCl on the average relative water content (RWC) of the two barley cultivars, Giza133 and Giza132, during two seasons, 2019/2020 and 2020/2021. Data represent the mean ± SE (n = 4). The same letter indicates no significant difference ($p \leq 0.05$).

3.2. Relative Water Content (RWC)

Under various levels of salt stress, the RWC of leaves declined significantly by around 10% and 15% in Giza133 and Giza132, respectively. Calcium lignosulfonate's treatments enhanced the RWC of stressed seedlings of both cultivars, which was particularly noticeable at the stress levels of 100 and 200 mM NaCl, more so than in the 300 mM NaCl level (Figure 4).

3.3. Electrolyte Leakage Rate (ELR)

Since salt stress elevated the abundance of free radicals in plants, the damage to membranes was examined by measuring the electrolyte leakage in both cultivars' leaf and root tissues (Figure 5). Electrolyte leakage was found in both cultivars' leaves, and according to the findings, was inhibited when the Ca-LIGN substance was added to all the studied salinity levels in both cultivars (Figure 5A). Although adding the Ca-LIGN substance to the salinity levels (100 mM and 200 mM NaCl) did not significantly reduce the leakage of root cells in Giza133, under the highest salinity level, 300 mM NaCl, Ca-LIGN maintained its root cell membrane stability by decreasing the ELR by 18.79% (Figure 5B). Additionally, application of Ca-LIGN led to alleviated salinity stress effects due to the decline in the ELR's root tissues by (8.18%, 11.32%, and 19.46%) under salinity levels of (100 mM, 200 mM, and 300 mM NaCl), respectively (Figure 5B).

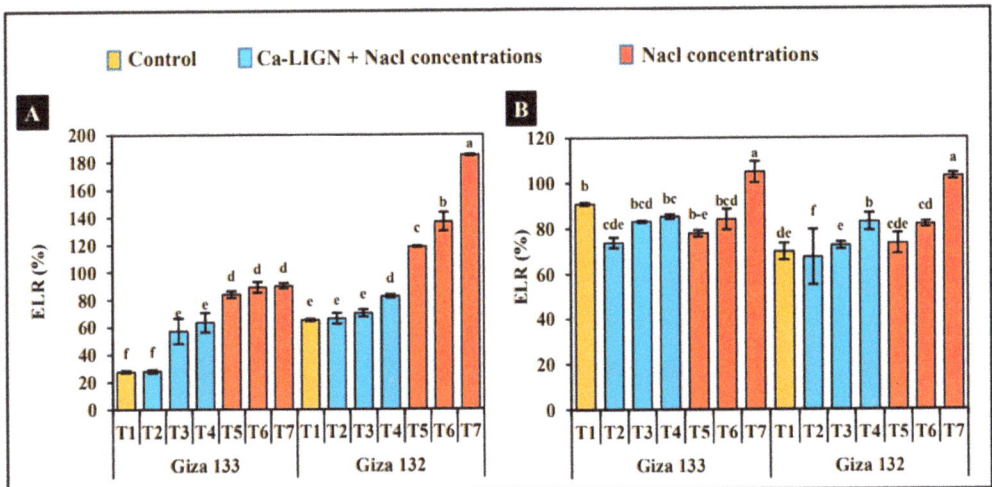

Figure 5. Effect of different treatments: (T1) control, (T2) 100 mM NaCl + Ca-LIGNO, (T3) 200 mM NaCl + Ca-LIGNO, (T4) 300 mM NaCl + Ca-LIGNO, (T5) 100 mM NaCl, (T6) 200 mM NaCl, and (T7) 300 mM NaCl on the average electrolyte leakage ratio (ELR) in (**A**) leaf and (**B**) root of the two barley cultivars, Giza133 and Giza132, during two seasons, 2019/2020 and 2020/2021. Data represent the mean ± SE (n = 4). The same letter indicates no significant difference ($p \leq 0.05$).

3.4. Peroxidase Enzyme Activity (POD)

The aforementioned features are supported by our findings on peroxidase (POD) activity, an enzyme implicated in antioxidative defense systems (Figure 6). The salinity stress levels increased the POD enzyme activity under the three levels (100 mM NaCl, 200 mM NaCl, and 300 mM NaCl) by (7.7-, 4.1-, 1.5-fold, and 4.2-, 2.5-, 1.4-fold) compared to under the control conditions in both cultivars of the leaves, Giza133 and Giza132, correspondingly (Figure 6A). Adding the Ca-LIGN substance to the different salinity levels did not alter the POD activity in either cultivar's leaves, except the Giza133 leaf under 300 mM NaCl-induced POD activity, which increased about 2.4 times compared to the control (Figure 6A). With regard to the root tissues, the activity of POD enzyme increased when the salinity level in the Giza133 cultivar was increased; however, its activity decreased when the salinity levels were increased in Giza132 cultivar root tissues (Figure 6B), and applying Ca-LIGN to salinity concentrations did not induce the POD activity in Giza133 root.

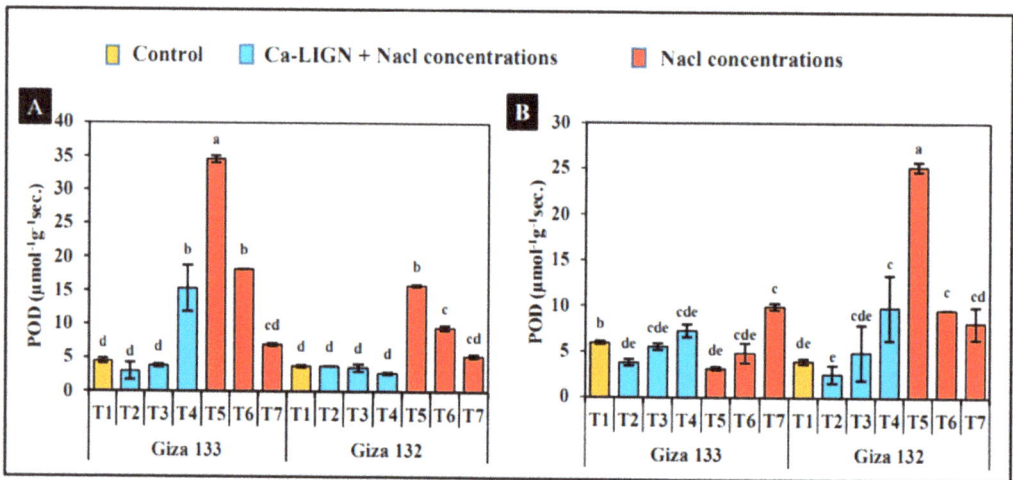

Figure 6. Effect of different treatments: (T1) control, (T2) 100 mM NaCl + Ca-LIGNO, (T3) 200 mM NaCl + Ca- LIGNO, (T4) 300 mM NaCl + Ca-LIGNO, (T5) 100 mM NaCl, (T6) 200 mM NaCl, and (T7) 300 mM NaCl on the average peroxide enzyme activity (POD) in (**A**) leaf and (**B**) root of the two barley cultivars, Giza133 and Giza132, during two seasons, 2019/2020 and 2020/2021. Data represent the mean ± SE (n = 4). The same letter indicates no significant difference ($p \leq 0.05$).

3.5. Different Photosynthetic Pigments

Salinity stress resulted in declined pigment production in the leaf tissues of both cultivars, Giza133 and Giza132, which was more pronounced in the Giza132 cultivar (Table 5). The different salinity levels (100 mM NaCl, 200 mM NaCl, and 300 mM NaCl) led to a significant decrease in the different pigments (Chl a, Chl b, total chlorophyll, and total carotenoids contents) by around (1.2, 3.38, and 3.8 times), respectively, compared with control conditions in Giza133. In the presence of sodium chloride stress, adding Ca-LIGN increased the Chl a, Chl b, total chlorophyll, and total carotenoids contents by gradually decreasing the salinity levels from 300 mM NaCl to 200 mM NaCl until reaching 100 mM NaCl in both cultivars (Table 5).

Table 5. Different pigments' concentrations (chlorophyll a, chlorophyll b, total chlorophyll content, total carotenoids) in the leaf tissues of Giza133 and Giza132 barley cultivars under various treatments (control, NaCl concentrations, and NaCl concentrations + Ca-LIGNO) during two seasons, 2019/2020 and 2020/2021, in combined analysis.

Cultivars	Treatments	Chlorophyll a (Chll a) ($\mu g\ mg^{-1}$ FW)	Chlorophyll b (Chll b) ($\mu g\ mg^{-1}$ FW)	Total Chlorophyll Content ($\mu g\ mg^{-1}$ FW)	Total Carotenoids Content ($\mu g\ mg^{-1}$ FW)
Giza133	Control	2.71 ± 0.00	2.67 ± 0.01	2.88 ± 0.01	2.56 ± 0.00
	100 mM NaCl + Ca-LIGNO	4.20 ± 0.00	4.28 ± 0.01	4.49 ± 0.01	4.05 ± 0.00
	200 mM NaCl + Ca-LIGNO	3.70 ± 0.00	3.81 ± 0.00	4.01 ± 0.00	3.54 ± 0.00
	300 mM NaCl + Ca-LIGNO	2.70 ± 0.00	2.74 ± 0.00	2.94 ± 0.01	2.55 ± 0.00
	100 mM NaCl	1.21 ± 0.00	1.25 ± 0.01	1.47 ± 0.01	1.06 ± 0.00
	200 mM NaCl	0.80 ± 0.00	0.86 ± 0.01	1.06 ± 0.01	0.66 ± 0.00
	300 mM NaCl	0.70 ± 0.00	0.72 ± 0.00	0.92 ± 0.00	0.56 ± 0.00

Table 5. Cont.

Cultivars	Treatments	Chlorophyll a (Chll a) (µg mg^{-1} FW)	Chlorophyll b (Chll b) (µg mg^{-1} FW)	Total Chlorophyll Content (µg mg^{-1} FW)	Total Carotenoids Content (µg mg^{-1} FW)
Giza132	Control	0.36 ± 0.00	0.34 ± 0.01	0.55 ± 0.01	0.21 ± 0.00
	100 mM NaCl + Ca-LIGNO	0.40 ± 0.00	0.42 ± 0.01	0.63 ± 0.02	0.26 ± 0.00
	200 mM NaCl + Ca-LIGNO	0.29 ± 0.00	0.21 ± 0.00	0.40 ± 0.02	0.15 ± 0.00
	300 mM NaCl + Ca-LIGNO	0.29 ± 0.00	0.29 ± 0.00	0.48 ± 0.00	0.16 ± 0.00
	100 mM NaCl	0.31 ± 0.00	0.29 ± 0.01	0.49 ± 0.01	0.16 ± 0.00
	200 mM NaCl	0.30 ± 0.00	0.32 ± 0.02	0.52 ± 0.00	0.16 ± 0.00
	300 mM NaCl	0.29 ± 0.00	0.28 ± 0.02	0.47 ± 0.02	0.16 ± 0.00
F. Test		**	**	**	**
L.S.D		0.0314	0.0316	0.0315	0.0445

L.S.D$_{0.05}$, least significant differences at 0.05 probability level and ** and N.S state the significant, highly significant and not significant respectively. Values are means ± SE of 4 replicates.

3.6. Na+ and K+ Concentrations

It is illustrated in (Figure 7A,B) that under salinity levels, the Giza133 cultivar accumulated less Na$^+$ content in its leaf and root tissues than the Giza132 cultivar. Simultaneously, Giza133 cultivar's K$^+$ content was higher than that of Giza132 in both leaf and root tissues (Figure 7C,D). Moreover, under the increase in NaCl levels, sodium content increased in the leaves and roots of both barley cultivars. The opposite results for K$^+$ content was obtained. Ca-LIGN application dropped dramatically the levels of Na$^+$ in the two organs of barley cultivar stressed seedlings (Figure 7A,B). Oppositely, applying Ca-LIGN increased the K$^+$ content in both tissues of both cultivars compared with the untreated tissues (Figure 7C,D)

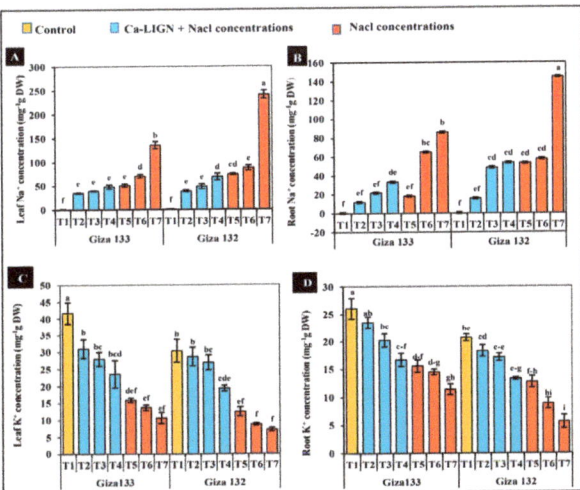

Figure 7. Effect of different treatments: (T1) control, (T2) 100 mM NaCl + Ca−LIGNO, (T3) 200 mM NaCl + Ca−LIGNO, (T4) 300 mM NaCl + Ca−LIGNO, (T5) 100 mM NaCl, (T6) 200 mM NaCl, and (T7) 300 mM NaCl on the average Na$^+$ concentrations in leaf (**A**) and root (**B**) and K$^+$ concentrations in leaf (**C**) and root (**D**) of the two barley cultivars, Giza133 and Giza132, during two seasons, 2019/2020 and 2020/2021.

3.7. Post-Harvest Characteristics

In both cultivars, irrigating barley with varied salt levels (100, 200, and 300 mM NaCl) dissolved in the nutrient solution reduced plant height (cm), spike length (cm), 1000-grain weight (gm), and grain yield (gm pot^{-1}) at harvest as compared to the control (Table 6). The results also reveal that watering the salinized barley plants with Ca-LIGN improved the values of those properties substantially. In the Giza133 cultivar, the reduction in plant height by salinity stressors was (5.6, 6.9, and 21.3%); however, this decrease was much higher in Giza132, for which these values declined by (11.18, 21.63 and 32.14%) under the salinity levels (100, 200, and 300 mM NaCl), respectively. Surprisingly, applying Ca-LIGN increased the plant height of both cultivars significantly under the three salinity levels, showing a gradual increase from the highest concentration to the lowest one, with plants being (8.6%, 3.1%) taller under these conditions than under control conditions in Giza133 and Giza132, respectively (Table 6). Overall, while the increasing salinity stress concentrations reduced grain yield and its components (spike length, no. grains spike-1, and 1000-grain weight), salt treatment with Ca-LIGN increased the aforesaid characteristics, particularly grain yield, in both cultivars. The effect of adding Ca-LIGN on grain yield was more evident in Giza133 than in Giza132, with increases of (61.46, 35.04, 29.21%, and 46.02, 24.16, 21.96%), respectively, at the corresponding salinity levels (100, 200, and 300 mM NaCl).

Table 6. Plant height, spike length, number of grains spike-1, grain yield pot-1, 1000-grain weight, and grain protein content of Giza133 and Giza132 barley cultivars under various treatments: (control, NaCl concentrations, and NaCl concentrations + Ca-LIGNO) during two seasons, 2019/2020 and 2020/2021, in combined analysis.

Cultivars	Treatments	Plant Height (cm)	Spike Lengths (cm)	No. Grains Spike^{-1}	1000-Grain Weight (gm)	Grain Yield Pot^{-1} (gm)	Grain Protein Content (%)
Giza133	Control	55.42 ± 2.05	8.46 ± 0.15	10.20 ± 0.25	42.50 ± 1.79	14.65 ± 0.51	9.48 ± 0.01
	100 mM NaCl + Ca-LIGNO	60.67 ± 1.77	9.88 ± 0.12	8.95 ± 0.09	50.00 ± 2.18	35.92 ± 2.15	11.36 ± 0.01
	200 mM NaCl + Ca-LIGNO	48.48 ± 0.97	8.80 ± 0.14	10.40 ± 0.19	33.38 ± 0.26	21.83 ± 1.18	12.13 ± 0.01
	300 mM NaCl + Ca-LIGNO	51.17 ± 0.75	7.15 ± 0.12	8.40 ± 0.14	23.38 ± 1.17	15.75 ± 0.77	13.82 ± 0.01
	100 mM NaCl	52.29 ± 0.46	7.61 ± 0.02	10.18 ± 0.27	26.68 ± 3.74	13.84 ± 0.08	9.65 ± 0.04
	200 mM NaCl	51.55 ± 0.53	8.13 ± 0.07	10.88 ± 0.17	26.13 ± 3.50	14.18 ± 0.33	10.23 ± 0.02
	300 mM NaCl	43.20 ± 0.68	7.33 ± 0.26	9.68 ± 0.06	29.86 ± 4.25	12.29 ± 0.44	10.88 ± 0.02
Giza132	Control	57.42 ± 1.40	9.77 ± 0.29	9.90 ± 0.30	32.00 ± 0.83	16.18 ± 0.67	9.27 ± 0.03
	100 mM NaCl + Ca-LIGNO	59.25 ± 0.44	9.95 ± 0.11	11.60 ± 0.15	38.50 ± 1.15	22.38 ± 1.04	11.04 ± 0.03
	200 mM NaCl + Ca-LIGNO	44.67 ± 0.61	9.45 ± 0.15	10.98 ± 0.21	26.00 ± 0.75	13.45 ± 0.21	11.66 ± 0.04
	300 mM NaCl + Ca-LIGNO	43.35 ± 0.38	8.00 ± 0.23	9.25 ± 0.35	20.75 ± 0.63	10.85 ± 0.99	12.30 ± 0.02
	100 mM NaCl	51.55 ± 1.20	9.38 ± 0.16	10.00 ± 0.02	22.00 ± 0.48	12.08 ± 0.10	9.44 ± 0.02
	200 mM NaCl	45.89 ± 1.03	9.20 ± 0.05	9.83 ± 0.28	16.88 ± 0.12	10.20 ± 0.15	9.95 ± 0.62

Table 6. Cont.

Cultivars	Treatments	Plant Height (cm)	Spike Lengths (cm)	No. Grains Spike^{-1}	1000-Grain Weight (gm)	Grain Yield Pot^{-1} (gm)	Grain Protein Content (%)
	300 mM NaCl	38.95 ± 0.40	8.54 ± 0.13	9.78 ± 0.21	17.28 ± 0.97	7.68 ± 0.42	10.46 ± 0.02
F. Test		*	**	**	*	**	**
L.S.D		4.135	0.5124	0.7568	5.861	3.345	0.3164

LSD$_{0.05}$, least significant differences at 0.05 probability level and *, ** and N.S state the significant, highly significant and not significant respectively. Values are means ± SE of 4 replicates.

3.8. Grain Protein Content

Similarly to grain protein content, which increased the quality of the grains, salinity levels induced steadily increasing protein content by increasing the concentration of NaCl in both cultivars compared to control conditions (Table 5). In addition, there was a significant increase in the grain protein content of both cultivars, Giza133 and Giza132, when the salinity stress levels with Ca-LIGN reached (21.27% and 14.95%) increases in the Giza133 cultivar and Giza132 cultivar, respectively, under the highest concentration, 300 mM NaCl (Table 6).

4. Discussion

Our results indicate that both cultivars clearly exhibited much tolerance to salinity stress levels when treated with calcium lignosulfonate (Ca-LIGN), as evidenced by increased dry mass, shoot and root length, LA, RWC, grain yield, and its components, with higher grain protein content as well as lower ELR and POD activity due to lower ROS production [36]. Otherwise, under salinity stress, the shot-gun approach, which includes the use of biostimulants as Ca-LIGN, would be a viable option for a variety of crops, including sunflower [47], barley [48], and corn [49]. The calcium element in calcium lignosulfonate (Ca-LIGN) is chemically active and can extrude sodium atoms that have been adsorbed to soil particles close to where roots are growing. Additionally, it contains an active alkyl group that binds to a sodium element and converts it to an organic form without harming the plant, similar to the results of Kang et al. [50]. The organic Ca-LIGN substance is utilized as a biostimulant for root growth and plant development since it also contains a sulfur element, which is crucial for protecting roots from salt stress.

For the above reasons, it was observed that salts in irrigation water can cause an inhibition in plant growth; however, applying Ca-LIGN was seen to alleviate diverse salt stress levels in both cultivars, Giza132 and Giza133.

When excessively salt enters the plant through the transpiration stream, it will harm the cells in the transpiring leaves, causing further growth reduction [7]. It was illustrated in our study that there was a negative effect of salinity stress levels on the plant DWs (Figure 1), shoot and root lengths (Figure 2), LA, and total TLA (Figure 3). This negative effect of salinity stress was recorded in many plants [51–53]; nevertheless, Ca-LIGN mitigates this negative impact on these growth characteristics. Lignin's efficient provision of the necessary ions required for plant growth results in such good impacts on plant growth. Our findings are also consistent with earlier research that has shown the benefits of lignin in stimulating in vitro plant development of rice [54]. It can be observed in Figure 4A that the LA was much higher in salt-stressed plants when Ca-LIGN was added than in the untreated stressed plants. Sulfur (S) is an essential element for the growth and development of crop plants; it is one of the main components in Ca-LIGN, which plays an important role in the normal functioning of plant chlorophyll and important proteins [55]. Chlorophyll a, chlorophyll b, total chlorophyll, and total carotenoids content in both cultivars on average for the two seasons increased significantly when Ca-LIGN was applied to NaCl concentrations compared with untreated stressed plants (Table 5). Sulfur in Ca-LIGN can help barley to cope with unfavorable environmental stressors; however, the effectiveness of Sulfur varies [56]. Our results indicate that Ca-LIGN acts as a source of sulfur, which improves the photosynthesis pigment content under salt stress levels (Table 5), leads to the plant's ability to store high dry mass (Figure 1), and increases the elongation of the leaf tissue, which was observed when expanding the LA under salinity levels (Figure 4).

Alternatively, by its function in lignin production, the peroxidase enzyme contributes to ROS detoxification under salinity stress in the tolerant cultivar Giza133 more than in the sensitive cultivar Giza132 under the different salt levels (Figure 6). However, the activity of the POD enzyme dramatically plummeted with Ca-LIGN application in both the leaf and root of both cultivars except under 300 mM NaCl, which means a low amount of ROS in plant cells and decreased oxidative stress are indicated by peroxidase activity in both leaf and root tissues. A decline in peroxidase activity when adding Ca-LIGN was found to significantly decrease the electrolyte leakage ratio (ELR) (Figure 5), which protects leaf and root tissues against the damage that often results from the stress. Adding Ca-LIGN during salinity stress helped both cultivars to maintain their leaf tissues' cell membrane stability (Figure 5A). According to the turgidity of leaves, we found the relative water content (RWC), which was exhibited by a massive increase in leaf turgor with the application of Ca-LIGN due to its ability to take up water in contrast to its lack of water uptake under salinity stress levels, especially under 300 mM NaCl and particularly in the sensitive cultivar Giza132 (Figure 4). The high salt stress level impact on RWC was observed to be much weaker in Giza133 (which retained as much as 100% of its leaf water on average for both seasons when grown in 100 mM NaCl) than in Giza132, in which it decreased to 95.25% (Figure 4). Earlier studies similarly established that the usage of exogenous biostimulants such as Ca-LIGN counteracted adverse salt-induced effects on RWC [57]. Nevertheless, our results indicate that this result is highly cultivar-dependent; for instance, Giza133 RWC seedlings were largely dependent on the salt concentration for the two seasons, independently of Ca-LIGN treatment; however, applying Ca-LIGN on different NaCl concentrations affected the leaf RWC of Giza132 seedlings, with by a noticeable upsurge. Salt stress tolerance in many crops is linked to a low Na^+ content in the shoots, particularly in leaves [58,59]. Ca-LIGN is a lignin-derived amorphous substance; it chelates the minerals from the soil and changes its charge so the plant does not absorb excessive harmful minerals such as Na^+ [60], which explained the declining Na^+ content in both leaf and root tissues in both cultivars treated with the Ca-LIGN substance (Figure 7A,B). Additionally, recent studies revealed that the intracellular K^+ plays an important role in the salinity tolerance mechanism. To sustain high intracellular K^+ concentrations, plasma membranes contain very negative internal potentials [9]. Excessive Na^+ influx under salinity stress depolarizes the membrane, allowing K^+ to be effluxed via depolarization-activated channels such NSCCs and GORK [61]. In contrast, the Ca-LIGN substance, which increases the efflux of Na^+ by connecting to the negatively charged alkali group of lignosulfonates, then changes into a neutral compound, and, when washed with the irrigated water in soil, it leads to an Na^+ exchange with a K^+ influx through the plant cells. This was crystal clear in both cultivars when Ca-LIGNO was applied under different salinity levels, which enhanced the K^+ content compared to untreated salinity levels and the control conditions (Figure 7B). Different previous studies concluded that salinity stress reduces the grain yield and its components of crop plants [62], which is in line with our results. Applying Ca-LIGNO stimulated the plant growth and caused an increase in the grain yield by increasing its components, which was obvious in our results (Table 6). Ca-LIGNO enhanced the plant height in both cultivars and increased their grain yield under 100mM NaCl more than under control conditions. Despite the fact that salinity stress increased the protein content in the grains, Ca-LIGN enhanced the protein content under salinity stress compared to under control conditions, in particular, for the Giza133 cultivar (Table 6). Applying Ca-LIGN helps plant to absorb many nutrients and increases nitrogen content, which is responsible for the protein in the grains under salinity levels.

5. Conclusions

In conclusion, our results indicate that superior tolerance was recorded for genotype Giza133 when Ca-LIGN was added, which enhanced its ability to tolerate the salinity stress levels. Moreover, applying Ca-LIGN improved cultivar Giza132's behavior under salinity stress levels. Calcium lignosulgonate application minimized the deleterious effects of salt stress in both cultivars. Physiological parameters analysis revealed that the difference in growth caused by adding Ca lignisulfonate was primarily due to higher antioxidant enzyme POD in the leaves and roots of barley cultivars Giza133 and Giza132 grown under control conditions and different NaCl stresses. The reluctance of extrusion would result in cytosolic Na^+ accumulation, which might injure plant tissues (as demonstrated in Giza132), producing in K^+ efflux and perhaps significant growth retardation under salt stress. Furthermore, adding Ca lignosulfonate to barley plants under various salinity levels boosted chlorophyll a,b content, relative water content, and the grain yield production as well as protein content; however, ELR was decreased in barley plants treated with Ca lignosulfonate. By using our findings in this work, we may be able to develop a breeding program for barely crop in the future. Overall, utilizing Ca-LIGN as an active and cost-effective component for farmers of Saigo-sal as a remedy for salt stress I'm plant crops improves development and allows them to tolerate salinity stress. Now, we are conducting new study on a variety of crops to determine how this compound may affect them.

Author Contributions: Conceptualization, H.I.A.E., A.M.M.M. and A.M.E.A.S.; methodology, H.I.A.E., K.A. (Khadiga Alharbi), A.M.M.M., A.U., M.A., L.A., K.A.A., K.A. (Khaled Abdelaal) and A.M.E.A.S.; software, H.I.A.E., K.A. (Khadiga Alharbi), A.M.M.M., A.U., M.A., L.A., K.A.A., K.A. (Khaled Abdelaal) and A.M.E.A.S.; validation, H.I.A.E., A.M.M.M., A.U. and A.M.E.A.S.; formal analysis, H.I.A.E., A.M.M.M., A.U., K.A.A., K.A. (Khaled Abdelaal) and A.M.E.A.S.; investigation, H.I.A.E., K.A. (Khadiga Alharbi), A.M.M.M., A.U., M.A., L.A., K.A.A., K.A. (Khaled Abdelaal) and A.M.E.A.S.; resources, H.I.A.E., A.M.M.M., A.U., K.A. (Khaled Abdelaal) and A.M.E.A.S.; data curation, H.I.A.E., A.M.M.M., A.U. and A.M.E.A.S.; writing—original draft preparation, H.I.A.E., K.A. (Khadiga Alharbi), A.M.M.M., A.U., K.A.A., K.A. (Khaled Abdelaal) and A.M.E.A.S.; writing—review and editing, H.I.A.E., K.A. (Khadiga Alharbi), A.M.M.M., A.U., M.A., L.A., K.A.A., K.A. (Khaled Abdelaal) and A.M.E.A.S.; supervision, H.I.A.E. and A.M.E.A.S.; funding acquisition, H.I.A.E., K.A. (Khadiga Alharbi), A.M.M.M., M.A., L.A., K.A.A., K.A. (Khaled Abdelaal) and A.M.E.A.S. All authors have read and agreed to the published version of the manuscript.

Funding: This research was funded by Princess Nourah bint Abdulrahman University Researchers Supporting Project Number (PNURSP2022R188), Princess Nourah bint Abdulrahman University, Riyadh, Saudi Arabia.

Institutional Review Board Statement: Not applicable.

Data Availability Statement: Not applicable.

Acknowledgments: The authors would like to thank Princess Nourah bint Abdulrahman University Researchers Supporting Project Number (PNURSP2022R188), Princess Nourah bint Abdulrahman University, Riyadh, Saudi Arabia. Additionally, the authors would like to thank Sherif Mohamed Abdeldayem, an associate professor at Kafrelshikh University, Egypt, for his support and for providing the chemicals and some materials for our experiment. Many thanks to Saigo chemical company for providing products and chemicals. Many thanks to the field crops research institute, Agricultural Research Center (ARC). Additionally, many thanks to all members of PPB Lab., and the EPCRS Excellence Centre (certified according to ISO/9001, ISO/14001, and OHSAS/18001), Department of Agricultural Botany, Faculty of Agriculture, Kafrelsheikh University, Kafr-Elsheikh, Egypt.

Conflicts of Interest: The authors declare no conflict of interest.

References

1. Abdelaal, K.A.; Mazrou, Y.S.; Hafez, Y.M. Silicon Foliar Application Mitigates Salt Stress in Sweet Pepper Plants by Enhancing Water Status, Photosynthesis, Antioxidant Enzyme Activity and Fruit Yield. *Plants* **2020**, *9*, 733. [CrossRef]
2. El-Shawa, G.M.R.; Rashwan, E.M.; Abdelaal, K.A.A. Mitigating salt stress effects by exogenous application of proline and yeast extract on morphophysiological, biochemical and anatomical characters of calendula plants. *Sci. J. Flowers Ornam. Plants* **2020**, *7*, 461–482. [CrossRef]
3. Abdelaal, K.A.A.; EL-Maghraby, L.M.; Elansary, H.; Hafez, Y.M.; Ibrahim, E.I.; El-Banna, M.; El-Esawi, M.; Elkelish, A. Treatment of Sweet Pepper with Stress Tolerance-Inducing Compounds Alleviates Salinity Stress Oxidative Damage by Mediating the Physio-Biochemical Activities and Antioxidant Systems. *Agronomy* **2020**, *10*, 26. [CrossRef]
4. El-Banna, M.F.; Abdelaal, K.A.A. Response of Strawberry Plants Grown in the Hydroponic System to Pretreatment with H2O2 before Exposure to Salinity Stress. *J. Plant Prot. Mansoura Univ.* **2018**, *9*, 989–1001. [CrossRef]
5. Shen, Q.; Fu, L.; Dai, F.; Jiang, L.; Zhang, G.; Wu, D. Multi-omics analysis reveals molecular mechanisms of shoot adaption to salt stress in Tibetan wild barley. *BMC Genom.* **2016**, *17*, 889. [CrossRef]
6. Akhilesh, K.; Singh, S.; Gaurav, A.K.; Srivastava, S. Plant Growth-Promoting Bacteria: Biological Tools for the Mitigation of Salinity Stress in Plants. *Front. Microbiol.* **2020**, *11*, 1216. [CrossRef]
7. El Nahhas, N.; AlKahtani, M.; Abdelaal, K.A.A.; Al Husnain, L.; AlGwaiz, H.; Hafez, Y.M.; Attia, K.; El-Esawi, M.; Ibrahim, M.; Elkelish, A. Biochar and jasmonic acid application attenuates antioxidative systems and improves growth, physiology, nutrient uptake and productivity of faba bean (*Vicia faba* L.) irrigated with saline water. *Plant Physiol. Biochem.* **2021**, *166*, 807–817. [CrossRef]
8. Shen, Q.; Fu, L.; Su, T.; Ye, L.; Huang, L.; Kuang, L.; Wu, L.; Wu, D.; Chen, Z.H.; Zhang, G. Calmodulin HvCaM1 negatively regulates salt tolerance via modulation of HvHKT1s and HvCAMTA41[OPEN]. *Plant Physiol.* **2020**, *183*, 1650–1662. [CrossRef]
9. Elsawy, H.I.A.; Mekawy, A.M.; Elhity, M.A.; Abdel-dayem, S.M.; Abdelaziz, M.N.; Assaha, D.V.M.; Ueda, A.; Saneoka, H. Differential responses of two Egyptian barley (*Hordeum vulgare* L.) cultivars to salt stress. *Plant Physiol. Biochem.* **2018**, *127*, 425–435. [CrossRef]
10. Munns, R. Genes and salt tolerance: Bringing them together. *New Phytol.* **2005**, *167*, 645–663. [CrossRef]
11. Munns, R.; Tester, M. Mechanisms of salinity tolerance. *Annu. Rev. Plant Biol.* **2008**, *59*, 651–681. [CrossRef] [PubMed]
12. Stadnik, B.; Tobiasz-Salach, R.; Mazurek, M. Physiological and Epigenetic Reaction of Barley (*Hordeum vulgare* L.) to the Foliar Application of Silicon under Soil Salinity Conditions. *Int. J. Mol. Sci.* **2022**, *23*, 1149. [CrossRef] [PubMed]
13. Khalil, S.R.M.; Ashoub, A.; Hussein, B.A.; Brüggemann, W.; Hussein, E.H.A.; Tawfik, M.S. Physiological and molecular evaluation of ten Egyptian barley cultivars under salt stress conditions. *J. Crop Sci. Biotechnol.* **2021**, *20*, 9669–9681. [CrossRef]
14. Abdelaal, K.; Mazrou, Y.; Hafez, Y. Effect of silcon and carrot extract on morphophysiological characters of pea (*Pisum sativum* L.) under salinity stress conditions. *Fresenius Environ. Bull.* **2022**, *31*, 608–615.
15. Alnusairi, G.S.H.; Mazrou, Y.S.A.; Qari, S.H.; Elkelish, A.A.; Soliman, M.H.; Eweis, M.; Abdelaal, K.; El-Samad, G.A.; Ibrahim, M.F.M.; ElNahhas, N. Exogenous Nitric Oxide Reinforces Photosynthetic Efficiency, Osmolyte, Mineral Uptake, Antioxidant, Expression of Stress-Responsive Genes and Ameliorates the Effects of Salinity Stress in Wheat. *Plants* **2021**, *10*, 1693. [CrossRef]
16. El-Flaah, R.F.; El-Said, R.A.R.; Nassar, M.A.; Hassan, M.; Abdelaal, K.A.A. Effect of rhizobium, nano silica and ascorbic acid on morpho-physiological characters and gene expression of POX and PPO in faba bean (*Vicia faba* L.) Under salinity stress conditions. *Fresenius Environ. Bull.* **2021**, *30*, 5751–5764.
17. Helaly, M.N.; Mohammed, Z.; El-Shaeery, N.I.; Abdelaal, K.A.; Nofal, I.E. Cucumber grafting onto pumpkin can represent an interesting tool to minimize salinity stress. Physiological and anatomical studies. *Middle East J. Agric. Res.* **2017**, *6*, 953–975.
18. Arzani, A. Improving salinity tolerance in crop plants: A biotechnological view. *In Vitro Cell. Dev. Biol. Plant* **2008**, *44*, 373–383. [CrossRef]
19. Hafez, Y.M.; Attia, K.A.; Alamery, S.; Ghazy, A.; Al-Dosse, A.; Ibrahim, E.; Rashwan, E.; El-Maghraby, L.; Awad, A.; Abdelaal, K.A.A. Beneficial Effects of Biochar and Chitosan on Antioxidative Capacity, Osmolytes Accumulation, and Anatomical Characters of Water-Stressed Barley Plants. *Agronomy* **2020**, *10*, 630. [CrossRef]
20. Abdelaal, K.A.A.; Attia, K.A.; Alamery, S.F.; El-Afry, M.M.; Ghazy, A.I.; Tantawy, D.S.; Al-Doss, A.A.; El-Shawy, E.-S.E.; Abu-Elsaoud, A.M.; Hafez, Y.M. Exogenous Application of Proline and Salicylic Acid can Mitigate the Injurious Impacts of Drought Stress on Barley Plants Associated with Physiological and Histological Characters. *Sustainability* **2020**, *12*, 1736. [CrossRef]
21. Abdelaal, K.A.A.; EL-Shawy, E.A.; Hafez, Y.M.; Abdel-Dayem, S.M.; Chidya, R.C.G.; Saneoka, H.; EL Sabagh, A. Nano-Silver and non-traditional compounds mitigate the adverse effects of net blotch disease of barley in correlation with up-regulation of antioxidant enzymes. *Pak. J. Bot.* **2020**, *52*, 1065–1072. [CrossRef]
22. El-Nashaar, F.; Hafez, Y.M.; Abdelaal, K.A.A.; Abdelfatah, A.; Badr, M.; El-Kady, S.; Yousef, A. Assessment of host reaction and yield losses of commercial barley cultivars to *Drechslera teres* the causal agent of net blotch disease in Egypt. *Fresenius Environ. Bull.* **2020**, *29*, 2371–2377.
23. Abdelaal, K.A.; Hafez, Y.M.; El-Afry, M.M.; Tantawy, D.S.; Alshaal, T. Effect of some osmoregulators on photosynthesis, lipid peroxidation, antioxidative capacity and productivity of barley (*Hordeum vulgare* L.) under water deficit stress. *Environ. Sci. Pollut. Res.* **2018**, *25*, 30199–30211. [CrossRef] [PubMed]
24. FAO. *The State of Food Agriculture Overcoming Water Challenges in Agriculture*; Food and Agriculture Organization: Rome, Italy, 2020.

25. Kanbar, A.; El drussi, I. Effect of Salinity Stress on Germination and Seedling Growth of Barley (*Hordeum vulgare* L.) Varieties. *Adv. Environ. Biol.* **2014**, 244–248. Available online: https://link.gale.com/apps/doc/A375184626/AONE?u=anon~||ed580df7&sid=googleScholar&xid=8bd31daa (accessed on 9 September 2022).
26. Porra, R.J.; Thompson, W.A.; Kriedemann, P.E. Determination of accurate extinction coefficients and simultaneous equations for assaying chlorophylls a and b extracted with four different solvents: Verification of the concentration of chlorophyll standards by atomic absorption spectrometry. *Biochem. Biophys. Acta* **1989**, *975*, 384–394. [CrossRef]
27. El-Wakeel, S.; Abdel-Azeem, A.; Mostafa, E. Assessment of salinity stress tolerance in some barley genotypes. *Alex. J. Agric. Sci.* **2019**, *64*, 195–206.
28. Maršálová, L.; Vítámvás, P.; Hynek, R.; Prášil, I.T.; Kosová, K. Proteomic Response of *Hordeum vulgare* cv. Tadmor and *Hordeum marinum* to Salinity Stress: Similarities and Differences between a Glycophyte and a Halophyte. *Front. Plant Sci.* **2016**, *7*, 1154. [CrossRef]
29. Docquier, S.; Kevers, C.; Lambé, P.; Gaspar, T.; Dommes, J. Beneficial use of lignosulfonates in in vitro plant cultures: Stimulation of growth, of multiplication and of rooting. *Plant Cell Tissue Organ Cult.* **2007**, *90*, 285–291. [CrossRef]
30. Gupta, B.; Huang, B. Mechanism of salinity tolerance in plants: Physiological, biochemical, and molecular characterization. *Int. J. Genom.* **2014**, *2014*, 701596. [CrossRef]
31. Gupta, S.; Manoj, G.; Kulkarni, J.; White, F.; Wendy, A.; Stirk, H.; Papenfus, B.; Karel, D.; Vince, Ö.; Jeffrey, N.; et al. Chapter 1—Categories of Various Plant Biostimulants—Mode of Application and Shelf-Life, Biostimulants for Crops from Seed Germination to Plant Development; Academic Press: Cambridge, MA, USA, 2021; pp. 1–60, ISBN 9780128230480.
32. Desoky, E.S.M.; Elrys, A.S.; Mansour, E.; Eid, R.S.M.; Selem, E.; Rady, M.M.; Ali, E.F.; Mersal, G.A.M.; Semida, W.M. Application of biostimulants promotes growth and productivity by fortifying the antioxidant machinery and suppressing oxidative stress in faba bean under various abiotic stresses. *Sci. Hortic.* **2021**, *288*, 110340. [CrossRef]
33. Van Oosten, M.J.; Pepe, O.; De Pascale, S.; Silletti, S.; Maggi, A. The Role of Biostimulants and Bioeffectors as Alleviators of Abiotic Stress in Crop Plants. *Chem. Biol. Technol. Agric.* **2017**, *4*, 5. [CrossRef]
34. Gul, S.; Yanni, S.F.; Whalen, J.K. Lignin controls on soil ecosystem services: Implications for biotechnological advances. *Biochem. Res. Trends Nova Sci. Publ.* **2014**, *978*, 375–416.
35. Ertani, A.; Nardi, S.; Francioso, O.; Pizzeghello, D.; Tinti, A. Metabolite-Targeted Analysis and Physiological Traits of *Zea mays* L. in Response to Application of a Products for Their Evaluation as Potential Biostimulants. *Agronomy* **2019**, *9*, 445. [CrossRef]
36. Kok, A.D.; Wan, W.M.A.N.; Tang, C.N. Sodium lignosulfonate improves shoot growth of *Oryza sativa* via enhancement of photosynthetic activity and reduced accumulation of reactive oxygen species. *Sci. Rep.* **2021**, *11*, 13226. [CrossRef] [PubMed]
37. Zörb, C.; Geilfus, C.M.; Dietz, K.J. Salinity and crop yield. *Plant Biol.* **2019**, *21*, 31–38. [CrossRef]
38. Porra, R.J. The chequered history of the development and use of simultaneous equations for the accurate determination of chlorophylls a and b. *Photosynth. Res.* **2002**, *73*, 149–156. [CrossRef]
39. Gonzalez, L.; Gonzalez-Vilar, M. Determination of relative water content. In *Handbook of Plant Ecophysiology Techniques*; Springer: Dordrecht, The Netherlands, 2001; pp. 207–212.
40. Murray, B.Y.M.B.; Cape, J.N.; Fowler, D. Quantification of frost damage in plant tissues by rates of electrolyte leakage. *New Phytol.* **1989**, *113*, 307–311. [CrossRef]
41. Takagi, H.; Yamada, M.; Takagi, H.; Yamada, S. Soil Science and Plant Nutrition Roles of enzymes in anti-oxidative response system on three species of chenopodiaceous halophytes under NaCl-stress condition Roles of enzymes in anti-oxidative response system on three species of chenopodiaceous halophyte. *Soil Sci. Plant Nutr.* **2013**, *59*, 603–611. [CrossRef]
42. Chance, B.; Meahly, A.C. Catalase Assay by Disappearance of Peroxide. *Methods Enzymol.* **1955**, *2*, 764–775.
43. Rahman, A.; Nahar, K.; Hasanuzzaman, M.; Fujita, M. Calcium supplementation improves Na$^+$/K$^+$ ratio, antioxidant defense and glyoxalase systems in salt-stressed rice seedlings. *Front. Plant Sci.* **2016**, *7*, 609. [CrossRef]
44. Sanful, R.E.; Darko, S. Utilization of soybean flour in the production of bread. *Pak. J. Nutr.* **2010**, *9*, 815–818. [CrossRef]
45. Gomez, K.A.; Gomez, A.A. *Statistical Procedures for Agricultural Research*; John Wiley and Sons, Inc.: New York, NY, USA, 1984.
46. Duncan, D.B. Multiple range and multiple F tset. *Biometrics* **1955**, *11*, 1–42. [CrossRef]
47. Noreen, S.; Faiz, O.; Akhter, M.S.; Shah, K.H. Influence of foliar application of osmoprotectants to ameliorate salt stress in sunflower (*Helianthus annuus* L.). *Sarhad J. Agric.* **2019**, *34*, 1316–1325. [CrossRef]
48. Fayez, K.A.; Bazaid, S.A. Improving drought and salinity tolerance in barley by application of salicylic acid and potassium nitrate. *J. Saudi Soc. Agric. Sci.* **2014**, *13*, 45–55. [CrossRef]
49. Fahad, S.; Bano, A. Effect of salicylic acid on physiological and biochemical characterization of maize grown in saline area. *Pak. J. Bot.* **2012**, *44*, 1433–1438.
50. Kang, F.; Lv, Q.l.; Liu, J.; Meng, Y.-S.; Wang, Z.-H.; Ren, X.-Q.; Hu, S.-W. Organic–inorganic calcium lignosulfonate compounds for soil acidity amelioration. *Environ. Sci Pollut Res.* **2022**. [CrossRef]
51. Hafez, Y.; Elkohby, W.; Mazrou, Y.S.A.; Ghazy, M.; Elgamal, A.; Abdelaal, K.A.A. Alleviating the detrimental impacts of salt stress on morpho-hpysiological and yield characters of rice plants (*Oryza sativa* L.) using actosol, Nano-Zn and Nano-Si. *Fresenius Environ. Bull.* **2020**, *29*, 6882–6897.
52. Abdelaal, K.A.A.; El-Afry, M.; Metwaly, M.; Zidan, M.; Rashwan, E. Salt tolerance activation in faba bean plants using proline and salicylic acid associated with physio-biochemical and yield characters' improvement. *Fresenius Environ. Bull.* **2021**, *30*, 3175–3186.

53. AlKahtani, M.D.F.; Hafez, Y.M.; Attia, K.; Al-Ateeq, T.; Ali, M.A.M.; Hasanuzzaman, M.; Abdelaal, K.A.A. *Bacillus thuringiensis* and Silicon Modulate Antioxidant Metabolism and Improve the Physiological Traits to Confer Salt Tolerance in Lettuce. *Plants* **2021**, *10*, 1025. [CrossRef]
54. Peanparkdee, M.; Iwamoto, S. Bioactive compounds from by-products of rice cultivation and rice processing: Extraction and application in the food and pharmaceutical industries. *Trends Food Sci. Technol.* **2019**, *86*, 109–117. [CrossRef]
55. Duncan, E.G.; O'Sullivan, C.A.; Roper, M.M.; Biggs, J.S.; Peoples, M.B. Influence of co-application of nitrogen with phosphorus, potassium and sulphur on the apparent efficiency of nitrogen fertilizer use, grain yield and protein content of wheat: Review. *Field Crops Res.* **2018**, *226*, 56–65. [CrossRef]
56. Liu, J.; Hou, H.; Zhao, L.; Sun, Z.; Li, H. Protective Effect of foliar application of sulfur on photosynthesis and antioxidative defense system of rice under the stress of Cd. *Sci. Total Environ.* **2020**, *710*, 136230. [CrossRef] [PubMed]
57. Torun, H. Arbuscular Mycorrhizal Fungi and K-Humate Combined as Biostimulants: Changes in Antioxidant Defense System and Radical Scavenging Capacity in *Elaeagnus angustifolia*. *J. Soil Sci. Plant Nutr.* **2020**, *20*, 2379–2393. [CrossRef]
58. Zhu, J.K. Plant salt tolerance. *Trends Plant Sci.* **2001**, *6*, 66–71. [CrossRef]
59. Assaha, D.V.M.; Mekawy, A.M.; Ueda, A.; Saneoka, H. Erratum: Salinity-induced expression of HKT may be crucial for Na^+ exclusion in the leaf blade of huckleberry (*Solanum scabrum* Mill.), but not of eggplant (*Solanum melongena* L.). *Biochem. Biophys. Res. Commun.* **2015**, *463*, 1342. [CrossRef]
60. Landlin, G.; Soundarya, M.K.; Bhuvaneshwari, S. Behavior of Lignosulphonate Amended Expansive Soil. In *Sustainable Practices and Innovations in Civil Engineering*; Ramanagopal, S., Gali, M., Venkataraman, K., Eds.; Springer: Singapore, 2021; Volume 79, pp. 151–162. [CrossRef]
61. Dave, A.; Agarwal, P.; Agarwal, P.K. Mechanism of high affinity potassium transporter (HKT) towards improved crop productivity in saline agricultural lands. *3 Biotech* **2022**, *12*, 51. [CrossRef]
62. Ikkonen, E.N.; Jurkevich, M.G. Effect of lignosulfonate application to sandy soil on plant nutrition and physiological traits. *IOP Conf. Ser. Earth Environ. Sci.* **2021**, *862*, 012079. [CrossRef]

Article

Reclamation of Saline Soil under Association between *Atriplex nummularia* L. and Glycophytes Plants

Monaliza Alves dos Santos [1,*], Maria Betânia Galvão Santos Freire [1], Fernando José Freire [1], Alexandre Tavares da Rocha [2], Pedro Gabriel de Lucena [1], Cinthya Mirella Pacheco Ladislau [1] and Hidelblandi Farias de Melo [3]

[1] Department of Agronomy, Federal Rural University of Pernambuco, Recife 52171-900, PE, Brazil; maria.freire@ufrpe.br (M.B.G.S.F.); fernando.freire@ufrpe.br (F.J.F.); pedro.lucena@ufrpe.br (P.G.d.L.); cinthya.m.pacheco@gmail.com (C.M.P.L.)
[2] Academic Unit of Garanhuns, Federal University of Agreste of Pernambuco, Garanhuns 55290-000, PE, Brazil; altarocha@gmail.com
[3] Department of Soils, Federal University of Viçosa, Viçosa 36570-900, MG, Brazil; hidelblandi@ufv.br
* Correspondence: alves.monaliza@yahoo.com.br

Abstract: Phytoremediation is an efficient technique for the reclamation of salt-affected soils by growing plants. The present study aims to evaluate the intercropping of halophyte *Atriplex nummularia* Lindl. with naturally occurring species (*Mimosa caesalpiniifolia* Benth, *Leucaena leucocephala* (Lam.) de Wit and *Azadirachta indica* A. Juss.) adapted to semiarid regions as a management capable of enhancing the phytoremediation capacity of these species. A field experiment was conducted in a randomized block and contained four replicates. Species were cultivated alone and in association with *A. nummularia* to evaluate their potential uses in the reclamation of soils. Exchangeable Ca^{2+}, Mg^{2+}, Na^+, and K^+, as well as salinity and sodicity variables, were evaluated. The evaluations were performed at 9 and 18 months of plant growth. The results indicated that *A. nummularia* individualized was the treatment most efficient; with reductions of 80%, 63%, and 84% in electrical conductivity, sodium adsorption ratio, and exchangeable sodium percentage values, respectively at 18 months compared to starting of the experiment. However, the use of *A. nummularia* and species adapted to the semiarid in association, or even alone, promoted beneficial effects on the soil quality after the establishment of the plants.

Keywords: phytoremediation; *Azadirachta indica*; *Leucaena leucocephala*; *Mimosa caesalpiniifolia*; salt-affected soils; soil reclaim

1. Introduction

Salinization is one of the main processes of soil degradation in the world, besides being one of the environmental factors limiting the productivity of agricultural crops [1,2]. Its occurrence is mainly associated with arid and semiarid regions around the world, occurring in practically all the continents and corresponds to 7% of the total world's surface area [3,4].

Salts in excess can lead to drastic changes in some soil's physical and chemical properties resulting in the development of an environment unsuitable for the growth of most crops [5,6]. The increase in the extent of degraded areas by salts is in the opposite direction of the necessary increase in food production (71%) between 2005 and 2050 [7,8]. In addition to agricultural impacts, the increase in soil degradation directly affects the maintenance of the hydrological cycle, the health of the terrestrial biosphere, favors pollution and eutrophication of water bodies, consequently, affecting the global and local economy [9].

On the other hand, the tendency of a given area to deteriorate can be reversed by restoring land use or by using appropriate management practices [7]. The reclamation of degraded areas by salinity stands out as an effective way of alleviating population pressure and contributing to food security for future generations [10,11].

Phytoremediation is an alternative and efficient technique that consists of the cultivation of tolerant vegetable species, with the capacity to extract expressive amounts of salts from the soil and to store them in their tissues throughout their life cycle [12,13]. This technique is presented as a low-cost technique and is more consistent with the socioeconomic and edaphoclimatic conditions of these regions [14,15].

Areas undergoing reclaim tend to have improved physical, chemical, and microbiological properties as the revegetation process evolves [16,17]. Among the main changes are the increase in soil fertility, organic matter accumulation in the soil, reduction of erodibility, increase in water retention in the soil, and reduction of salinity [5].

Studies with plants with proven effectiveness in reclamation saline soils are still scarce and have been carried out with halophytes, especially plants of the Atriplex genus. However, these plants are not known by some rural populations and, therefore, are not well accepted. Other species, such as Sabiá (*Mimosa caesalpiniifolia* Benth) and Leucaena (*Leucaena leucocephala* (Lam.) de Wit), already quite adapted to the semiarid region could be tested for adaptability to saline and sodic soils. Or even less known ones, but which have been highlighted recently, such as Neem (*Azadirachta indica* A. Juss.) which has the capacity to adapt to numerous climatic and edaphic factors.

The adaptation of these plants to salinity in field conditions is not well understood and their cultivation in association with a halophyte could improve their establishment. The use of plant species with different root systems can promote a more effective action of these in the reclamation, leading to improvements in the quality of the soils under revegetation, and increasing the extraction of salts from the soil. Moreover, *Atriplex nummularia* Lindl. may improve soil properties so that other plant species can develop on degraded soils, protecting soil and water in semiarid environments, and contributing to environmental quality. The objective of this study was to evaluate the use of tree species alone and in association with *Atriplex nummularia* L. in the revegetation of degraded soil in the semi-arid region, increasing reclamation.

2. Materials and Methods

2.1. Study Area

The experiment was conducted in Cachoeira II Irrigated Perimeter, Municipality of Serra Talhada, Pernambuco, semiarid region of Brazil (7°58′54″ to 8°01′36″ S and 38°18′24″ to 38°21′21″ W) (Figure 1). The soil in the area is Fluvisol [18], which has flat relief, with small gradients and serious problems of water infiltration. For a long time (around 30 years), the area was used for banana cropping under furrow irrigation. Thereafter, due to soil degradation, the area was left without agricultural use for eight years.

The climate in the area is BSh type (semiarid of low latitude and altitude), with a dry period of nine months and rainfall concentrated from February to April [19]. Temperature ranges from 64.4 to 98.6 °F (average 80.6 °F) and 720 mm of average annual rainfall.

2.2. Species Selection

To test the adaptability of species to saline-sodic soils, as well as to enable their cultivation as an alternative practice in the management of soil reclamation, we selected three plant species normally found in the semi-arid region of Brazil: Sabiá (*Mimosa caesalpiniifolia* Benth), Leucaena [*Leucaena leucocephala* (Lam.) de Wit] and Neem (*Azadirachta indica* A. Juss). These plants were compared to Atriplex (*Atriplex nummularia* Lindl.), which has already been used in other degraded areas and with proven effectiveness in the reclamation of salt-affected soils, given the characteristics of vegetative growth and the high concentration of salts in the tissues [14].

The *A. indica*, *M. caesalpiniifolia*, and *L. leucocephala* are plant species adapted to the semiarid region and could be tested regarding the potential of reclamation saline and sodic soils, for sustainable management and use of products from plants.

An additional consideration for the species selection deals with the root effect, which provides channels for the percolating soil solution. An added advantage relates to the better availability of some macro and micronutrients after soil amelioration that involves ions leaching.

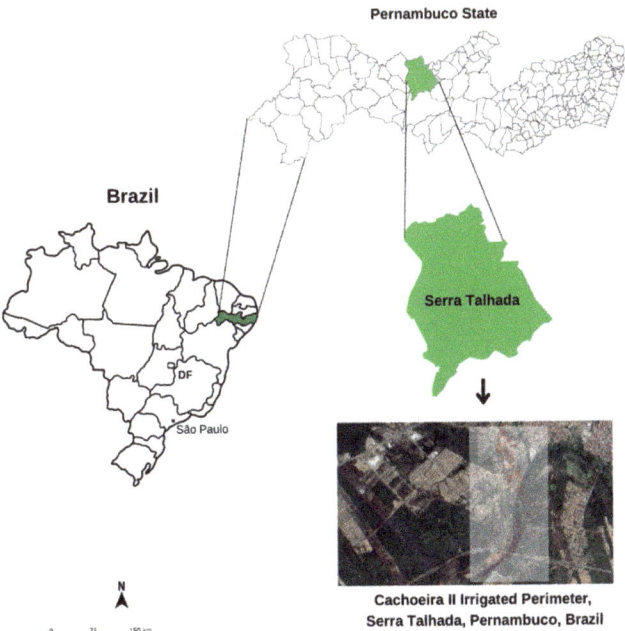

Figure 1. Study area location, Irrigated Perimeter Cachoeira II, Serra Talhada, State of Pernambuco, Brazil.

2.3. Seedling Production

The seedlings of each species were cultivated in a greenhouse at the Federal Rural University of Pernambuco, on an organic substrate. It was cropped 85-day-old *A. nummularia* seedlings that were produced from the cuttings of a single mother plant to maintain genetic uniformity. The seedlings of *A. indica*, *L. leucocephala*, and *M. caesalpinifolia* were produced from the seeds and were cropped 90-day-old seedlings.

The seedlings were transplanted to the field when they were 30 cm long (Atriplex, Neem, and Sabiá) or 40 cm (Leucaena). Manually transplanting one plant per hole (0.3 × 0.3 × 0.3 m), without the addition of organic or chemical compost, only seedling substrate.

2.4. Treatments and Experimental Design

The experimental area was divided into four randomized blocks, with the eight tested treatments, totaling 32 experimental plots. Each plot was dimensioned in 8 × 8 m (64 m^2) and the useful plot was 4 × 4 m (16 m^2). These eight studied treatments were: four individualized treatments where the selected species were cultivated alone (Sabiá, Leucaena, Neem, Atriplex, one in each plot); three treatments associating Atriplex and one of the other species (Atriplex/Sabiá, Atriplex/Leucaena, and Atriplex/Neem); and one control treatment, without cultivation of any plant species.

The control plots were not cultivated and were kept without plants, by manual weeding the naturally occurring species (self-sown plants) once a month, using a hoe.

The planting spacing was 2 × 2 m, totaling 16 plants per plot and 4 plants in the useful plot. The seedlings were transplanted to the field in November 2013, with a height varying between 0.3–0.4 m. No addition of fertilizer, organic, or chemical was performed during the experiment. To ensure the establishment of seedlings, weekly irrigations were carried out for 60 days after transplanting. The water used for irrigation was pumped directly from the Pajeú River (Table 1).

Table 1. Water characteristics: pH, electrical conductivity (EC), soluble cations (Na^+, K^+, Ca^{2+}, and Mg^{2+}), sodium adsorption ratio (SAR), and Pajeú river water classification for salinity and sodicity risk.

pH	EC	Na^+	K^+	Ca^{2+}	Mg^{2+}	SAR	Risk [1]
	dS m^{-1}	\multicolumn{4}{c}{mmol$_c$ L^{-1}}		($mmol_c$ L^{-1})$^{0.5}$	
7.89	1.28	19.25	0.25	2.30	0.23	17.11	C3S2

[1] USSL Staff [19].

2.5. Soil Sampling

For the characterization and evaluation of soil chemical quality, four simple samples were collected at the center of each plot to constitute composite samples, one for each plot. The samples were taken at 0.5 m from the stem of the four central plants in the useful area. In the control treatment, the center of the useful area was used as a reference for sampling.

Three soil samplings were performed. In the sampling for chemical characterization of the soil in the experiment set up, thirty-two samples were collected in the 0–30 cm layer. Whereas in the two samplings to evaluate changes in soil chemical quality, performed at 9 and eighteen months after setting up the experiment, the samples were always collected at depths of 0–10, 10–30, and 30–60 cm layers.

After each sampling, the soil samples were air dried and sieved in a 2 mm mesh, and reserved for chemical analysis.

2.6. Soil Chemical Analyzes

Exchangeable cations Ca^{2+}, Mg^{2+}, Na^+ and K^+ were extracted by 1 mol L^{-1} ammonium acetate at pH 7.0 [20], in which Na^+ and K^+ were quantified by flame photometry, and Ca^{2+} and Mg^{2+} by atomic absorption spectrophotometry [21]. Cation exchangeable capacity was determined by the cation index method [19]. Sodium adsorption ratio (SAR) and exchangeable sodium percentage (ESP) were calculated according to USSL Staff [19].

Soluble cations were quantified in the saturation extract, obtained by the preparation of the saturated paste, and extracted under vacuum according to the saturation paste method [19]. In the extract was measured electrical conductivity (EC 25 °C), soluble cations Ca^{2+} and Mg^{2+} were determined by atomic absorption spectrophotometry, and Na^+ and K^+ by flame emission photometry. Soil pH was measured in water in the proportion of 1:2.5 (soil:water).

Particle size distribution was performed using the pipetting method modified by Ruiz [22]. To each 10-g sample was added 50 mL of NaOH solution (0.1 mol L^{-1}) and 150 mL of deionized water; the mixture was stirred with a glass rod and left to settle overnight. The mixtures were then dispersed by shaking at 12,000 rpm for 15 min, after which the suspension was passed through a 0.053-mm sieve to quantify the total sand, and then this total was passed through a 0.2-mm sieve to separate and quantify the fine sand and coarse sand fractions.

Silt and clay fractions were collected in a 500-mL graduated cylinder and shaken again. Immediately afterward, we collected 25 mL of the silt + clay suspension and allowed it to settle for the time calculated by Stokes' Law for the ambient temperature. Then we collected another 25 mL of the suspension from 5 cm below the surface (clay fraction). All fractions were oven-dried at 100 °C and weighed to calculate the percentages of coarse sand, fine sand, silt, and clay. Soil chemical and physical characterization are in Table 2.

Table 2. Soil chemical characteristics and particle size distribution of the soil collected at the experiment set up (0–30 cm layer).

EC [2] dS m^{-1}	pH 1:2.5	Exchange Complex [1]				CEC [3]	ESP [4]	Soil Particle Size [5]			
		Na$^+$	K$^+$	Ca^{2+}	Mg^{2+}			Coarse Sand	Fine Sand	Silt	Clay
		cmol$_c$ kg^{-1}					%	g kg^{-1}			
5.48	7.23	5.99	1.05	1.59	0.68	9.51	64	51.0	712.0	75.0	162.0

[1] USSL Staff [19]. [2] Electrical conductivity; [3] Cation exchange capacity; [4] Exchangeable sodium percentage; [5] Ruiz [22]. Clay (≤2 μm); silt (2–53 μm); fine sand (53–200 μm); coarse sand (200–2000 μm). Average values of thirty-two soil samples collected in the 0–30 cm layer.

In the experiment set up, the soil prior to treatments (0–30 cm depth) was saline-sodic, EC = 6.48 dS m^{-1} and ESP = 64% (Table 2). Despite the sandy loam texture, the exchangeable cations are maintained in the soil exchange complex in the sequence Na$^+$ > Ca^{2+} > K$^+$ > Mg^{2+}. And pH is slightly alkaline.

2.7. Statistical Analysis

First, we applied the normality test to the evaluated variables, using the Kolmogorov-Smirnov test. Subsequently, an analysis of variance was performed using Fisher's F test, both with 5% significance. For significant variables, the means obtained were submitted to the Student's T test, comparing the effects of orthogonal contrasts for treatments, compared the effects at 5% probability. The Skott-Knott test was applied at 10% probability for the means of salinity and sodicity variables (pH, EC, SAR, and ESP). The results were evaluated according to the treatments applied at different sampling times.

3. Results

The growth rate of A. nummularia was uniform until eight months after transplantation (MAT), after this period, the increase was less pronounced until 18 MAT. Although, the growth rate of A. indica and L. leucocephala the increase was less pronounced after 12 MAT and after 10 MAT to M. caesalpinifolia.

The height of Atriplex, Leucaena, and Neem plants showed higher mean values observed in the individualized treatments compared to the associated treatments, possibly due to competition between species. However, Atriplex plant development was observed regardless of soil salinity in the area. In contrast, inhibition of the growth dynamic of Sabiá plants was observed both when cultivated alone and in association with Atriplex.

The A. nummularia and L. leucocephala showed no signs of stress. The adaptive responses of M. caesalpinifolia were more evident visually, expressed by the yellowing of leaves and early foliar senescence, resulting in a reduced number of leaves, which may be associated with the toxicity caused by saline ions. It is also possible to state that as a criterion of adaptability to stress conditions, there was a halt in the growth of M. caesalpinifolia, with a loss of biomass production. Signs of stress were also observed in A. indica, with a number and position of branches frequent in the lower parts of the plants. This is a common mechanism in stressed neem plants to reduce the effects of soil salt concentrations.

3.1. Exchangeable Cations

There was no difference for exchangeable Na$^+$ and K$^+$ in soil between applied treatments at 9 months of growth at 0–10, 10–30, and 30–60 cm layers (Figure 2). However, changes occurred in these exchangeable cations at 18 months of cultivation. In general, the cultivation of plants promoted a decrease in exchangeable Na$^+$ and K$^+$ contents in the soil.

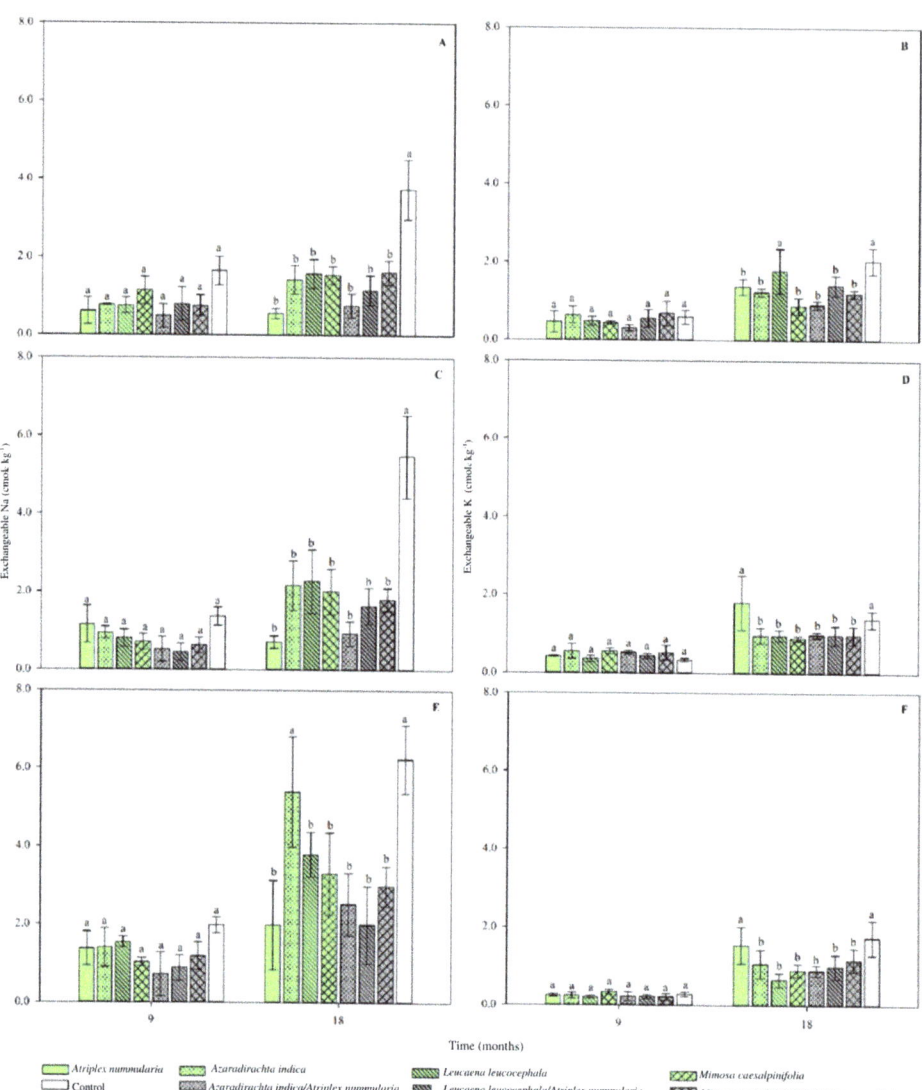

Figure 2. Exchangeable Na⁺ [(**A**) 0–10 cm; (**C**) 10–30 cm; (**E**) 30–60 cm] and K⁺ [(**B**) 0–10 cm; (**D**) 10–30 cm; (**F**) 30–60 cm] at 9 and 18 months as result of the applied treatments (Averages of four replications). Averages followed by the same letter have no difference in the same layer and time of sampling by the Skott-Knott test ($p \leq 0.05$).

These Na⁺ values influenced the results of contrasts at 9 months, a significant difference was observed between the groups for the variables of sodicity (SAR and ESP), were significant for SAR at 0–10 cm and 10–30 cm layers; and for ESP at 0–10 cm and 30–60 cm layers. At the same sampling time, the contrasts (All treatments × control), (Associations × control), and (Isolated cultures × associations) were significant for pH at 30–60 cm layer; and the contrast (No leguminous × leguminous) at 10–30 cm layer. For EC, only contrasts (Isolated cultures × control) and (Leguminous × control) were significant at the 30–60 cm layer.

The effect of growing plants on mean values of SAR and ESP was verified, ESP was decreased and the soil did not remain as sodic at 9 months in the evaluated layers, except for *A. indica* treatment in the first layer and control treatment at 0–10 cm and 30–60 cm layers (Table 3). Initial changes in SAR and ESP were observed under growing plants in some soil layers.

Table 3. Orthogonal contrasts of chemical attributes pH (potential of hydrogen), EC (electrical conductivity), SAR (sodium adsorption ratio), and ESP (exchangeable sodium percentage) in the soil at 0–10, 10–30, and 30–60 cm depth as a function of applied treatments at 9 months.

Contrast	pH			EC			SAR			ESP		
	0–10 cm	10–30 cm	30–60 cm	0–10 cm	10–30 cm	30–60 cm	0–10 cm	10–30 cm	30–60 cm	0–10 cm	10–30 cm	30–60 cm
All treatments × control	0.253	0.695	2.169 *	−2.003	−1.242	−2.075	−0.095	−0.950	−0.439	−1.139	−1.261	−1.219
Isolated cultures × control	−0.247	0.582	1.414	−1.860	−1.239	−2.198 *	−0.146	−0.789	−0.393	−0.840	−0.814	−0.873
Associations × control	0.867	0.751	2.860 **	−1.926	−1.084	−1.644	−0.018	−1.034	−0.441	−1.377	−1.674	−1.506
Leguminous × control	−0.710	−0.531	1.463	−1.631	−1.534	−2.286 *	−0.084	−0.908	0.168	−0.888	−1.239	−0.905
No leguminous × control	0.258	1.593	1.119	−1.765	−0.728	−1.728	−0.183	−0.532	−0.886	−0.646	−0.248	−0.688
Isolated cultures × associations	−1.672	−0.284	−2.254 *	0.189	−0.174	−0.732	−0.187	0.408	0.092	0.852	1.338	1.000
No leguminous × leguminous	1.185	2.601 *	−0.422	−0.165	0.986	0.684	−0.121	0.460	−1.291	0.296	1.214	0.266
Atriplex × all treatments	−0.463	0.602	−0.683	−0.044	−0.628	−0.768	−1.389	−0.774	−1.044	−0.378	1.740	1.000

* Significant at 0.05 probability; ** Significant at 0.01 probability.

At 18 months of cultivation, exchangeable Na^+ and K^+ contents of the soil increased in relation to those recorded at 9 months (Figure 2), although the Na^+ values remained below those found in the initial characterization of the soil (Table 2).

These Na^+ values changes influenced the results of variables of sodicity SAR and ESP at 18 months, a significant difference was observed at any of the layers evaluated (Table 4).

Table 4. Chemical attributes pH (hydrogen potential), EC (electrical conductivity), SAR (sodium adsorption ratio), and ESP (exchangeable sodium percentage) in the soil at 0–10, 10–30, and 30–60 cm depth as a function of treatments at 9 and 18 months (mean of four replicates).

Treatment	pH 1:2.5			EC [1] dS m^{-1}			SAR [2] (mmol$_c$ L^{-1})$^{0.5}$			ESP [3] %		
	\multicolumn{12}{c}{9 Months}											
	0–10 cm	10–30 cm	30–60 cm	0–10 cm	10–30 cm	30–60 cm	0–10 cm	10–30 cm	30–60 cm	0–10 cm	10–30 cm	30–60 cm
Atriplex nummularia	7.15 a	7.31 a	7.30 a	1.23 a	1.32 a	3.23 a	5.72 b	6.52 b	6.22 a	9.33 b	12.61 a	14.41 a
Azaridachta indica	7.28 a	7.48 a	7.21 a	1.24 a	2.94 a	4.50 a	13.38 a	12.59 a	7.37 a	17.73 a	9.54 a	11.38 a
Leucaena leucocephala	7.00 a	7.10 a	7.34 a	0.82 a	1.30 a	2.54 a	10.37 a	7.53 b	8.87 a	8.43 b	7.81 a	13.57 a
Mimosa caesalpinifolia	7.09 a	7.10 a	7.23 a	1.88 a	1.39 a	4.16 a	9.53 a	9.03 b	8.99 a	12.00 b	7.42 a	10.54 a
A. indica/A. nummularia	7.32 a	7.16 a	7.48 a	1.19 a	2.07 a	5.26 a	13.34 a	6.43 b	8.59 a	8.19 b	6.59 a	7.43 b
L. leucocephala/A. nummularia	7.22 a	7.34 a	7.37 a	2.11 a	2.01 a	1.85 a	4.16 b	5.20 b	4.48 a	9.78 b	5.27 a	9.72 a
M. caesalpinifolia/A. nummularia	7.46 a	7.29 a	7.58 a	1.32 a	1.35 a	4.19 a	13.15 a	12.43 a	10.16 a	9.44 b	6.73 a	11.75 a
Control	7.15 a	7.18 a	7.10 a	2.54 a	2.85 a	5.51 a	10.30 a	11.39 a	8.58 a	17.80 a	13.01 a	17.83 a
CV (%)	5.41%	11.43%	11.83%	28.01%	32.81%	34.94%	35.28%	36.47%	55.74%	34.60%	66.08%	38.25%
	\multicolumn{12}{c}{18 Months}											
	0–10 cm	10–30 cm	30–60 cm	0–10 cm	10–30 cm	30–60 cm	0–10 cm	10–30 cm	30–60 cm	0–10 cm	10–30 cm	30–60 cm
Atriplex nummularia	6.77 a	6.59 a	6.57 a	0.97 b	1.08 a	1.34 b	8.55 b	9.36 b	15.84 a	8.09 c	10.39 c	10.59 c
Azaridachta indica	6.36 a	6.64 a	6.41 a	2.76 a	3.06 a	3.23 a	18.92 a	13.08 b	18.24 a	15.52 b	22.82 b	44.81 a
Leucaena leucocephala	6.10 a	6.20 a	6.46 a	2.56 a	2.97 a	2.67 a	10.02 b	7.87 b	6.95 c	17.65 b	19.41 b	32.75 a
Mimosa caesalpinifolia	6.23 a	6.21 a	6.36 a	3.78 a	4.92 a	4.75 a	9.77 b	7.78 b	11.11 b	20.35 b	26.69 b	28.26 b
A. indica/A. nummularia	6.06 a	6.10 a	6.43 a	1.96 a	1.91 a	2.12 a	9.96 b	12.73 b	13.57 b	10.92 c	19.43 b	28.55 b
L. leucocephala/A. nummularia	6.30 a	6.37 a	6.73 a	1.71 a	1.49 a	1.71 b	13.02 b	12.31 b	17.02 a	14.13 b	18.52 b	19.52 b
M. caesalpinifolia/A. nummularia	6.35 a	6.49 a	6.53 a	2.23 a	1.94 a	2.58 a	9.14 b	6.87 b	9.39 b	19.78 b	20.12 b	27.90 b
Control	6.52 a	6.64 a	6.64 a	4.54 a	3.85 a	5.51 a	23.28 a	29.58 a	21.29 a	41.03 a	48.07 a	48.53 a
CV (%)	8.61%	8.32%	8.57%	22.19%	39.25%	27.14%	27.16%	29.82%	26.28%	35.60%	32.22%	35.12%

[1] Electrical conductivity; [2] Sodium adsorption ratio; [3] Exchangeable sodium percentage. (Averages of four replications). Averages followed by the same letter have no difference in the same layer and time of sampling by the Skott-Knott test ($p \leq 0.10$).

At 18 months, an effect of treatments with the use of the studied species was observed on EC compared to the control treatment. The influence on salinity reduction (EC values) by the Atriplex treatment in isolated culture was significantly observed ($p \leq 0.10$) at the 0–10 cm and 30–60 cm layers (Table 4).

In the 0–10 and 10–30 cm layers, the exchangeable Na^+ of the soil in all treatments had a difference in relation to the control, showing the efficiency of plant cultivation in not allowing the increase in Na^+ contents in the exchange complex of soil (Figure 2). However, in the last evaluated layer (30–60 cm), isolated *A. indica* cultivation was not able to reduce the content of this element in relation to the soil without plants. The other tested treatments, with the use of isolated or associated intercropped plants, promoted a decrease in exchangeable Na^+ contents of the soil.

There was an increase in the levels of exchangeable Ca^{2+} in the soil at 9 months (Figure 3) in relation to the initial values (Table 2); the contents were higher than 2 $cmol_c\ kg^{-1}$ in the soil of all treatments in the three layers. Nevertheless, the soil exchangeable Ca^{2+} contents were reduced at 18 months of cultivation, with no difference between treatments.

Exchangeable Mg^{2+} contents remained stable, not differing between the treatments, neither in relation to the control for any of the layers evaluated at 9 or 18 months of plant growth (Figure 3).

3.2. Soluble Cations

No difference among treatments was recorded for concentrations of soluble cations Na^+ and K^+ at 9 months at either depth (Figure 4). The same was observed for Na^+ soluble at 18 months in the superficial layer (0–10 cm). However, differences between treatments were recorded for soluble Na^+ in the subsurface (Figure 4). Plant cultivation reduced soluble Na^+ contents to almost half of that registered in the control treatment in the soil at 10–30 cm. In the 30–60 cm layer, only the isolated cultures and association *M. caesalpiniifolia*/*A. nummularia* were efficient in reducing the soluble Na^+ contents in soil.

For soluble K^+, a significant difference was observed among the treatments only on the surface for an isolated culture of the *M. caesalpiniifolia* and association *A. indica*/*A. nummularia* and *M. caesalpiniifolia*/*A. nummularia*. At layers of 10–30 and 30–60 cm, no differences were found among treatments (Figure 4).

At 9 months of growing plants, there was no difference between treatments for soluble Ca^{2+} contents at 0–10 and 10–30 cm layers. Only at the layer of 30–60 cm, the isolated cultivation treatments of *A. indica*, *L. leucocephala*, and *M. caesalpiniifolia*, and the association *A. indica*/*A. nummularia* and *M. caesalpiniifolia*/*A. nummularia* differed from the control (Figure 5).

For soluble Mg^{2+}, differences to control were observed for the isolated cultivation of *L. leucocephala*, *M. caesalpiniifolia*, and for association *A. indica*/*A. nummularia* and *M. caesalpiniifolia*/*A. nummularia* at 0–10 cm layer at 9 months (Figure 5). In subsurface (10–30 cm), the increase of Mg^{2+} was observed for association treatment *A. indica*/*A. nummularia* with respect to the control and other plant treatments. While in the 30–60 cm layer only isolated cultures of *L. leucocephala* and *M. caesalpiniifolia* and associated *L. leucocephala*/*A. nummularia* and *M. caesalpiniifolia*/*A. nummularia* differed in relation to the control (Figure 5). The soluble Ca^{2+} and Mg^{2+} values at 18 months after plant cultivation showed no significant difference in any of the layers evaluated (Figure 5).

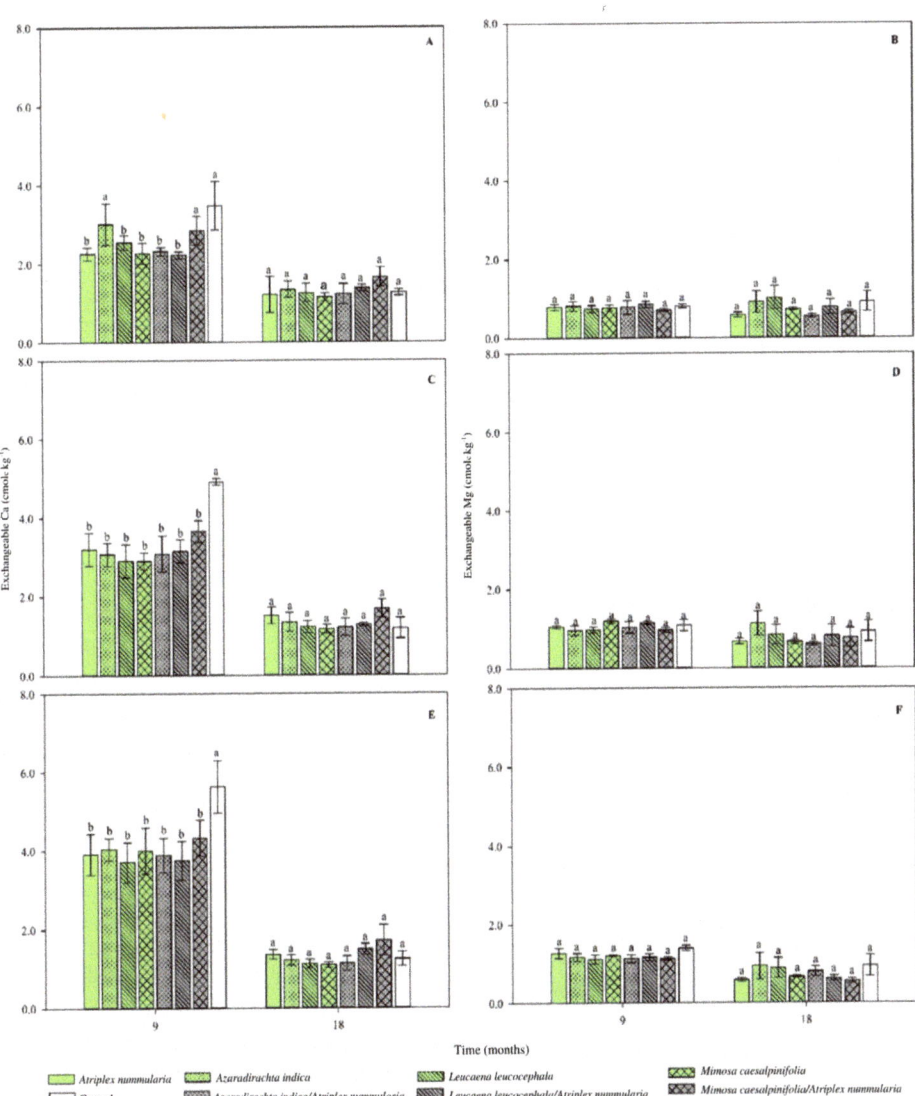

Figure 3. Exchangeable Ca^{2+} [(**A**) 0–10 cm; (**C**) 10–30 cm; (**E**) 30–60 cm] and Mg^{2+} [(**B**) 0–10 cm; (**D**) 10–30 cm; (**F**) 30–60 cm] at 9 and 18 months as result of the applied treatments. (Averages of four replications). Averages followed by the same letter have no difference in the same layer and time of sampling by the Skott-Knott test ($p \leq 0.05$).

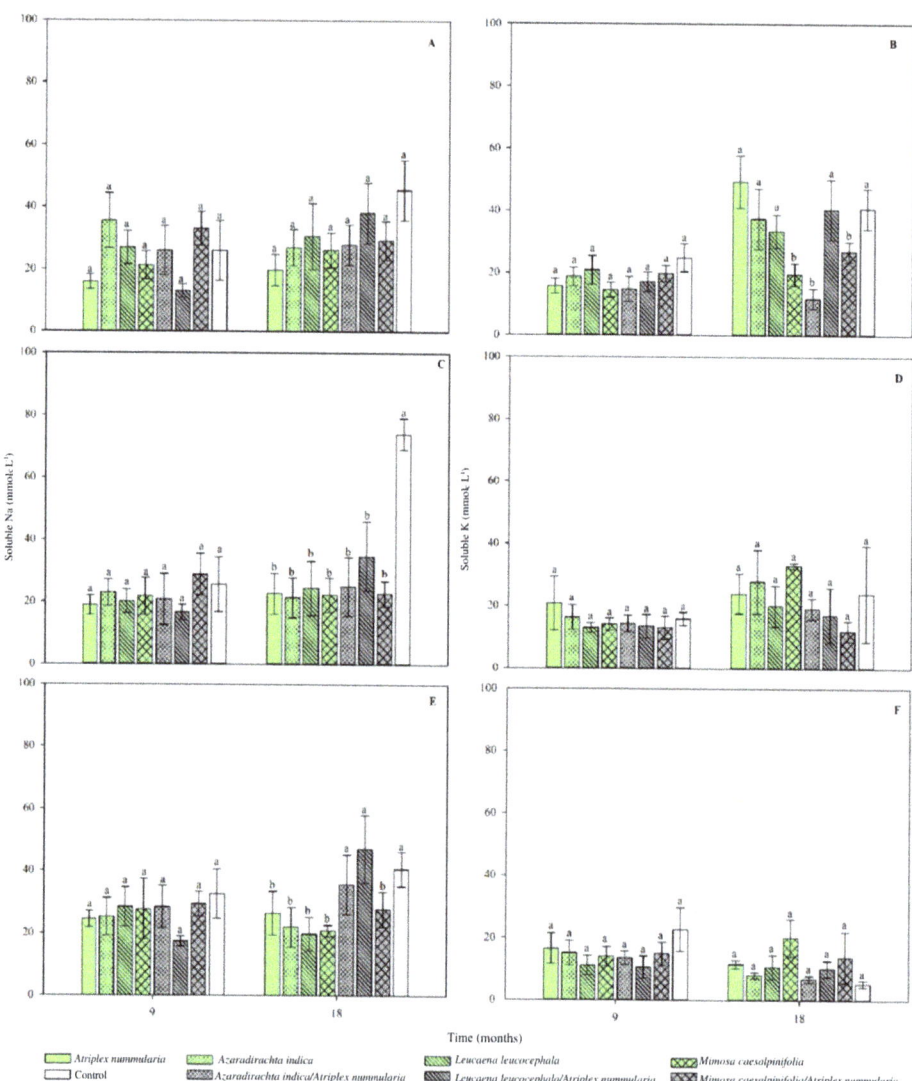

Figure 4. Soluble Na$^+$ [(**A**) 0–10 cm; (**C**) 10–30 cm; (**E**) 30–60 cm] and soluble K$^+$ [(**B**) 0–10 cm; (**D**) 10–30 cm; (**F**) 30–60 cm] at 9 and 18 months as result of the applied treatments. (Averages of four replications). Averages followed by the same letter have no difference in the same layer and time of sampling by the Skott-Knott test ($p \leq 0.05$).

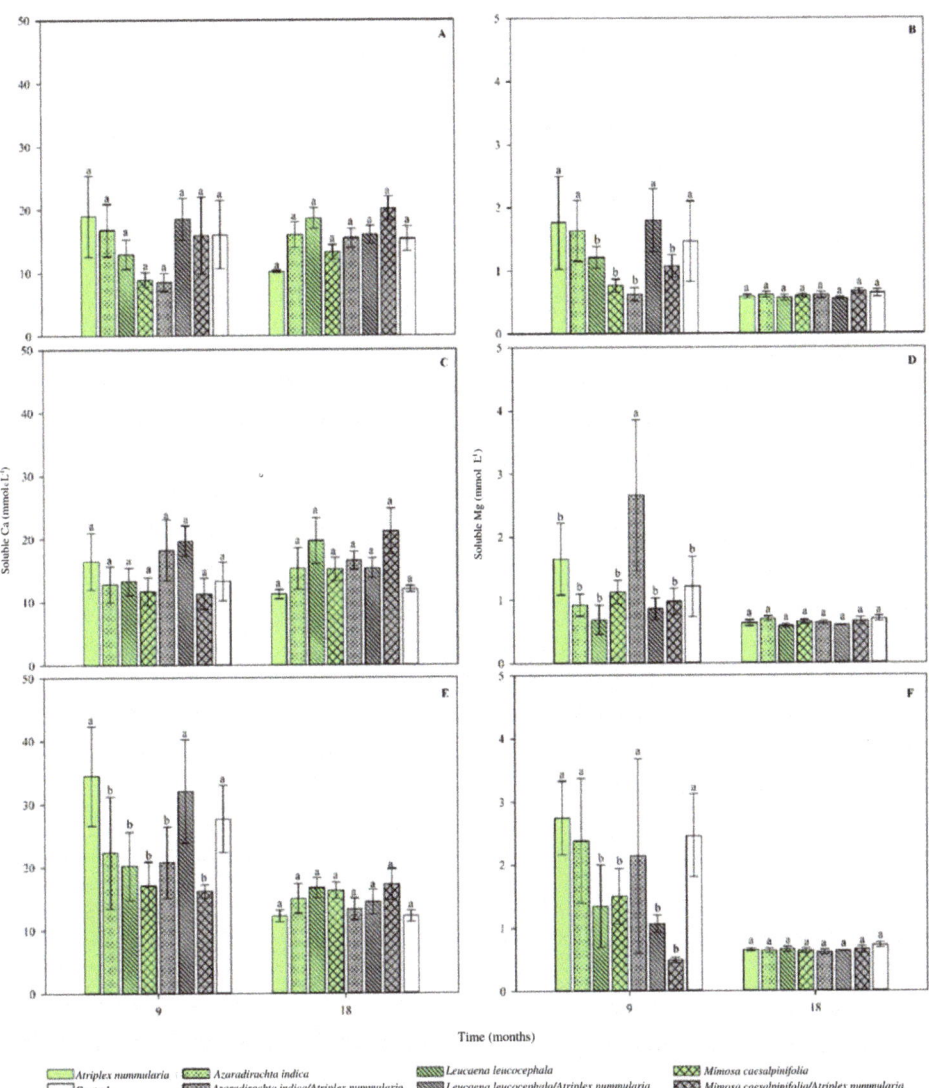

Figure 5. Soluble Ca^{2+} [(**A**) 0–10 cm; (**C**) 10–30 cm; (**E**) 30–60 cm] and soluble Mg^{2+} [(**B**) 0–10 cm; (**D**) 10–30 cm; (**F**) 30–60 cm] at 9 and 18 months as result of the applied treatments. treatments (Averages of four replications). Averages followed by the same letter have no difference in the same layer and time of sampling by the Skott-Knott test ($p \leq 0.05$).

3.3. Soil Salinity and Sodicity Variables

The orthogonal contrasts performed for pH, EC, SAR, and ESP results at 9 months after plant cultivation did not present a significant difference for any of the sodicity variables (SAR and ESP) at the layers evaluated (Table 3).

Changes were observed in EC at 9 months in the layer of 30–60 cm between the group of isolated cultivation plants and the control treatment, and the group of legumes in comparison to the control. At the same depth, pH recorded differed between the group of all plant treatments and the control, the association between plant treatments in relation to the control, and isolated cultures in relation to plant associations (Table 3).

At 18 months, the orthogonal contrasts were significant for all the arrangements confronting plants and control (all treatments × control, isolated cultures × control, plants associations × control, leguminous × control, and non-legume × control) in all evaluated layers (Table 5). In addition, there was a difference between isolated A. nummularia in relation to all treatments for the 30–60 cm layer.

Table 5. Orthogonal contrasts of the chemical attributes pH (hydrogen potential), EC (electrical conductivity), SAR (sodium adsorption ratio), and ESP (exchangeable sodium percentage) in the soil at 0–10, 10–30, and 30–60 cm depth as a function of treatments at 18 months.

Contrast	pH			EC			SAR			ESP		
	0–10 cm	10–30 cm	30–60 cm	0–10 cm	10–30 cm	30–60 cm	0–10 cm	10–30 cm	30–60 cm	0–10 cm	10–30 cm	30–60 cm
All treatments × control	−1.020	−1.319	−1.031	−2.382 *	−1.493	−6.059 **	−2.359 *	−5.306 **	−1.404	−3.635 **	−3.635 **	−2.838 *
Isolated cultures × control	−0.694	−1.072	−1.258	−2.600 *	−1.272	−6.010 **	−2.165 *	−5.201 **	−1.361	−3.451 **	−3.451 **	−2.504 *
Associations × control	−1.307	−1.466	−0.603	−1.790	−1.584	−5.330 **	−2.300 *	−4.748 **	−1.275	−3.397 **	−3.397 **	−2.898 *
Leguminous × control	−1.479	−1.900	−1.378	−2.096 *	−0.570	−5.868 **	−2.306 *	−5.152 **	−1.845	−2.791 *	−2.791 *	−2.120 *
No leguminous × control	0.211	−0.058	−0.919	−2.650 *	−1.751	−5.104 **	−1.647	−4.344 **	−0.639	−3.510 **	−3.510 **	−2.450 *
Isolated cultures × consortium	0.960	0.646	−0.929	−1.099	0.533	−0.739	0.308	−0.434	−0.064	0.083	0.083	0.716
No leguminous × leguminous	2.070	2.257 *	0.563	−0.678	−1.447	0.935	0.807	0.990	1.477	−0.881	−0.881	−0.404
Atriplex × all treatments	2.715 *	1.371	0.496	−1.854	−2.694 *	−1.965	−0.637	−0.203	0.536	−1.365	−1.365	−2.630 *

* Significant at 0.05 probability; ** Significant at 0.01 probability.

The same was observed on SAR data in relation to the same groups, differing only in the layers. Differences were significant at 0–10 and 10–30 cm layers, except for the non-legume group in relation to the control, which did not show differences at the 0–10 cm layer (Table 5).

Similarly, differences were recorded on the EC for the contrasts between the same groups, except for the depth of 10–30 cm, and for the contrast between plants associations × control group at 0–10 cm layer, which did not present a significant difference (Table 5).

4. Discussion

The present study provides new insights into reclaim of saline soils strategies. In brief, this study first reported the phytoremediation potential of the association between Atriplex plants with M. caesalpiniifolia, L. leucocephala, and A. indica, and under isolated cultivation of them in salt-affected soil.

Most of the evaluated variables were not modified after only 9 months of cultivation, however, the effect of the plants was more consistent at 18 months. This indicates that saline soil reclamation studies in the field should be conducted for a longer period of time. In field conditions, even the plants acting in the extraction of salts continue to be supplied to the soil.

The dynamics of groundwater in semi-arid environments have a direct influence on soil salinization in the edaphoclimatic conditions of these regions [23]. Although not measured, this fact may have influenced the increase in exchangeable levels of Na^+ and K^+ at 18 months, as a result of the probable exposure of the area to fluctuations in the water table depth and consequent entry of salts in conditions of insufficient drainage [24]. On the other hand, the reduction in exchangeable Na^+ and K^+ in the soil of the cultivated areas in relation to the control demonstrates the influence of the root system of these species on the drainage of the area [15,25].

In addition to the tolerance required to withstand the environmental conditions of salinity and sodicity [15], the extraction of salts from the environment is an essential characteristic for reclamation and contributes to the reduction of the levels of these ions in the soil [14]. It would be necessary to cut the plants to quantify the biomass production and its composition, and to estimate the extraction of salts. However, as our proposal is the improvement of the soil by revegetation, we chose the permanence of the plants in the area.

The balance between bivalent and monovalent cations in the soil exchange complex directly interferes with the proportions of these in the soluble form and in their absorption by plants. Under semiarid conditions, the constant evaporation of the water stored in the soil increases the concentration of the ions in solution with subsequent precipitation of the less soluble ions. Ions such as Ca^{2+} and Mg^{2+} precipitate first while Na^+ remains in solution, leading to its passage to the exchangeable phase. This would explain the increase in exchangeable Na^+ (Figure 2) and the reduction of exchangeable Ca^{2+} (Figure 3) at 18 months of plant growth.

The importance of plant utilization in the reclamation of salt-affected areas has been highlighted [26]. Although the total reclaim of the evaluated chemical attributes was not recorded, treatments with both isolated and associated plants, except for *A. indica*, were efficient in keeping Na^+ contents low in the exchange complex at all depths even with the reduction of exchangeable Ca^{2+}.

The differences found between the cultivation of the evaluated plants and the association of these with the *A. nummularia* are important for the soils degraded by salts reclamation. It has increased the need to use species adapted to the climatic conditions of the arid and semiarid regions, where the processes of salinization and sodification of soils occur [27], mainly in the context of the reclaim of salt-affected areas and the control of the degradation process. Therefore, the use of halophyte plants that remove salt from the vicinity of roots of crop plants has more potential for alleviating saline soil conditions in the future [28].

When looking at the salt resistance of *A. nummularia*, we have to consider that its phytoextraction potential is associated with the fact that it stores significant amounts of salts in its tissues [17], due to the existence of the major mechanisms, the secretion of ions by salt glands and exclusion of accumulated toxic ions via bladder hairs that is a common strategy in *Atriplex* [29].

Another metabolic response to hyperosmotic salinity is the physiological mechanism because we maintain high photosynthetic rates with a low stomatal aperture and low transpiration simultaneously, and higher water use efficiency [30]. Finally, the biochemical mechanisms, as a consequence of synthesis and high cellular accumulation of organic and inorganic solutes adjust osmotically [31].

Even with the satisfactory establishment of the species, the reductions in soluble ion contents are not necessarily due to the accumulation of these in the plant tissue. In a reclamation study conducted in a semi-arid region of Tunisia, the reductions observed by the authors in the salinity and sodicity variables were attributed not only to the accumulation of salts in the aerial part of the species [*Tecticornia indica* (Willd.) Subsp. *Indica* and *Suaeda fruticosa* Forssk.], but also to the leaching due to physical-water improvements of the soil after the establishment of the plants.

Similar behavior in relation to Na^+ and K^+ was explained by Mahmoodabadi et al. [31], which demonstrated the lower mobility of bivalent cations. A factor that may have restricted the performance of the species implanted in the area over the soluble Ca^{2+} and Mg^{2+} contents along the profile. However, the permanence of these bivalent cations in solution compared to the significant reductions found for Na^+ and K^+ is an important factor in the ion balance of this soil. Cations adsorbed to a charged colloid surface (interface) can be replaced, by mass action, by other cations with greater activity in the soil solution [32].

Thus, soluble Ca^{2+} and Mg^{2+} can displace the exchangeable Na^+, reducing the sodicity characteristics of this soil [31].

Changes in EC are mainly due to the removal of soluble salts by leaching and accumulation in plant tissues, without being sufficient to cause greater changes in the proportions of basic cations in the soil exchange complex [23,33].

In addition, with respect to salt removal through plant species as a contributing factor in controlling soil salinity in saline soil, the process involves different mechanisms, such as: (1) the increase in the partial pressure of carbon dioxide (PCO_2) in the root zone, (2) release of protons in the rhizosphere of legumes, such as *L. leucocephala* and *M. caesalpinifolia* (3) can promote soil aggregation (4) root effect: the physical action of plant roots improves the soil structure and provides channels for infiltrating water, with the leaching of salts out of the root zone and (5) absorption of salts and removal of salts by aerial plant parts [34,35].

An additional consideration is that the aerial plant part also contributes toward a decrease in soil salinity, by providing shade to the soil. Furthermore, lowers soil temperature and decreases evaporation from the soil surface compared to a non-cropped surface [35].

Thus, plants that thrive under saline conditions became an option for the remediation of salt-affected soils. Especially, the halophytes, such as *A. nummularia*, that keep toxic ions in their vacuoles, accumulate compatible solutes in their cytoplasm and activate genes for salt tolerance that confer salt resistance [36].

In summary, our results suggested that reclamation was not evident at 9 months after planting, and significant results to decrease of soil salinity (EC) were verified at 18 months. Soil reclamation studies in the field should be longer so that the results be really of environmental importance. Both the effects of salt extraction from the soil are greater and the increase in salt leaching in rain events due to the action of the plant root system.

The isolated cultivation of saltbush provided environmentally better conditions for other plants, promoting coverage and protection of the soil, maintaining its humidity, and reducing its temperature. Thus, the implantation of salt-tolerant cultures, such as *A. nummularia*, offers a more sustainable system.

These results also reflect the *A. nummularia* potential to occupy areas degraded by salts where other plants would not have growing conditions and can be considered an economically viable alternative to claim the productive capacity in semiarid soils degraded by salinity.

Therefore, in soils with sodicity problems, the practice of reclamation with these species need to be managed over a period of more than nine months for significant changes in soil solution (SAR) and the percentage of sodium on the soil exchange complex (ESP). From 18 months of treatment, in similar environments to the study, a significant alteration in ESP can be observed when cultivated with *A. nummularia*, *A. indica*, *L. leucocephala*, and *M. caesalpinifolia*, in consortium or isolated culture. These changes were observed involving depths of 60 cm in relation to the surface, but these plant roots can go beyond, along with associated microorganisms. These changes can only be achieved when the modifications in the concentrations of the soluble cations are enough to cause displacement of the Na^+ ions in the exchange complex [32,37].

Although with less expressive results, the intercropping management can be indicated for reclaim of saline soil and, in parallel, it offers an alternative source of biomass production, as well as increasing the diversity of plant species in the environment, with environmental significance. It is even possible that other species associations may promote greater increases in the effects of crops on improving the properties of soils degraded by salinity.

Results entre o beginning and end of the experiment indicated that salinity and sodicity variables levels were increased in the control treatment. These fluctuations levels are associated with problems in the physical properties of the soil, causing restrictions to water and air movement in the soil profile.

An additional consideration for this behavior deals with the fact that the dynamics of rainfall and the ground-water level. In the period between November 2013 and May 2015. In 2013, 5.0 mm of rainfall was recorded until November (first soil sampling—0 times); In 2014, 740.0 mm was registered between the first and second samplings (0 and 9 months) and 253.50 mm occurred between the second and third samplings (9 to 18 months).

It was not possible to infer a significant decrease in soil ($p \leq 0.10$) salt levels (EC) during the experimental period of 9 months as a result of plant cultivation due to high rainfall events, low evapotranspiration rates, and periodic changes in groundwater level.

In addition, due to rainfall events, high evapotranspiration rates, and periodic changes in groundwater levels, it's possible to infer a significative decrease ($p \leq 0.10$) in salinity and sodicity variables levels (EC, SAR, and ESP) during the experimental period 18 months as a result of plants cultivation, especially to *A. nummularia* individualized treatment. These plant species contribute to the improvement of soil structure by the formation of biopores, that work as alternative routes and increase the movement of water in the soil and, consequently, increase the leaching of saline ions.

Moreover, as crops include species that respond differently to salinity soil, comparative studies concerning the salt stress effects on the reclamation ability of these plants should be conducted in order to assess their potential as crops in the predicted world of climate change.

When reclamation is used, these changes in the concentrations may take longer than the conventional recovery techniques, because they depend on the ion extraction rate of the culture used and the improvements in the physical-water properties caused by the presence of the roots [15]. In this work, we demonstrated that reclamation is an alternative technique, and it could be successfully used for the removal of salts from the soil. Such a scenario would fit poorly drained soils with sodium accumulation, indicating their potential for future revegetation projects of salt-affected soils with environmental benefits.

Additionally, one of the ecological benefits of newly theses created plant cultures is the fact that they seem to be particularly suited for long-term CO_2 sequestration which counteracts the greenhouse effect.

5. Conclusions

The *Atriplex nummularia* individualized cultivation can be effective in reducing the sodicity and salinity of saline-sodic soil. However, when assessing the efficiency for the reclamation of salt-affected soils with the cultivation of *Azadirachta indica*, *Leucaena leucocephala* and *Mimosa caesalpiniifolia* individualized or even associated with *A. nummularia* contribute to improvement in the soil chemical quality due to root effect, promoting leaching of salts.

At 9 months, we were unable to detect a reduction in the electrical conductivity of the saturation extract in the soil as a result of plant species cultivation due to the variation in rainfall in the study area. However, at 18 months after planting, the results of this study showed decreases in the electrical conductivity, sodium adsorption ratio, and exchangeable sodium percentage due to cultivation.

As an important consideration is highlighted: that the degraded soil reclamation with species adapted to the semiarid would be suitable agronomic practices as well as to improve soil quality and sustainability, contributing to increased water uptake by infiltration and carbon sequestration in soils with no vegetation cover.

Author Contributions: Conceptualization, M.A.d.S. and M.B.G.S.F.; methodology, M.A.d.S.; software, M.A.d.S. and F.J.F.; validation, M.B.G.S.F.; formal analysis, M.A.d.S. and P.G.d.L.; investigation, M.A.d.S.; resources, M.B.G.S.F.; data curation, M.A.d.S. and M.B.G.S.F.; writing—original draft preparation, M.A.d.S., M.B.G.S.F., F.J.F., C.M.P.L. and H.F.d.M.; writing—review and editing, M.A.d.S. and M.B.G.S.F.; visualization, M.A.d.S.; supervision, M.B.G.S.F. and A.T.d.R.; project administration, M.A.d.S., M.B.G.S.F. and A.T.d.R.; funding acquisition, M.B.G.S.F. All authors have read and agreed to the published version of the manuscript.

Funding: CNPq (National Council for Scientific and Technological Development), CAPES (Coordination for the Improvement of Higher Education Personnel) and FACEPE (Foundation of Science and Technology Support of Pernambuco).

Acknowledgments: We thank the whole Group of Soil Salinity from the Federal Rural University of Pernambuco (Brazil) for their dedication to research. The authors are grateful for the financial support of the institutions: CNPq, CAPES and FACEPE. We appreciate the reviews by the editor and anonymous reviewers that improved the manuscript quality.

Conflicts of Interest: The authors declare no conflict of interest.

References

1. Cuevas, J.; Daliakopoulos, I.N.; del Moral, F.; Hueso, J.J.; Tsanis, I.K. A Review of Soil-Improving Cropping Systems for Soil Salinization. *Agron* **2019**, *9*, 295. [CrossRef]
2. Leal, L.Y.C.; de Souza, E.R.; Santos Júnior, J.A.; Santos, M.A. Comparison of soil and hydroponic cultivation systems for spinach irrigated with brackish water. *Sci. Hortic.* **2020**, *274*, 109616. [CrossRef]
3. Shahid, S.A.; Zaman, M.; Heng, L. Soil Salinity: Historical Perspectives and a World Overview of the Problem. In *Guideline for Salinity Assessment, Mitigation and Adaptation Using Nuclear and Related Techniques*; Springer: Cham, Switzerland, 2018. [CrossRef]
4. Bouaziz, M.; Hihi, S.; Chtourou, M.; Osunmadewa, B. Soil Salinity Detection in Semi-Arid Region Using Spectral Unmixing, Remote Sensing and Ground Truth Measurements. *J. Geogr. Inf. Syst.* **2020**, *12*, 372–386. [CrossRef]
5. Zhang, J.B.; Yang, J.S.; Yao, R.J.; Yu, S.P.; Li, F.R.; Hou, X.J. The effects of farmyard manure and mulch on soil physical properties in a reclaimed coastal tidal flat salt-affected soil. *J. Integr. Agric.* **2014**, *13*, 1782–1790. [CrossRef]
6. Besser, H.; Mokadem, N.; Redhouania, B.; Rhimi, N.; Khlifi, F.; Ayadi, Y.; Omar, Z.; Bouajila, A.; Hamed, Y. GIS-based evaluation of groundwater quality and estimation of soil salinization and land degradation risks in an arid Mediterranean site (SW Tunisia). *Arab. J. Geosci.* **2017**, *10*, 350. [CrossRef]
7. Lal, R. Restoring soil quality to mitigate soil degradation. *Sustainability* **2015**, *7*, 5875–5895. [CrossRef]
8. Kibria, M.; Hoque, M. A Review on Plant Responses to Soil Salinity and Amelioration Strategies. *Open J. Soil Sci.* **2019**, *9*, 219–231. [CrossRef]
9. Xu, Z.; Shao, T.; Lv, Z.; Yue, Y.; Liu, A.; Long, X.; Zhou, Z.; Gao, X.; Rengel, Z. The mechanisms of improving coastal saline soils by planting rice. *Sci. Total Environ.* **2020**, *703*, 135529. [CrossRef]
10. Li, J.; Pu, L.; Zhua, M.; Zhang, J.; Li, P.; Dai, X.; Xua, Y.; Liu, L. Evolution of soil properties following reclamation in coastal areas: A review. *Geoderma* **2014**, *226–227*, 130–139. [CrossRef]
11. Li, X.; Zhang, X.; Wang, X.; Yang, X.; Cui, Z. Bioaugmentation-assisted phytoremediation of lead and salinity co-contaminated soil by *Suaeda salsa* and *Trichoderma asperellum*. *Chemosphere* **2019**, *224*, 716–725. [CrossRef] [PubMed]
12. Lam, E.J.; Cánovas, M.; Gálvez, M.E.; Montofré, Í.L.; Keith, B.F.; Faz, Á. Evaluation of the phytoremediation potential of native plants growing on a copper mine tailing in northern Chile. *J. Geochem. Explor.* **2017**, *182*, 210–217. [CrossRef]
13. Moura, E.S.R.; Cosme, C.R.; Leite, T.S.; Dias, N.D.; Fernandes, C.S.; Sousa Neto, O.N.; Sousa Junior, F.S.; Rebouças, T.C. Phytoextraction of salts by *Atriplex nummularia* Lindl. irrigated with reject brine under varying water availability. *Int. J. Phytoremediat.* **2019**, *21*, 892–898. [CrossRef] [PubMed]
14. Souza, E.R.; Freire, M.B.G.S.; Melo, D.V.M.; Montenegro, A.A.A. Management of *Atriplex nummularia* Lindl. in a salt affected soil in a semi-arid region of Brazil. *Int. J. Phytoremediat.* **2014**, *16*, 73–85. [CrossRef] [PubMed]
15. Jesus, J.M.; Danko, A.S.; Fiúza, A.; Borges, M.T. Phytoremediation of salt-affected soils: A review of processes, applicability, and the impact of climate change. *Environ. Sci. Pollut. Res.* **2015**, *22*, 6511–6525. [CrossRef]
16. Nouri, H.; Chavoshi Borujeni, S.; Nirola, R.; Hassanli, A.; Beecham, S.; Alaghmand, S.; Saint, C.; Mulcahy, D. Application of green remediation on soil salinity treatment: A review on halophytoremediation. *Process Saf. Environ. Prot.* **2017**, *107*, 94–107. [CrossRef]
17. Leite, M.C.B.S.; Freire, M.B.G.S.; Queiroz, J.V.J.; Maia, L.C.; Duda, G.P.; Medeiros, E.V. Mycorrhizal *Atriplex nummularia* promote revegetation and shifts in microbial properties in saline Brazilian soil. *Appl. Soil Ecol.* **2020**, *153*, 103574. [CrossRef]
18. IUSS Working Group WRB. *World Reference Base for Soil Resources 2014, Update 2015. International Soil Classification System for Naming Soil and Creating Legends for Soil Maps*. No. 106; World Soil Resources Reports; FAO: Rome, Italy, 2015; p. 192.
19. Alvares, C.A.; Stape, J.S.; Sentelhas, P.C.; de Moraes Gonçalves, J.L.; Sparovek, G. Köppen's climate classification map for Brazil. *Meteorol. Z.* **2014**, *22*, 711–728. [CrossRef]
20. Thomas, G.W. Exchangeable cations. In *Methods of Soil Analysis. Part-2 Chemical Methods*, 1st ed.; Page, A.L., Ed.; American Society of Agronomy: Madison, WI, USA, 1982; pp. 159–165.
21. USSL Staff–United States Salinity Laboratory. *Diagnosis and Improvement of Saline and Alkali Soils*; Richards, L.A., Ed.; US Government Printing Office: Washington, DC, USA, 1954.
22. Ruiz, H.A. Incremento da exatidão da análise granulométrica do solo por meio da coleta da suspensão (silte+argila). *Rev. Bras. Cienc. Solo* **2005**, *29*, 297–300. [CrossRef]
23. Bouksila, F.; Bahri, A.; Berndtsson, R.; Persson, M.; Rozema, J.; Zee, S.E.A. Assessment of soil salinization risks under irrigation with brackish water in semiarid Tunisia. *Environ. Exp. Bot.* **2013**, *92*, 176–185. [CrossRef]
24. Valipour, M. Drainage, waterlogging, and salinity. *Arch. Agron. Soil Sci.* **2014**, *1*, 1–16. [CrossRef]
25. Baquero, J.E.; Ralisch, R.; Medina, C.C.; Tavares Filho, J.; Guimarães, M.F. Soil physical properties and sugarcane root growth in a Red Oxisol. *Rev. Bras. Cienc. Solo* **2012**, *36*, 63–70. [CrossRef]
26. Gairola, S.; Bhatt, A.; El-Keblawy, A. A perspective on potential use of halophytes for reclamation of salt-affected lands. *Wulfenia* **2015**, *22*, 88–97.
27. Gharaibeh, M.A.; Rusan, M.J.; Eltaif, N.I.; Shunnar, O.F. Reclamation of highly calcareous saline-sodic soil using low quality water and phosphogypsum. *Appl. Water Sci.* **2014**, *4*, 223–230. [CrossRef]
28. Karakas, S.; Çullu, M.A.; Dikilitas, M. Comparison of two halophyte species (*Salsola soda* and *Portulaca oleracea*) for salt removal potential under different soil salinity conditions. *Turk J. Agric. For.* **2017**, *41*, 183–190. [CrossRef]

29. Paulino, M.K.S.S.; Souza, E.R.; Lins, C.M.T.; Dourado, P.R.M.; Leal, L.Y.C.; Monteiro, D.M.; Rego Júnior, F.E.A.; Silva, C.U.C. Influence of vesicular trichomes of *Atriplex nummularia* on photosynthesis, osmotic adjustment, cell wall elasticity and enzymatic activity. *Plant Physiol. Biochem.* **2020**, *155*, 177–186. [CrossRef]
30. Geissler, N.; Hussin, S.; El-Far, M.M.M.; Koyro, H.W. Elevated atmospheric CO_2 concentration leads to different salt resistance mechanisms in a C3 (*Chenopodium quinoa*) and a C4 (*Atriplex nummularia*) halophyte. *Environ. Exp. Bot.* **2015**, *118*, 67–77. [CrossRef]
31. Mahmoodabadi, M.; Yazdanpanah, N.; Sinobas, L.R.; Pazira, E.; Neshat, A. Reclamation of calcareous saline sodic soil with different amendments (I): Redistribution of soluble cations within the soil profile. *Agric. Water Manag.* **2013**, *120*, 30–38. [CrossRef]
32. Kharel, T.P.; Clay, D.E.; Reese, C.; DeSutter, T.; Malo, D.; Clay, S. Do Precision Chemical Amendment Applications Impact Sodium Movement in Dryland Semiarid Saline Sodic Soils? *Agron. J.* **2018**, *110*, 1103–1110. [CrossRef]
33. Hussin, S.; Geissler, N.; Koyro, H.W. Effect of NaCl salinity on *Atriplex nummularia* (L.) with special emphasis on carbon and nitrogen metabolism. *Acta Physiol. Plant* **2013**, *35*, 1025–1038. [CrossRef]
34. Qadir, M.; Tubeileh, A.; Akhtar, J.; Larbi, A.; Minhas, P.S.; Khan, M.A. Productivity enhancement of salt-affected environments through crop diversification. *Land Degrad. Dev.* **2008**, *19*, 429–453. [CrossRef]
35. Qadir, M.; Ghafoor, A.; Murtaza, G. Amelioration strategies for saline soils: A review. *Land Degrad. Dev.* **2000**, *11*, 501–521. [CrossRef]
36. Nikalje, G.C.; Srivastava, A.K.; Pandey, G.K.; Suprasanna, P. Halophytes in biosaline agriculture: Mechanism, utilization, and value addition. *Land Degrad. Dev.* **2018**, *29*, 1081–1095. [CrossRef]
37. Han, L.; Liu, H.; Yu, S.; Wang, W.; Liu, J. Potential application of oat for phytoremediation of salt ions in coastal saline-alkali soil. *Ecol. Eng.* **2013**, *61*, 274–281. [CrossRef]

Communication

Stomatal Regulation and Osmotic Adjustment in Sorghum in Response to Salinity

Pablo Rugero Magalhães Dourado [1], Edivan Rodrigues de Souza [1,*], Monaliza Alves dos Santos [1], Cintia Maria Teixeira Lins [1], Danilo Rodrigues Monteiro [1], Martha Katharinne Silva Souza Paulino [1] and Bruce Schaffer [2]

[1] Agronomy Department, Av. Dom Manuel de Medeiros, Dois Irmãos, Recife CEP 52171-900, Brazil; rugerodm@hotmail.com (P.R.M.D.); alves.monaliza@yahoo.com.br (M.A.d.S.); cintia_lins2@hotmail.com (C.M.T.L.); danilor.monteiro1@gmail.com (D.R.M.); marthakatharinne@gmail.com (M.K.S.S.P.)

[2] Tropical Research and Education Center, Department of Horticultural Sciences, Institute of Food and Agricultural Sciences, University of Florida, 18905 S.W. 280 Street, Homestead, FL 33031, USA; bas56@ufl.edu

* Correspondence: edivan.rodrigues@ufrpe.br

Abstract: *Sorghum bicolor* (L.) Moench, one of the most important dryland cereal crops, is moderately tolerant of soil salinity, a rapidly increasing agricultural problem due to inappropriate irrigation management and salt water intrusion into crop lands as a result of climate change. The mechanisms for sorghum's tolerance of high soil salinity have not been elucidated. This study tested whether sorghum plants adapt to salinity stress via stomatal regulation or osmotic adjustment. Sorghum plants were treated with one of seven concentrations of NaCl (0, 20, 40, 60, 80, or 100 mM). Leaf gas exchange (net CO_2 assimilation (A), transpiration (Tr); stomatal conductance of water vapor (gs), intrinsic water use efficiency (WUE)), and water (Ψw), osmotic (Ψo), and turgor Ψt potentials were evaluated at 40 days after the imposition of salinity treatments. Plants exhibited decreased A, gs, and Tr with increasing salinity, whereas WUE was not affected by NaCl treatment. Additionally, plants exhibited osmotic adjustment to increasing salinity. Thus, sorghum appears to adapt to high soil salinity via both osmotic adjustment and stomatal regulation.

Keywords: stomatal conductance; transpiration; net CO_2 assimilation; water and osmotic potentials; salt tolerance

1. Introduction

Sorghum bicolor (L.) Moench, one of the most important dryland cereal crops [1], is used for food, animal feed, and fuel. In addition to its resistance to water stress [2–4], this species with a C_4 photosynthetic pathway, is moderately tolerant to saline soil conditions, and therefore has the potential for cultivation in areas prone to salt water intrusion or high salinity of the irrigation water [5,6].

Soil salinity negatively affects the productivity of agricultural crops, hindering plant development through osmotic and ionic effects [7–9]. The adverse effects caused by soil salinity range from metabolic changes, ionic toxicity, and osmotic stress to biochemical and physiological disturbances [10]. Osmotic stress, as a result of a plant's exposure to salinity, has an immediate negative impact on water and nutrient absorption due to stomatal closure, which not only limits transpiration, but also inhibits photosynthesis [11–13]. High soil salinity also causes reductions in the leaf water potential (Ψw), which further reduces osmotic (Ψo) and turgor (Ψt) potentials, hindering many physiological processes, and causing the accumulation of toxic ions and an increase in the amount of reactive oxygen species (ROS) in exposed plants [13,14].

Salt tolerance in sorghum, as in other crops, is not due to one trait but involves several traits including morphological, physiological, biochemical, and molecular markers [15].

These include maintenance of ionic homeostasis, transport and ion uptake, osmotic adjustment, and production of antioxidant enzymes [16]. Among these coping mechanisms, one of the most common is osmotic adjustment, which is characterized by the synthesis of compatible osmolytes that stabilize the structure of cells and proteins, maintaining the osmotic potential of the cell under osmotic stress [17,18]. Some of the salinity tolerance mechanisms reported in *Sorghum bicolor*, include proline accumulation, protection of photosynthetic enzymes and antioxidants [19–21], increased root hydraulic conductance [11], retention of plant water status, maintenance of the photosynthetic rate; increased concentrations of phenolic compounds [10] and turning on of genes associated with the detection and signaling and transport of Na+ in salt-specific QTL [6].

Stomatal conductance is often negatively impacted by soil salinity levels [22,23]. The low soil–water potential imposed by salinity can cause a marked decline in stomatal conductance (gs); the physiological rationale behind this reduction is the plant's attempt to minimize water loss under conditions of reduced water availability ("physiological drought") imposed by salinity. This reduction in gs often results in a reduction in net CO_2 assimilation, and therefore a reduction in plant growth [24]. To better understand the adaptive response of sorghum to high soil salinity, it is important to understanding the relative contribution of stomata and the relative cost to CO_2 assimilation and growth by determining stomatal conductance and net CO_2 assimilation [25].

Drought and salinity are two major abiotic stresses that severely limit agricultural production worldwide [26]. Plant response to salinity follows a biphasic model, wherein an early phase shows a similarity to drought (osmotic stress), and in the long term induces ion toxicity [27]. In response to drought stress, plants are classified as either isohydric, whereby plants reduce stress by closing their stomata, or anisohydric, whereby plants osmotically adjust to stress. Sorghum is classified as anisohydric because it adapts to drought stress by osmotic adjustment [28]. The objectives of this study were to determine if sorghum adapts to salinity stress in a similar manner as it does to drought stress via osmotic adjustment, or is stomatal regulation involved. Our hypothesis was that moderate tolerance of sorghum to soil salinity is solely due to osmotic adjustment. To test this hypothesis, we exposed sorghum plants to increasing soil salinity concentrations and we measured leaf gas exchange and osmotic adjustment at different salinity levels.

2. Materials and Methods
2.1. Experimental Design and Treatments

The experiment was conducted for 55 days in a greenhouse at the Federal Rural University of Pernambuco, Recife, Brazil. During the experiment, the average temperature and relative humidity in the greenhouse were 28.59 °C and 70%, respectively. Sorghum seeds (cv. IPA 2502) were sown in 10-L cylindrical plastic pots filled with Fluvic Neosol (Fluvisol) non-saline soil obtained from Pesqueira, Pernambuco, Brazil (8°34′11″ lat. and 37°48′54″ long). Initial soil chemical characteristics are shown in Table 1. Treatments consisted of irrigating plants with different salinity levels by adding differing concentrations of NaCl to the irrigation water beginning 15 after planting. Treatments were: 0, 10, 20, 40, 60, 80, or 100 mmol L^{-1} of NaCl. The experiment was arranged as a randomized complete block design with seven treatments (salinity levels) and five single-plant replicates per treatment.

The bulk density, particle density, soil total porosity, sand, silt, and clay were 1.37 mg m^{-3}, 2.63 mg m^{-3}, 47.91 %, 433 g kg^{-1}, 466 g kg^{-1}, and 101 g kg^{-1}, respectively. The soil was maintained at 65% field capacity, at a moisture content of 0.19 g g^{-1}, equivalent to a matric potential of −0.01 Mpa (field capacity). Water lost by evapotranspiration was measured daily by weighing each pot in late afternoon. Each plant was then irrigated to bring each pot to 65% field capacity.

Table 1. Chemical characteristics of the of the Fluvic Neosol (Fluvisol) soil used in this study.

Exchangeable Complex	Mean Value *	Soil Solution	Mean Value *
pH (1:2.5)	6.75	ECse (dS m^{-1})	3.36
Ca^{2+} (cmol$_c$ kg^{-1})	4.35	Ca^{2+} (mmol L^{-1})	9.12
Mg^{2+} (cmol$_c$ kg^{-1})	3.14	Mg^{2+} (mmol L^{-1})	8.63
Na$^+$ (cmol$_c$ kg^{-1})	1.65	Na$^+$ (mmol L^{-1})	13.51
K$^+$ (cmol$_c$ kg^{-1})	1.20	K$^+$ (mmol L^{-1})	2.13
H$^+$ (cmol$_c$ kg^{-1})	1.54	Cl$^-$ (mmol L^{-1})	25.47
ESP (%)	15.96	SAR [(mmoles L^{-1})$^{0.5}$]	4.54

ESP: exchangeable sodium Percentage; ECse: electrical conductivity of the saturation paste extract; SAR: sodium adsorption relation Data are expressed as means. * n = 10 samples.

2.2. Osmotic Potential, Water Potential, Turgor Potential, and Osmotic Adjustment

Fifty-five days after sowing (DAS) or forty days after the imposition of salinity treatments, five leaflets were collected from fully expanded leaves in the middle third of the canopy of each plant. Leaf water potential (Ψw) was determined in each leaf sample with a Scholander pressure chamber (Model 1515D; PMS Instrument Company, Albany, OR, USA). The osmotic potential (Ψo) in the leaf was quantified after freezing the same leaf sample used for Ψw determination, then thawing it and extracting the cell sap by macerating the leaf and filtering the extract through fine nylon mesh with the aid of a syringe. A drop of cell sap was placed on a filter paper disc and Ψo was measured with a vapor pressure osmometer (Vapro model 5600, Wescor, Inc., Logan, UT, USA). Osmometer readings (mmol kg^{-1}) were converted to -MPa and Ψo was calculated using the Van't Hoff equation [29]:

$$\Psi o = -RTC \qquad (1)$$

where C is the solute concentration; R is the gas constant; and T is the absolute temperature.

The turgor potential (Ψt) was calculated as the difference between Ψo and Ψw. The osmotic adjustment ability was defined as the net increase in the solute concentration when the leaf was fully turgid in plants treated with NaCl compared to plants in the control treatment [30] and calculated by the equation:

$$OA = \Psi oc^{100} - \Psi os^{100} \qquad (2)$$

where OA is the total osmotic adjustment, Ψoc^{100} is the osmotic potential of the control plants at full turgor and Ψos^{100} is the osmotic potential of the stressed plants at full turgor.

2.3. Leaf Gas Exchange

Leaf gas exchange (net CO$_2$ assimilation (A), transpiration (Tr), and stomatal conductance of water vapor (gs)) was measured 40 days after the imposition of salinity treatments, between 09:00 and 14:00 h in the first fully expanded leaf below the apex of the canopy. Leaf gas exchange was measured with a portable gas exchange system (model LI-6400XT, LI-COR Biosciences, Lincoln, NE, USA). In the leaf cuvette, the light intensity was maintained at 1800 mmol mol^{-1}, the ambient CO$_2$ concentration at 400 μmol mol^{-1}, and the air temperature at 25 °C. Intrinsic water use efficiency (WUE) was calculated as A/Tr.

2.4. Leaf Fresh Weight, Leaf Dry Weight, and Plant Height

At 40 days after the imposition of salinity treatments (55 days after sowing) plant height was measured and the leaves were collected for fresh and dry weight determinations. Leaves were oven dried at 65 °C prior to dry weight determination.

2.5. Statistical Analyses

The data were analyzed by linear regression using R Statistical Software [31].

3. Results

3.1. Xylem Osmotic, Water, and Turgor Potentials, and Osmotic Adjustment Ability

There was a strong inverse linear relationship between Ψw ($R^2 = 0.99$) or Ψo ($R^2 = 0.99$) and NaCl concentration (Figure 1). The Ψw decreased from −0.10 MPa in the control treatment to −0.90 MPa in the 100 mM treatment, and the Ψo decreased from −0.80 MPa in the control treatment to −1.5 MPa in the 100 mM NaCl treatment (Figure 1). Although there was also a significant inverse linear relationship between Ψt and NaCl concentration ($R^2 = 0.84$), the decrease was more gradual than for Ψw or Ψo as indicated by a lower slope of the regression line (Figure 1) for Ψt versus NaCl concentration compared to the Ψw or Ψo regression lines. After 55 days, electrical conductivity of saturated paste extracts from each treatment was 2.9, 5.4, 9, 16.4, 20.7, 24.6, and 33.8 dS m^{-1} for the 0, 10, 20, 40, 60, 80, and 100 mM of NaCl treatments, respectively.

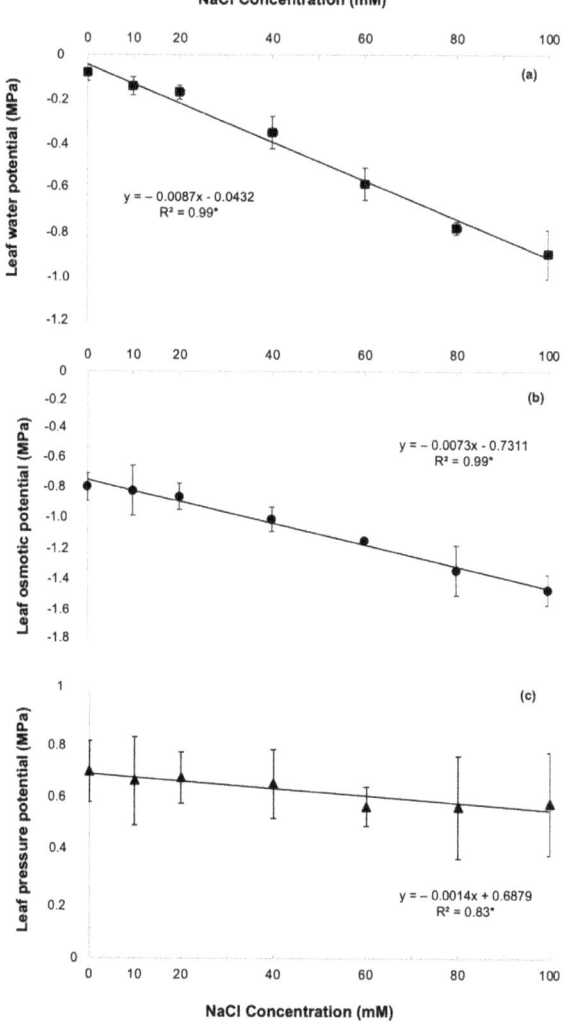

Figure 1. (a) Water, (b) osmotic, and (c) turgor potentials in sorghum leaves, 40 days after NaCl treatments were imposed. Symbols represent the means of each treatment and error bars indicate ± 1 std. dev. * ($p < 0.05$).

There was a strong positive linear relationship ($R^2 = 0.97$) between NaCl concentration and OA (Figure 2). The osmotic adjustment increased from 0.23 in the control treatment to 0.7 in the 100 mM NaCl treatment.

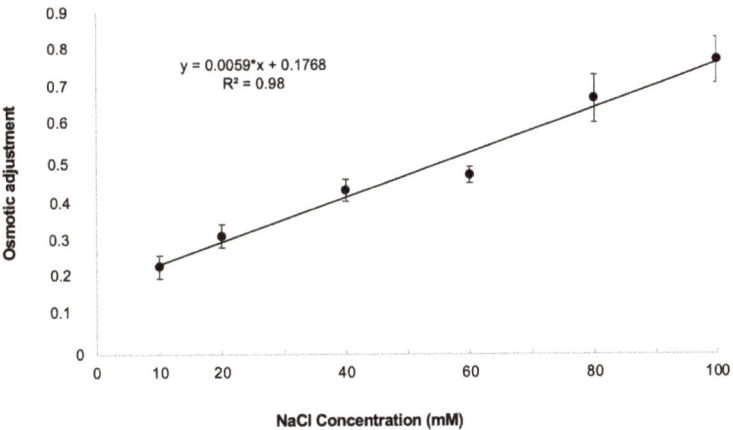

Figure 2. Osmotic adjustment in sorghum plants 40 days after NaCl treatments were imposed. Symbols represent means of each treatment. Symbols represent the means of each treatment and error bars indicate ± 1 std. dev. * ($p < 0.05$).

3.2. Leaf Gas Exchange

There was a strong inverse linearly relationship ($R^2 = 0.97$) between NaCl concentration and A (Figure 3). Net CO_2 assimilation decreased by 0.204 per 1 mM of increase in NaCl concentration.

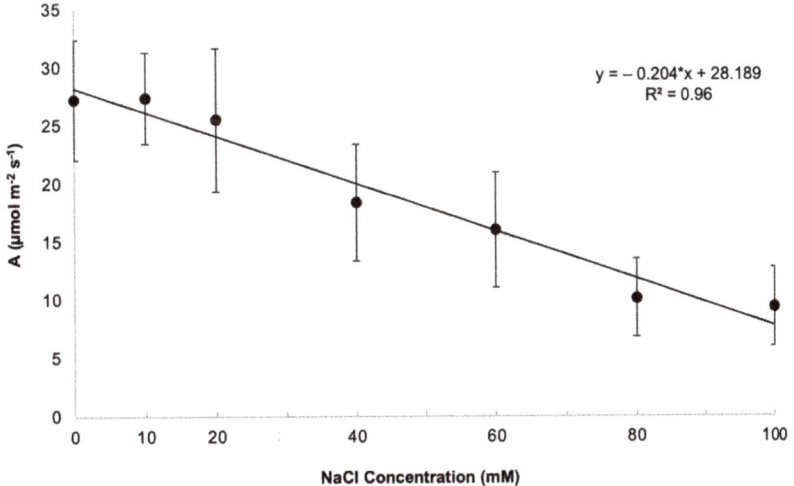

Figure 3. Net CO_2 assimilation (A) of sorghum plants 40 days after NaCL treatments were imposed. Symbols represent means of each treatment. Symbols represent the means of each treatment and error bars indicate ± 1 std. dev. * ($p < 0.05$).

Similar to A, there was a significant linear decrease in gs ($R^2 = 0.95$) and Tr ($R^2 = 0.97$) as NaCl concentration increased, whereas WUE was not affect by NaCl concentration and was similar for all treatments (Figure 4).

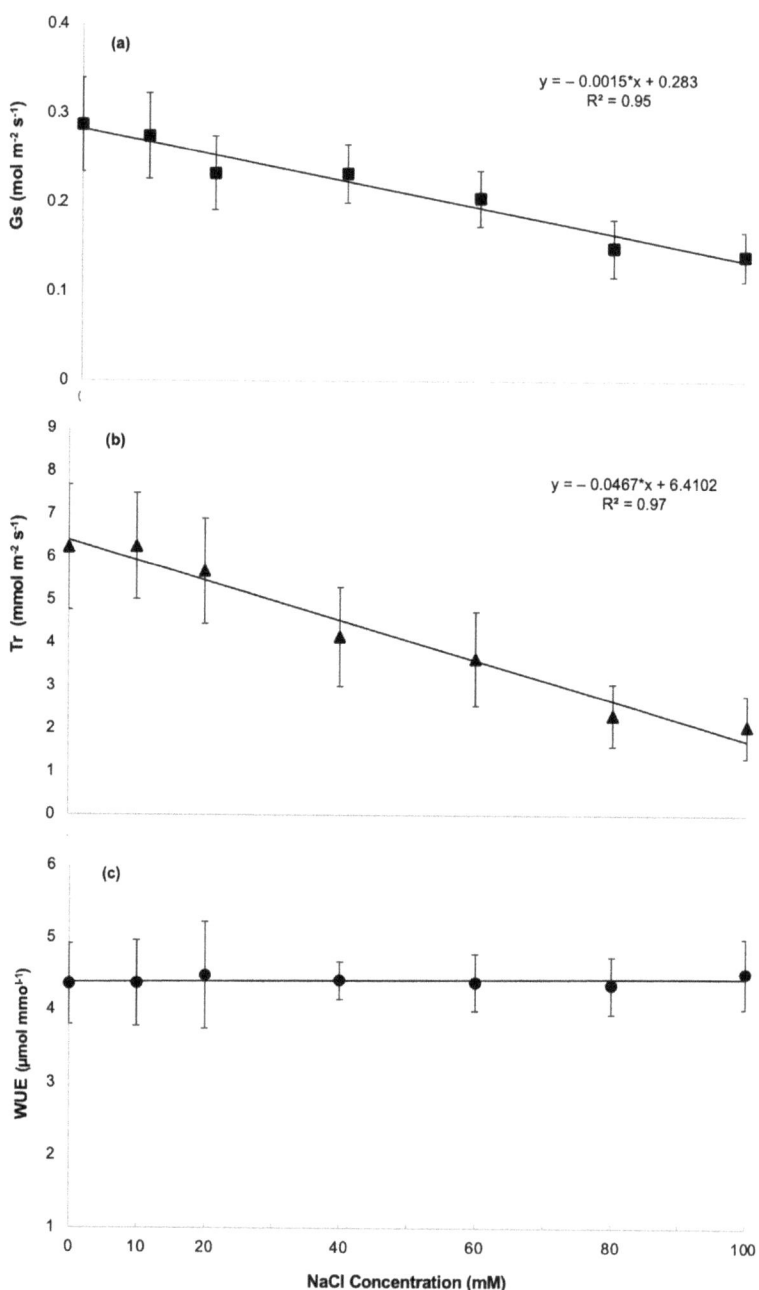

Figure 4. (a) Stomatal conductance (gs), (b) transpiration, and (c) and intrinsic water use efficiency (WUE) (D) of sorghum seedlings, 40 days after NaCl treatments were imposed. Symbols represent means of each treatment. Symbols represent the means of each treatment and error bars indicate ± 1 std. dev. * ($p < 0.05$).

There was a strong linear correlation between A ($R^2 = 0.88$) or Tr ($R^2 = 0.89$) and gs (Figure 5). For both variables, plants in the control and lower NaCl treatments were grouped at the top of the regression line and plants in the highest NaCl treatments grouped at the bottom of the regression line (Figure 5), indicating that A and Tr decreased as a result of decreased gs in response to increasing soil salinity.

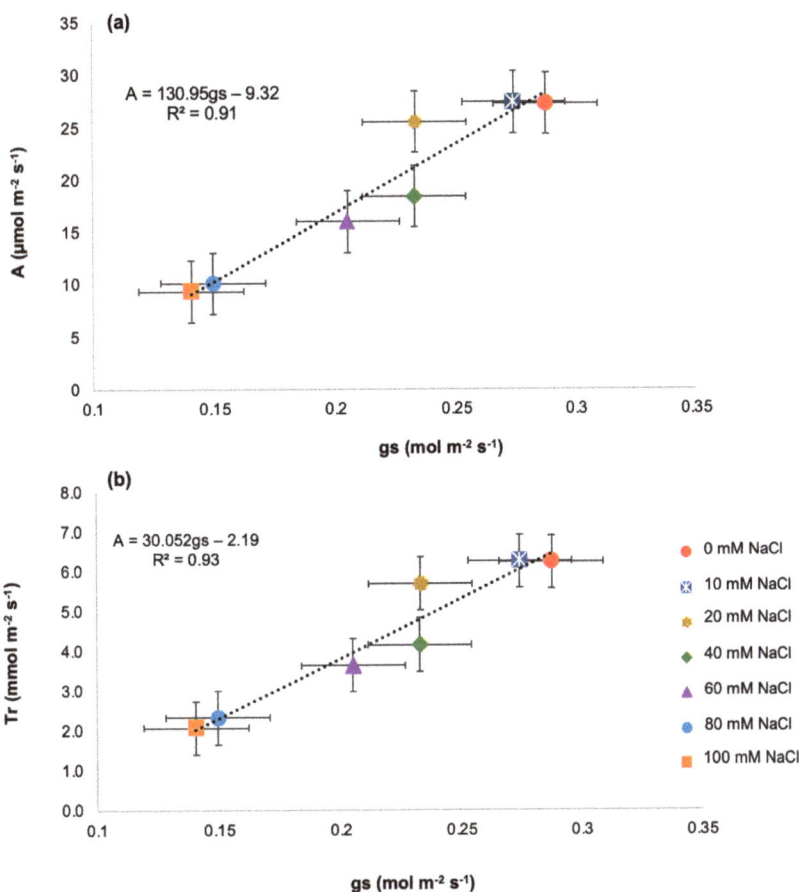

Figure 5. (a) Relationship between on a net CO_2 assimilation (A) and stomatal conductance (gs), and (b) transpiration (Tr) and (gs) in sorghum plants in different NaCl treatments, 40 days after NaCl treatments were imposed. Symbols represent the means of each treatment and error bars indicate ± 1 std. dev.

3.3. Leaf Fresh Weight, Leaf Dry Weight, and Plant Height

There was a linear decrease in plant height, leaf fresh weight, and leaf dry weight as salinity increased (Figure 6). For the highest salinity treatment (100 mM of NaCl) the reductions in plant height, leaf fresh weight, and leaf dry weight were 27% (154 cm to 112 cm), 48% (99 to 51 g plant^{-1}), and 33% (18 to 12 g plant^{-1}), respectively, compared to the control treatment (0 mM of NaCl).

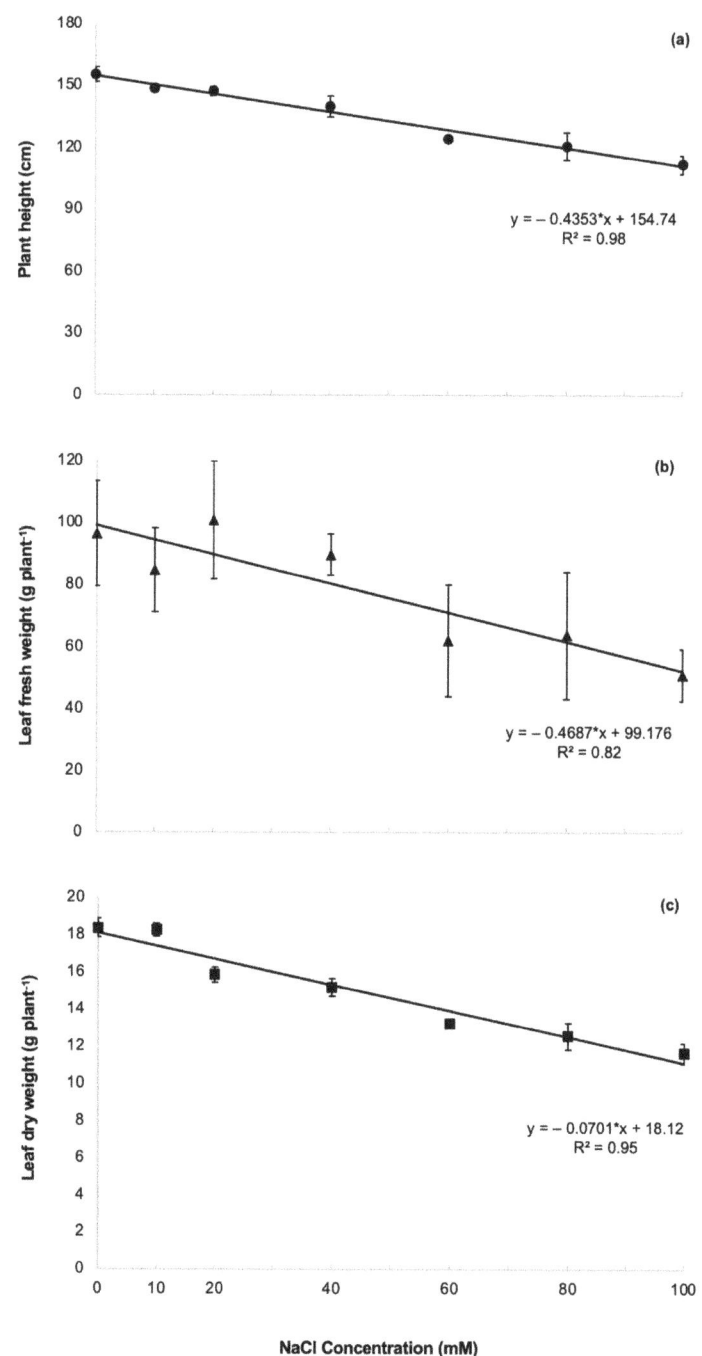

Figure 6. (a) Plant height, (b) leaf fresh weight, and (c) leaf dry weight in sorghum plants in different NaCl treatments, 40 days after NaCl treatments were imposed. Symbols represent the means of each treatment and error bars indicate ± 1 std. dev. * ($p < 0.05$).

4. Discussion

The observation that there was less of a decrease in Ψt with increasing NaCl concentration compared Ψo or Ψw suggests that there is the capacity for osmotic adjustment in sorghum. This was confirmed by OA measurements, which indicated that the values of Ψw, Ψo, and Ψt could be used to assess osmotic adjustment in the absence of direct determinations of OA. Monteiro et al. [32] evaluated the same cultivar of sorghum evaluated in the present study and found Ψw values ranged from −0.119 MPa (0 dS m^{-1}) to −0.875 MPa (7.5 dS m^{-1} – 75 mM of NaCl), which were similar to values observed in the present study. In saline soil conditions, many plants osmotically adjust by accumulating solutes which function to regulate Ψo or Ψw, allowing plants to maintain water uptake and/or Ψt [33], thereby decreasing stress. Inorganic solutes, such as potassium, magnesium, chloride, and nitrate have all been shown to contribute to as much as 52% of osmotic adjustment in sorghum plants, while organic solutes contribute to approximately 30% of the osmotic adjustment [34]. In a study of different varieties of sorghum, Bafeel [35] suggested that sorghum plants survive in saline conditions due to the osmotic adjustment involving accumulation of inorganic salts in the vacuole and accumulation of organic solutes in the cytoplasm. Negrão, Schmöckel, and Tester [8] observed compartmentalization of toxic ions into specific tissues, cells, and subcellular organelles as one the key strategies of plant adaptation to salt stress. A similar situation may occur in sorghum plants under high salinity conditions.

Lacerda et al. [36] tested the quantitative and qualitative aspects of leaf and root osmotic adjustment in two genotypes of sorghum cultivated in NaCl concentrations of 0, 50 and 100 mmol L^{-1}. Our results from leaf osmotic potential in plants treatment with 0 and 100 mmol L^{-1} of NaCl (−0.77 and −1.47 MPa, respectively) were similar to the values found in the previous study for the salt-tolerant genotype (−0.752 and −1.204 MPa for 0 and 100 mmol L^{-1} treatments, respectively). It is important to note that the genotype (IPA 2502) we tested is recommended for semiarid regions affected by abiotic stress such as salinity and drought in northeastern Brazilian. A relevant discussion about increasing osmotic adjustment is related to the balance of Na and Cl versus compatible solutes. According to Lacerda et al. [36], the higher decrease in Ψs in the salt sensitive genotype was due to a higher Na$^+$ and Cl$^-$ accumulation and suggested the importance of evaluating the osmotic adjustment quality.

The decrease in A with increasing NaCl concentrations observed in the present study was also observed by Nabati et al. [37], who found that after 21 days of exposure of sorghum to high NaCl concentrations (electrical conductivity of 10.5 and 23.1 dS m^{-1}), A decreased by 18 and 26%, respectively, compared to a treatment with a lower electrical conductivity of 5.2 dS m^{-1} (−52 mM NaCl). The negative effect of high salinity on A is related to a decrease in the osmotic potential of the soil solution, which limits water uptake by the roots [38], resulting in stomatal closure to conserve water. As a result of stomatal closure, gs and Tr are reduced and there is a limitation of CO$_2$ diffusion into the leaf thereby limiting A [39]. This is supported by our observation that the concomitant reductions of A and Tr offset each other, resulting in no significant effect of NaCl concentration on WUE. Plants growing in saline soils often adapt to high salinity by minimizing water loss because growth depends on the ability to maintain A, while reducing water loss [8]. This was not the case with sorghum. Although we observed that sorghum was able to maintain WUE when gs was reduced, the decrease in A at high salinity levels inhibited plants growth under high soil salinity (Figure 6). This may partially explain why sorghum is considered only moderately tolerant of saline soil conditions.

Salinity effects on photosynthesis are often associated with inhibition of electron transport proteins in chloroplasts [40–42]. Wang et al. [43] determined that in response to high soil salinity, there was a reduction in a complex of three proteins in *Ricinus communis* that negatively influenced the initial stage of CO$_2$ fixation, compromising CO$_2$ uptake and fixation dye to decreased by Rubp-carboxylase/oygenase (RuBisCO) activity. Thus, the decreasing A in sorghum as salinity increased in the present study may have not only been due to physical factors such as changes in water potentials, by may also have been affected by biochemical factors such as reduced enzyme activity. It was also reported that several photosynthetic proteins involved in the stability of PSII and photosynthetic electron transport from photosystem II to photosystem I are affected by salt stress [42]. The authors also observed that a NaCl-induced reduction in enzymes involved in the Calvin cycle and the first step of carbon fixation, such as carbonate dehydratases, are potentially regulated by salinity stress.

Calone et al. [44] compared the growth of sorghum to three salinity levels (0, 3, or 6 dS m^{-1}) with leaching (water applied to above water holding capacity, of the soil) and without leaching (irrigated below water holding capacity of the soil). When comparing the 0 to the 6 dS m^{-1} treatments, they observed an 87% and 42% reduction in dry weight without and with leaching, respectively. In our study, where there was no leaching, and plant growth differences between highest salinity level and the control treatment were less than those observed by Calone et al. [44]. Growth differences between the present study and those observed by Calone et al. [44] may have been due to the difference in physical and/or chemical qualities of the soils, source of salt and/or genotype tested, which have a significant impact on results [25]. The present study provides new information about salinity effects on an important sorghum genotype that is grown commercially in areas of Northeast Brazil that are prone to high salinity levels.

5. Conclusions

Our hypothesis that sorghum's ability to moderately tolerate high soil salinity is due to osmotic adjustment (similar to their tolerance to drought stress), was only partially true. Our data showed that the sorghum plants respond to increasing soil salinity by both osmotic adjustment and by stomatal regulation, as indicated by reductions in gs with increasing salt concentrations. However, there was a metabolic cost when soil salinity was high due to A being limited by reduced gs under these conditions. The concomitant decreased in Tr with decreasing A as soil salinity increased, resulted in maintenance of WUE even at high salinity, allowing sorghum to tolerate high soil salinity (100 mM of NaCl = -10 dS m^{-1}). Sorghum is not a halophytic species and soil salinity reached 33.8 dS m^{-1} at 55 days after exposure to the 100 mM of NaCl treatment, suggesting salinity tolerance, which is also supported by low growth reductions, such as only 27% and 33% to plant height and dry weight, respectively, in the 100 mM of NaCl treatment compared to the control treatment.

Author Contributions: Conceptualization, P.R.M.D. and E.R.d.S.; methodology-investigation, P.R.M.D., E.R.d.S., M.A.d.S., C.M.T.L., D.R.M. and M.K.S.S.P.; writing—original draft preparation, P.R.M.D., E.R.d.S., M.K.S.S.P., B.S.; writing—review and editing, P.R.M.D., E.R.d.S., M.A.d.S. and B.S.; funding acquisition, E.R.d.S., B.S. All authors have read and agreed to the published version of the manuscript.

Funding: This research was funded by the National Council for the Improvement of Higher Education (CAPES) and the National Council for Scientific and Technological Development for granting scholarships and funding the research project (CNPq n° 308530/2015-2 and 420723/2016-1) and Foundation for Science and Technology Development of the State of Pernambuco.

Institutional Review Board Statement: Not applicable.

Informed Consent Statement: Not applicable.

Data Availability Statement: Available upon request.

Conflicts of Interest: The authors declare no conflict of interest.

References

1. Nagaraju, M.; Reddy, P.S.; Kumar, S.A.; Kumar, A.; Rajasheker, G.; Rao, D.M.; Kishor, P.K. Genome-wide identification and transcriptional profiling of small heat shock protein gene family under diverse abiotic stress conditions in Sorghum bicolor (L.). *Int. J. Biol. Macromol.* **2020**, *142*, 822–834. [CrossRef] [PubMed]
2. Sutka, M.R.; Manzur, M.E.; Vitali, V.A.; Micheletto, S.; Amodeo, G. Evidence for the involvement of hydraulic root or shoot adjustments as mechanisms underlying water deficit tolerance in two Sorghum bicolor genotypes. *J. Plant Physiol.* **2016**, *192*, 13–20. [CrossRef] [PubMed]
3. Mccormick, R.F.; Truong, S.K.; Sreedasyam, A.; Jenkins, J.; Shu, S.; Sims, D.; Mullet, J.E. The Sorghum bicolor reference genome: Improved assembly, gene annotations, a transcriptome atlas, and signatures of genome organization. *Plant J.* **2018**, *93*, 338–354. [CrossRef] [PubMed]
4. Devnarain, N.; Crampton, B.G.; Olivier, N.; Van der Westhuyzen, C.; Becker, J.V.; O'Kennedy, M.M. Transcriptomic analysis of a Sorghum bicolor landrace identifies a role for beta-alanine betaine biosynthesis in drought tolerance. *S. Afr. J. Bot.* **2019**, *127*, 244–255. [CrossRef]
5. Guimarães, M.J.M.; Simões, W.L.; Tabosa, J.N.; Santos, J.E.; Willadino, L. Cultivation of forage sorghum varieties irrigated with saline effluent from fish-farming under semiarid conditions. *Rev. Bras. Eng. Agríc. Ambien.* **2016**, *20*, 461–465. [CrossRef]
6. Hostetler, A.N.; Govindarajulu, R.; Hawkins, J.S. QTL mapping in an interspecific sorghum population uncovers candidate regulators of salinity tolerance. *Plant Stress* **2021**, *2*, 100024. [CrossRef]
7. Munns, R.; Tester, M. Mechanisms of salinity tolerance. *Annu. Rev. Plant Biol.* **2018**, *59*, 651–681. [CrossRef]
8. Negrão, S.; Schmöckel, S.M.; Tester, M. Evaluating physiological responses of plants to salinity stress. *Ann. Bot.* **2017**, *119*, 1–11. [CrossRef]
9. Morton, M.J.; Awlia, M.; Al-Tamimi, N.; Saade, S.; Pailles, Y.; Negrão, S.; Tester, M. Salt stress under the scalpel–dissecting the genetics of salt tolerance. *Plant J.* **2019**, *97*, 148–163. [CrossRef]
10. Punia, H.; Tokas, J.; Malik, A.; Singh, S.; Phogat, D.S.; Bhuker, A.; Sheokand, R.N. Discerning morpho-physiological and quality traits contributing to salinity tolerance acquisition in sorghum [Sorghum bicolor (L.) Moench]. *S. Afr. J. Bot.* **2021**, *140*, 409–418. [CrossRef]
11. Liu, P.; Yin, L.; Wang, S.; Zhang, M.; Deng, X.; Zhang, S.; Tanaka, K. Enhanced root hydraulic conductance by aquaporin regulation accounts for silicon alleviated salt-induced osmotic stress in Sorghum bicolor L. *Environ. Exp. Bot.* **2015**, *111*, 42–45. [CrossRef]
12. Betzen, B.M.; Smart, C.M.; Maricle, K.L.; Maricle, B.R. Effects of increasing salinity on photosynthesis and plant water potential in Kansas salt marsh species. *Trans. Kansas Acad. Sci.* **2019**, *122*, 49–58. [CrossRef]
13. Arif, Y.; Singh, P.; Siddiqui, H.; Bajguz, A.; Hayat, S. Salinity induced physiological and biochemical changes in plants: An omic approach towards salt stress tolerance. *Plant Physiol. Biochem.* **2020**, *156*, 64–77. [CrossRef]
14. Methenni, K.; Abdallah, M.B.; Nouairi, I.; Smaoui, A.; Zarrouk, M.; Youssef, N.B. Salicylic acid and calcium pretreatments alleviate the toxic effect of salinity in the Oueslati olive variety. *Sci. Hortic.* **2018**, *233*, 349–358. [CrossRef]
15. Mansour, M.M.F.; Emam, M.M.; Salama, K.H.A.; Morsy, A.A. Sorghum under saline conditions: Responses, tolerance mechanisms, and management strategies. *Planta* **2021**, *254*, 24. [CrossRef] [PubMed]
16. Zhang, Q.; Dai, W. Plant response to salinity stress. In *Stress Physiology of Woody Plants*; CRC Press: Boca Raton, FL, USA, 2019; pp. 155–173.
17. Yang, Y.; Guo, Y. Unraveling salt stress signaling in plants. *J. Integr. Plant Biol.* **2018**, *60*, 796–804. [CrossRef] [PubMed]
18. Hussain, S.; Hussain, S.; Ali, B.; Ren, X.; Chen, X.; Li, Q.; Ahmad, N. Recent progress in understanding salinity tolerance in plants: Story of Na+/K+ balance and beyond. *Plant Physiol. Biochem.* **2021**, *160*, 239–256. [CrossRef] [PubMed]
19. Zörb, C.; Geilfus, C.-M.; Dietz, K.-J. Salinity and crop yield. *Plant Biol.* **2019**, *32* (Suppl. 1), 31–38. [CrossRef]
20. Swami, A.K.; Alam, S.I.; Sengupta, N.; Sarin, R. Differential proteomic analysis of salt stress response in Sorghum bicolor leaves. *Environ. Exp. Bot.* **2011**, *71*, 321–328. [CrossRef]
21. Reddy, P.S.; Jogeswar, G.; Rasineni, G.K.; Maheswari, M.; Reddy, A.R.; Varshney, R.K.; Kishor, P.K. Proline over-accumulation alleviates salt stress and protects photosynthetic and antioxidant enzymactivities in transgenic sorghum [Sorghum bicolor (L.) Moench]. *Environ. Exp. Bot.* **2015**, *94*, 104–113. [CrossRef]
22. Soltabayeva, A.; Ongaltay, A.; Omondi, J.O.; Srivastava, S. Morphological, physiological and molecular markers for salt-stressed plants. *Plants* **2021**, *10*, 243. [CrossRef] [PubMed]
23. Rasouli, F.; Kiani-Pouya, A.; Shabala, L.; Li, L.; Tahir, A.; Yu, M.; Hedrich, R.; Chen, Z.; Wilson, R.; Zhang, H.; et al. Salinity effects on guard cell Proteome in Chenopodium quinoa. *Int. J. Mol. Sci.* **2021**, *22*, 428. [CrossRef] [PubMed]
24. Zhao, C.; Zhang, H.; Song, C.; Zhu, J.-K.; Shabala, S. Mechanisms of plant responses and adaptation to soil salinity. *Innovation* **2020**, *1*, 1. [CrossRef] [PubMed]
25. Amombo, E.; Ashilenje, D.; Hirich, A.; Kouisni, L.; Oukarroum, A.; Ghoulam, C.; Gharous, M.E.; Nilahyane, A. Exploring the correlation between salt tolerance and yield: Research advances and perspectives for salt-tolerant forage sorghum selection and genetic improvement. *Planta* **2022**, *255*, 71. [CrossRef]
26. Fan, Y.; Shabala, S.; Ma, Y.; Xu, R.; Zhou, M. Using QTL mapping to investigate the relationships between abiotic stress tolerance (drought and salinity) and agronomic and physiological traits. *BMC Genom.* **2015**, *16*, 43. [CrossRef]
27. Ma, Y.; Dias, M.C.; Freitas, H. Drought and Salinity Stress Responses and Microbe-Induced Tolerance in Plants. *Front. Plant Sci.* **2020**, *13*. [CrossRef]

28. Jones, H.G.; Tardieu, F. Modelling water relations of horticultural crops: A review. *Sci. Hortic.* **1998**, *74*, 21–46. [CrossRef]
29. Silveira, J.A.G.; Araújo, S.A.M.; Lima, J.P.M.S.; Viegas, R.A. Roots and leaves display contrasting osmotic adjustment mechanisms in response to NaCl-salinity in *Atriplex nummularia*. *Environ. Exp. Bot.* **2009**, *66*, 1–8. [CrossRef]
30. Lins, C.M.T.; Souza, E.R.; Melo, H.F.; Paulino, M.K.S.S.; Dourado, P.R.M.; Leal, L.Y.C.; Santos, H. R B. Pressure-volume (P-V) curves in *Atriplex nummularia* Lindl. for evaluation of osmotic adjustment and water status under saline conditions. *Plant Physiol. Biochem.* **2018**, *124*, 55–159. [CrossRef]
31. R Core Team. *R: A Language and Environment for Statistical Computing*; R Foundation for Statistical Computing: Vienna, Austria, 2020; Available online: https://www.R-project.org/ (accessed on 20 June 2021).
32. Monteiro, D.R.; Souza, E.R.; Dourado, P.R.M.; Melo, H.; Santos, H.; Dos Santos, M.A. Soil Water Potentials and Sweet Sorghum under Salinity. *Commun. Soil Sci. Plant Anal.* **2021**, *52*, 1149–1160. [CrossRef]
33. Pan, Y.-Q.; Guo, H.; Wang, S.-M.; Zhao, B.; Zhang, J.-L.; Ma, Q.; Yin, H.-J.; Bao, A.-K. The photosynthesis, NaC/KC homeostasis and osmotic adjustment of *Atriplex canescens* in response to salinity. *Front. Plant Sci.* **2016**, *7*, 848. [CrossRef] [PubMed]
34. Turner, N.C. Turgor maintenance by osmotic adjustment: 40 years of progress. *J. Exp. Bot.* **2018**, *69*, 3223–3233. [CrossRef] [PubMed]
35. Bafeel, S.O. Physiological parameters of salt tolerance during germination and seedling growth of *Sorghum bicolor* cultivars of the same subtropical origin. *Saudi J. Biol. Sci.* **2014**, *21*, 300–304. [CrossRef] [PubMed]
36. Lacerda, C.F.; Cambraia, J.; Oliva, M.A.O.; Ruiz, H.A. Osmotic adjustment in roots and leaves of two sorghum genotypes under NaCl stress. *Braz. J. Plant Physiol.* **2003**, *15*, 113–118. [CrossRef]
37. Nabati, J.; Kafi, M.; Masoumi, A.; Zare, M. Effect of salinity and silicon application on photosynthetic characteristics of sorghum (*Sorghum bicolor* L.). *Internat. J. Agric. Sci.* **2013**, *3*, 483–492.
38. Meng, Y.; Yin, C.; Zhou, Z.; Meng, F. Increased salinity triggers significant changes in the functional proteins of *Anammox* bacteria within a biofilm community. *Chemosphere* **2018**, *207*, 655–664. [CrossRef]
39. Oliveira, W.J.; de Souza, E.R.; Cunha, J.C.; Silva, E.F.F.; Velos, V.L. Leaf gas exchange in cowpea and CO_2 efflux in soil irrigated with saline water. *Rev. Bras. Eng. Agric. Ambiental.* **2017**, *21*, 32–37. [CrossRef]
40. Ashraf, M.; Harris, P.J.C. Photosynthesis under stressful environments: An overview. *Photosynthetica* **2013**, *51*, 163–190. [CrossRef]
41. Maswada, H.F.; Djanaguiraman, M.; Prasad, P.V.V. Seed treatment with nano-iron (III) oxide enhances germination, seeding growth and salinity tolerance of sorghum. *J. Agron. Crop Sci.* **2018**, *204*, 577–587. [CrossRef]
42. Monteiro, D.R.; Melo, H.F.; de Lins, C.M.T.; Dourado, P.R.M.; Santos, H.R.B.; Souza, E.R. Chlorophyll a fluorescence in saccharine sorghum irrigated with saline water. *Rev. Braz. Eng. Agríc. Ambient.* **2018**, *22*, 673–678. [CrossRef]
43. Wang, Y.; Penga, X.; Salvato, F.; Wang, Y.; Yan, X.; Zhou, Z.; Lin, J. Salt-adaptive strategies in oil seed crop *Ricinus communis* early seedlings (cotyledon vs. true leaf) revealed from proteomics analysis. *Ecotoxicol. Environ. Saf.* **2019**, *171*, 12–25. [CrossRef] [PubMed]
44. Calone, R.; Sanoubar, R.; Lambertini, C.; Speranza, M.; Antisari, L.V.; Vianello, G.; Barbanti, L. Salt tolerance and Na allocation in *Sorghum bicolor* under variable soil and water salinity. *Plants* **2020**, *9*, 561. [CrossRef] [PubMed]

Article

Supplemental Irrigation with Brackish Water Improves Carbon Assimilation and Water Use Efficiency in Maize under Tropical Dryland Conditions

Eduardo Santos Cavalcante [1,*], Claudivan Feitosa Lacerda [1], Rosilene Oliveira Mesquita [2], Alberto Soares de Melo [3], Jorge Freire da Silva Ferreira [4], Adunias dos Santos Teixeira [1], Silvio Carlos Ribeiro Vieira Lima [5], Jonnathan Richeds da Silva Sales [1], Johny de Souza Silva [2] and Hans Raj Gheyi [6]

1. Agricultural Engineering Department, Federal University of Ceará, Fortaleza 60356-001, Brazil; cfeitosa@ufc.br (C.F.L.); adunias@ufc.br (A.d.S.T.); salesjrs@alu.ufc.br (J.R.d.S.S.)
2. Department of Agronomy, Federal University of Ceará, Fortaleza 60356-001, Brazil; rosilenemesquita@ufc.br (R.O.M.); johny.ufca@alu.ufc.br (J.d.S.S.)
3. Biological Sciences Center and Health, State University of Paraiba, Campina Grande 58429-600, Brazil; alberto@uepb.edu.br
4. United States Salinity Laboratory USDA-ARS, Agricultural Water Efficiency and Salinity Research Unit, Riverside, CA 92507, USA; jorge.ferreira@usda.gov
5. Secretariat of Economic Development and Labor of the State of Ceará, Fortaleza 60356-001, Brazil; silvio.carlos@sedet.ce.gov.br
6. Agricultural Engineering Department, Federal University of Campina Grande, Campina Grande 58428-830, Brazil; hans.gheyi@ufcg.edu.br
* Correspondence: educavalcantes@alu.ufc.br

Citation: Cavalcante, E.S.; Lacerda, C.F.; Mesquita, R.O.; de Melo, A.S.; da Silva Ferreira, J.F.; dos Santos Teixeira, A.; Lima, S.C.R.V.; da Silva Sales, J.R.; de Souza Silva, J.; Gheyi, H.R. Supplemental Irrigation with Brackish Water Improves Carbon Assimilation and Water Use Efficiency in Maize under Tropical Dryland Conditions. *Agriculture* 2022, 12, 544. https://doi.org/10.3390/agriculture12040544

Academic Editor: Guodong Liu

Received: 1 March 2022
Accepted: 5 April 2022
Published: 11 April 2022

Publisher's Note: MDPI stays neutral with regard to jurisdictional claims in published maps and institutional affiliations.

Copyright: © 2022 by the authors. Licensee MDPI, Basel, Switzerland. This article is an open access article distributed under the terms and conditions of the Creative Commons Attribution (CC BY) license (https://creativecommons.org/licenses/by/4.0/).

Abstract: Dry spells in rainfed agriculture lead to a significant reduction in crop yield or to total loss. Supplemental irrigation (SI) with brackish water can reduce the negative impacts of dry spells on net CO_2 assimilation in rainfed farming in semi-arid tropical regions and maintain crop productivity. Thus, the objective of this study was to evaluate the net carbon assimilation rates, indexes for water use efficiency, and indicators of salt and water stress in maize plants under different water scenarios, with and without supplemental irrigation with brackish water. The experiment followed a randomized block design in a split-plot design with four replications. The main plots simulated four water scenarios found in the Brazilian semi-arid region (Rainy, Normal, Drought, and Severe Drought), while the subplots were with or without supplemental irrigation using brackish water with an electrical conductivity of 4.5 dS m^{-1}. The dry spells reduced the photosynthetic capacity of maize, especially under the Drought (70% reduction) and Severe Drought scenarios (79% reduction), due to stomatal and nonstomatal effects. Supplemental irrigation with brackish water reduced plant water stress, averted the excessive accumulation of salts in the soil and sodium in the leaves, and improved CO_2 assimilation rates. The supplemental irrigation with brackish water also promoted an increase in the physical water productivity, reaching values 1.34, 1.91, and 3.03 times higher than treatment without SI for Normal, Drought, and Severe Drought scenarios, respectively. Thus, the use of brackish water represents an important strategy that can be employed in biosaline agriculture for tropical semi-arid regions, which are increasingly impacted by water shortage. Future studies are required to evaluate this strategy in other important crop systems under nonsimulated conditions, as well as the long-term effects of salts on different soil types in this region.

Keywords: tropical semi-arid; saline water; biosaline agriculture; complementary irrigation; water stress; photosynthesis

1. Introduction

Irrigated agriculture is essential for crop production for human and animal consumption in semi-arid regions. However, the expansion of irrigation in these regions is generally limited by the scarcity of water resources, especially in years of drought [1]. On the other hand, rainfed agriculture is a high-risk activity in semi-arid environments, as observed in Northeastern Brazil, due to the high interannual variability and poor distribution of rainfall in space and time [2]. These risks can be minimized with the use of supplemental irrigation [3], a promising climate-smart practice for dryland agriculture [4] even when saline water sources are used [5–7].

The salinity of water and soil is a problem present on all continents, impacting ecosystems and agricultural activities, notably in arid and semi-arid regions [8]. However, the growing demand for food, the scarcity of water resources, and the overuse of groundwater under the ongoing scenario of global climate change have created the need to tap into saltwater resources to maintain food production and generate jobs and income for farmers in drylands. Therefore, it is necessary to use appropriate management techniques and salt-tolerant species, both aspects being part of biosaline agriculture [9,10].

Many studies have been carried out on the use of brackish water applied in a mixed or cyclic way as supplemental irrigation in rainfed farming [5,6,11,12]. The use of brackish water in irrigation can be beneficial for the Brazilian semi-arid region, given the large number of wells containing water with a moderate salt concentration [13]. Under such conditions, supplemental irrigation with brackish water may increase the possibilities of plant cultivation, especially on small farms [7], which predominate in the Brazilian semi-arid region.

From the point of view of plant physiology, it is known that leaf gas exchange, i.e., the loss of water vapor during the transpiration process and the CO_2 intake in the photosynthetic pathways, plays a fundamental role in crop yield. The reduction of leaf gas exchange under water shortages limits plant growth and contributes to low maize yields under rainfed farming in tropical semi-arid regions. In these cropping systems, photosynthesis decreases at different rates depending on the duration of dry spells, with the effects being more intense in years of drought and severe drought [14].

Supplemental irrigation can reduce the negative effects of dry spells on leaf gas exchange. However, the use of brackish water as supplemental irrigation can also impact photosynthetic rates, given both the osmotic and the toxic effects of salts accumulated in the soil [15]. These effects can be minimized by the fact that the salts applied during dry spells tend to be leached down during the rainy season, especially in soils with good natural drainage, thus having a negligible impact even on moderately salt-sensitive crops such as maize [7]. Thus, supplemental irrigation can also increase water productivity and allow the use of an alternative source of water (brackish water) with improved efficiency of irrigation management.

In this context, this research applied the hypothesis that the water stress associated with dry spells is more deleterious to the physiology and water use efficiency of maize plants than the salt stress associated with the use of brackish water in supplemental irrigation. Thus, the objective of the present study was to evaluate the net carbon assimilation rates and the indexes for water use efficiency in maize cultivated under different water scenarios in the Brazilian semi-arid region, with and without supplemental irrigation employing brackish water.

2. Material and Methods

The experiment was conducted in Fortaleza (3°74′ S, 38°58′ W and altitude of 19 m), Ceará, Brazil in two cycles during the dry season: from 31 August to 21 November 2018, and from 28 September to 18 December 2019. During the experiment, the average, minimum and maximum air temperatures were, respectively, 27.6, 22.7, 30.6 °C (2018) and 27.2, 25.3, 30.1 °C (2019). The average air relative humidity was 70.7% for 2018 and 68.3% for 2019. The soil in the area is classified as a Ultisol, with a sandy loam texture in the 0–20 cm layer, pH 6.3, electrical conductivity of the saturation extract of soil 0.20 dS m^{-1}, and an exchangeable sodium percentage of 4.4.

The experiment followed a randomized block design in a split-plot arrangement with four replications. The main plots were designed to simulate the water supply in the soil corresponding to four rainfall scenarios—Rainy, Normal, Drought, and Severe Drought (simulations based on historical series of precipitation data for the rainfed cropping season in the Brazilian semi-arid region). The subplots were assigned to the use or lack of supplemental irrigation with brackish water (electrical conductivity of water—ECw = 4.5 dS m^{-1}). The sub-subplots were assigned to the sampling dates (27, 47, 49, 56, 60, and 67 days after planting). Each experimental plot consisted of six 10 m long rows and each subplot had three 10 m long rows, with a spacing of 0.80 × 0.20 m between rows and plants, respectively.

The water scenarios were defined according to rainfall and dry spell data for the semi-arid region of Vale do Curu, Ceará, Brazil, for February to May (the period of rainfed farming in the region). The climate of this region is very hot and semi-arid according to the Köppen classification, with annual rainfall and potential evapotranspiration of 800 mm and 1700 mm, respectively. For definition of the scenarios, historical series data for 30 years (1989 to 2019) of precipitation in this region were provided by the Foundation for Meteorology and Water Resources of Ceará (FUNCEME). Based on rainfall data during the rainy season (February to May) and the characterization of rainfall patterns in the region [16], the following scenarios were defined: Rainy (803 to 1040 mm), Normal (456 to 590 mm), Drought (383 to 443 mm), and Severe Drought (158 to 338 mm).

Supplemental applications of brackish water were estimated for dry spell periods, adding a leaching fraction of 0.20 in each irrigation event. The applied water depths were estimated based on the values of crop evapotranspiration [17]. In the periods without dry spells, irrigations were carried out using well water of low salinity ((pH 7.1, ECw = 0.9 dS m^{-1}, (SAR 3.9 mmol L^{-1})$^{0.5}$). The brackish water (ECw = 4.5 dS m^{-1}) was prepared by adding NaCl, CaCl$_2$.2H$_2$O, and MgCl$_2$.6H$_2$O to the well water in the equivalent proportion of 7:2:1. This salt ratio is representative of the chemical composition of brackish waters in the Brazilian semi-arid region [18].

For each crop cycle, the total water depths applied, without and with supplemental irrigation, were as follows: 745 and 796 mm (Rainy), 465 and 567 mm (Normal), 345 and 517 mm (Drought), and 240 and 500 mm (Severe Drought). The dry spells were distributed in the water scenarios and their quantity and duration varied according to the rainfall patterns. Dry spells were defined as periods of at least five continuous days without rain [14]. Detailed information on water scenarios and supplemental irrigation was reported previously [7].

We used seeds of Hybrid BRS 2022 maize, a double hybrid with moderate resistance to diseases. To represent the reality of traditional family farming in the northeastern region of Brazil, sowing was carried out after applying a 30 mm water depth of low-salinity water. Fertilization with nitrogen (70 kg ha^{-1}), phosphorus (40 kg ha^{-1} of P$_2$O$_5$), and potassium (20 kg ha^{-1} of K$_2$O) was carried out according to the recommendations for rainfed maize cultivation in the State of Ceará [19]. The phosphorus dose (as simple superphosphate) was applied at planting and the nitrogen (as urea) and potassium (as potassium chloride) doses were split into three applications: one at planting and two as topdressing. Irrigation was performed by drip, using drip tapes with a flow rate of 2.7 L h^{-1}, with self-compensating emitters spaced 0.4 m apart.

Soil moisture was determined using the gravimetric method [20], with soil samples from the 0–20 cm layer collected at 47, 56, 60, and 67 days after planting in both cycles. For the determination of soil salinity, samples were collected from each subplot at the end of each crop cycle, from the 0–20 cm layer at three points of the central row: beginning, middle, and end. Soil electrical conductivity was measured in a soil:water suspension of 1:1 (v/v) and expressed in dS m^{-1}.

Measurements of leaf gas exchange were performed at 27, 47, 49, 56, 60, and 67 days after planting (Figure 1) using the third fully expanded leaf from the apex of the plant. The net photosynthesis rate A (µmol CO$_2$ m^{-2} s^{-1}), stomatal conductance gs (mol m^{-2} s^{-1}), transpiration rate E (mmol m^{-2} s^{-1}), and internal CO$_2$ concentration Ci (µmol mol^{-1}) were measured using an infrared gas analyzer (Li–6400XT, LICOR, USA) under the following conditions: ambient air temperature, CO_2 concentration of 400 ppm, and photosynthetically active radiation of 1800 µmol m^{-2} s^{-1}. The instantaneous water use efficiency (WUEi) was estimated using the photosynthesis and transpiration rate data.

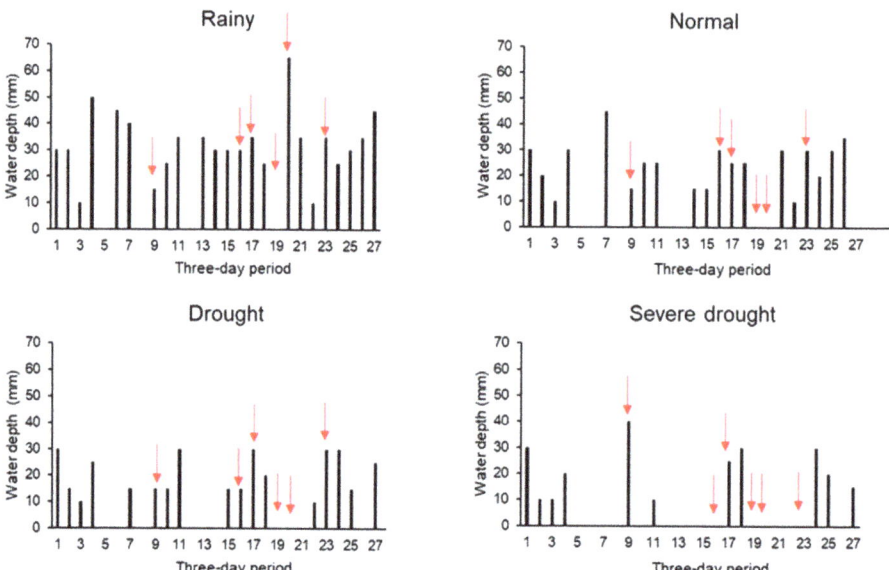

Figure 1. Water depths applied at three-day intervals in two maize crop cycles for different simulated water scenarios, without supplemental irrigation. The red arrows indicate the moments of leaf gas exchange measurements.

Samples of fully expanded leaf blades were collected at 47, 56, and 60 days after planting in both cycles for the determination of sodium and proline concentrations. The material was lyophilized and ground to obtain the extract, according to the method described by [21]. Sodium concentration was determined using a flame photometer. Free proline levels were determined according to a previously established method [22]. Proline readings were performed using a spectrophotometer (model UV−1650PC, Shimadzu, Japan).

At 82 days after planting, 15 plants were collected per subplot, and the production of dry ear biomass and total biomass was determined. Dry biomass productivity per hectare (ears and total) was estimated, taking into account the planting density and final stand. The physical water productivity (PWP, kg m^{-3}) was estimated using the relationship between the production of ears (PWP$_{ear}$) or total dry biomass (PWP$_{biomass}$) and the total volume of water applied (simulated rainfall plus supplementary irrigation), according to Equations (1) and (2) [1]:

$$\text{PWP}_{ear} = \frac{\text{Biomass of ears } (\text{kg ha}^{-1})}{\text{Total water appied } (\text{m}^3 \text{ ha}^{-1})} \quad (1)$$

$$\text{PWP}_{biomass} = \frac{\text{Total biomass of plants } (\text{kg ha}^{-1})}{\text{Total water applied } (\text{m}^3 \text{ ha}^{-1})} \quad (2)$$

The efficiency of supplemental irrigation (WUE$_{SI}$) was estimated using the ratio between the increment of biomass (ears and total) and the volume of supplemental water applied, according to Equation (3):

$$\text{WUE}_{SI} = \frac{Y_{SI} - Y \ (\text{kg ha}^{-1})}{\text{Suplemental irrigation } (m^3 \ \text{ha}^{-1})} \quad (3)$$

where Y_{SI} and Y represent the yields of plots with and without supplemental irrigation, respectively.

The data were submitted to the analysis of variance (F-test) after passing the Kolmogorov–Smirnov normality test. When the F-test determined statistical significance, the means were compared using the Tukey test ($p < 0.05$). Statistical analyses were performed using the Sisvar software version 5.6 [23].

3. Results and Discussion

3.1. Soil Moisture and Salinity

Soil moisture content was high throughout the crop cycle in the Rainy scenario, with a small difference between treatments with and without supplemental irrigation only on the last sampling date (Figure 2A). Moisture contents were close to the soil field capacity value (7.21%) on most sampling dates, regardless of supplemental irrigation. The excess water supplied to the soil in the Rainy scenario resulted in water accumulation in the soil profile, with part lost to runoff and part percolated [24–26]. The storage of water in the soil in this scenario favored the maintenance of moisture with minimal need for supplementation, since torrential rains are usually interspersed with medium and low-intensity rains under tropical semi-arid conditions.

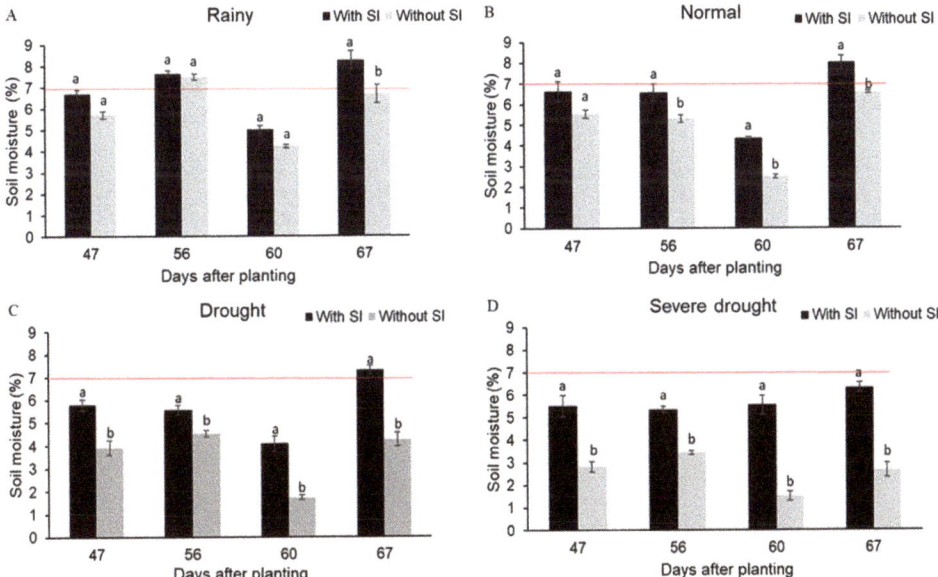

Figure 2. Soil moisture (layer from 0 to 20 cm) in areas cultivated with maize for the Rainy (**A**), Normal (**B**), Drought (**C**) and Severe drought (**D**) scenarios, showing sampling date and presence or absence of supplemental irrigation (SI) with brackish water (ECw = 4.5 dS m^{-1}). For each sampling date, means followed by the same letters do not differ ($p \geq 0.05$) from each other according to the Tukey test. Error bars represent the standard error of the mean ($n = 4$). The red line in each figure represents the field capacity of the soil.

For the Normal scenario, differences in soil moisture were observed on almost all sampling dates except for the first one, with higher values always recorded in the treatment with supplementation (Figure 2B). However, these differences were intensified in the Drought and Severe Drought scenarios (Figure 2C,D), taking as a reference the value of the soil field capacity. In the Severe Drought treatment, soil moisture was always above 5.0% with supplemental irrigation, while the values without supplementation ranged from 1.0 to 3.0%. Supplementation with brackish water in this scenario practically doubled the soil moisture content at 60 and 67 days after planting compared to the treatment without supplementation. These differences reflect the distribution of soil water application during maize crop cycles (Figure 1), which indicate situations of increasing water shortage for both Drought and Severe Drought scenarios.

The mean soil electrical conductivity ($EC_{1:1}$) increased significantly only with supplemental irrigation with brackish water for the Severe Drought scenario, reaching a mean value of 1.3 dS m^{-1} towards the ends of the crop cycles (Figure 3). However, the values were relatively low compared to those obtained with continuous irrigation with similar water salinity [27–30]. These differences occurred because in supplemental irrigation with brackish water, part of the salt content was leached after the application of water of lower salinity, simulating rain events. These results also reflect the weighted salinity values of the irrigation water, which were 1.1, 1.5, 2.1, and 2.8 dS m^{-1}, respectively, for the Rainy, Normal, Drought, and Severe Drought scenarios. According to [28], only irrigation water with electrical conductivities higher than 2.2 dS m^{-1} impacts the biomass production of maize plants under tropical conditions, indicating that brackish water supplementation may have little or no effect on maize crops.

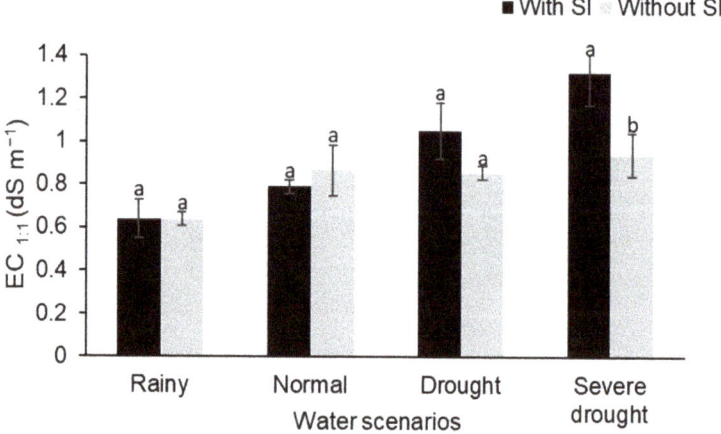

Figure 3. Soil electrical conductivity (soil:water extract 1:1) cultivated with maize under different water scenarios, with or without supplemental irrigation (SI) with brackish water (ECw = 4.5 dS m^{-1}). For each water scenario, means followed by the same letters do not differ ($p \geq 0.05$) from each other according to the Tukey test. Error bars represent the standard error of the mean ($n = 4$).

It is worth noting that the well water used in this study had an electrical conductivity of 0.9 dS m^{-1}, much higher than the salinity of rainwater (less than 0.1 dS m^{-1}). This indicates that under real conditions, the effects of supplemental irrigation on the accumulation of salts in the soil may be even smaller than those observed in the present study. These results suggest that brackish water supplementation presents a small risk of soil salinization if the soil has good natural drainage. The adoption of practices such as leaching fractions and the application of amendments may enable the mitigation of negative effects of the application of brackish water on the soil under Severe Drought scenarios, thus preventing soil degradation [26,31–33].

3.2. Leaf Gas Exchange

Maize photosynthetic rates reached values above 40 µmol m^{-2} s^{-1} (Figure 4), which are compatible with the photosynthetic metabolism of C4 species [34–36]. In the Rainy scenario, no difference was observed in photosynthetic rates between treatments with and without supplementation in the six evaluations, indicating the absence of water deficit and that supplemental irrigation would not be necessary in this scenario (Figure 4A). However, in the Normal scenario (Figure 4B), the need for supplemental irrigation during the dry periods, and in the reproductive stage (56 and 60 days after planting), was evident, based on the significant increase (2 to 3.5 times) observed for the rate of photosynthesis when compared to the treatment without supplementation.

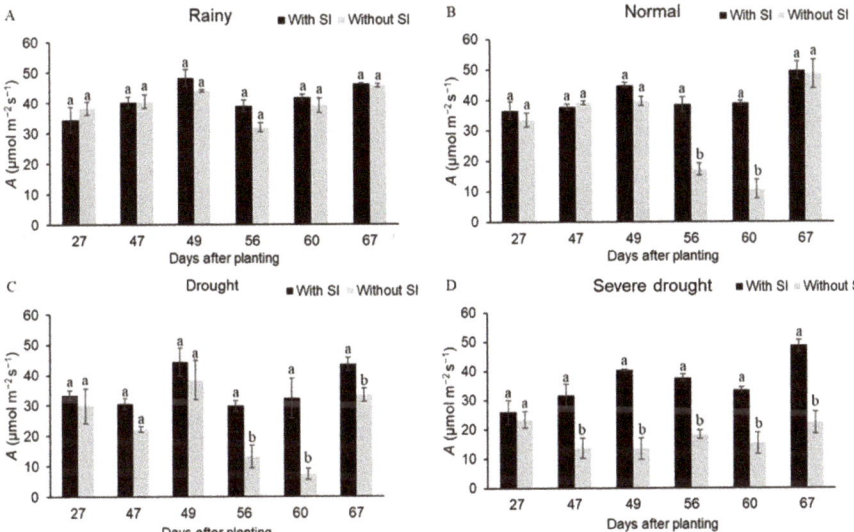

Figure 4. Net photosynthetic rate of maize leaves under Rainy (**A**), Normal (**B**), Drought (**C**) and Severe drought (**D**) scenarios, showing sampling date and presence or absence of supplemental irrigation (SI) with brackish water. For each sampling date, means followed by the same letters do not differ ($p \geq 0.05$) from each other according to the Tukey test. Error bars represent the standard error of the mean ($n = 4$).

In the Drought scenario, supplemental irrigation was important in maintaining the photosynthetic rate on three (56, 60, and 67 days after planting) out of six sampling dates (Figure 4C), while in the Severe Drought scenario this was observed for five dates (47, 49, 56, 60, and 67 days after planting) (Figure 4D), reflecting the distribution of dry spells (Figure 1) and soil moisture (Figure 2). It should be noted that a reduction in the rate of photosynthesis can be caused by several biotic and abiotic constraints. From the physiological point of view, this reduction can be explained by stomatal limitations, reduction in chlorophyll concentration, photochemical damage, and inhibition of enzymatic activities [37–39].

The water deficit during the dry spells caused a significant reduction in stomatal conductance (Figure 5). Such a reduction in stomatal opening limits the influx of CO_2 during the photosynthetic process [37], decreasing the net carbon assimilation capacity, as observed in the present study (Figure 4). However, on some sampling dates, there was a concomitant reduction in the photosynthetic rate (Figure 4) and an increase in the internal CO_2 concentration (Figure 6), especially for treatments without supplemental irrigation (Drought and Severe Drought scenarios). This indicates that longer dry spells induced more severe water stress, which affected the photosynthetic metabolism. According to [33], even if there was enough substrate (CO_2) in the mesophyll in a situation of severe water shortage, there was a high degree of photosynthetic restriction in response to the deterioration of the photochemical and/or biochemical apparatus of the carbon assimilation process in chloroplasts.

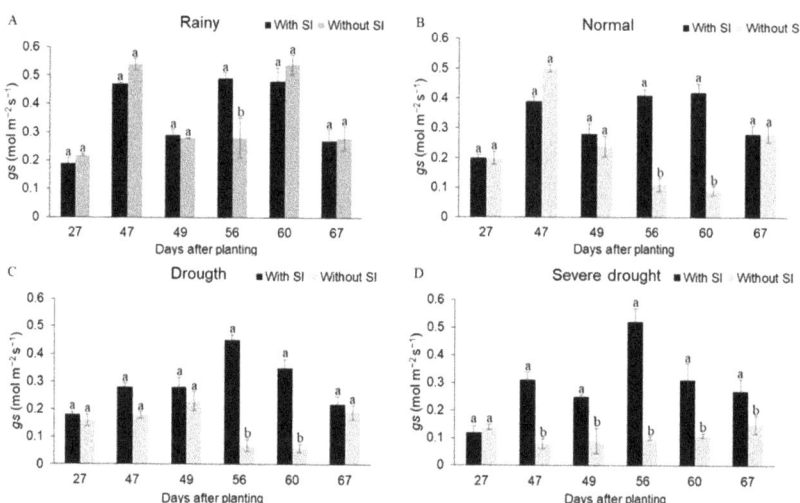

Figure 5. Stomatal conductance of maize leaves under Rainy (**A**), Normal (**B**), Drought (**C**) and Severe drought (**D**) scenarios, showing sampling date and presence or absence of supplemental irrigation (SI) with brackish water (ECw = 4.5 dS m^{-1}). For each sampling date, means followed by the same letters do not differ ($p \geq 0.05$) from each other according to the Tukey test. Error bars represent the standard error of the mean ($n = 4$).

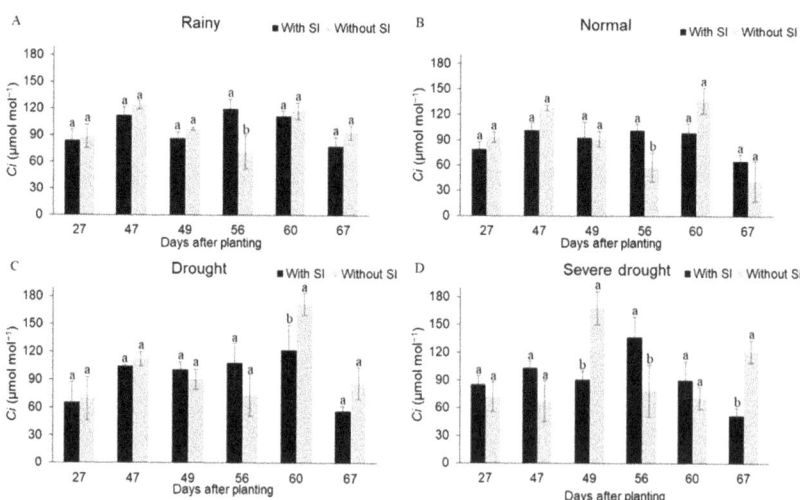

Figure 6. Internal CO_2 concentration of maize leaves under Rainy (**A**), Normal (**B**), Drought (**C**) and Severe drought (**D**) scenarios, showing sampling date and presence or absence of supplemental irrigation (SI) with brackish water (ECw = 4.5 dS m^{-1}). For each sampling date, means followed by the same letters do not differ ($p \geq 0.05$) from each other according to the Tukey test. Error bars represent the standard error of the mean ($n = 4$).

The transpiration rate data (Figure 7) show, in most cases, a similar behavior for the net photosynthetic rate (Figure 4), demonstrating strong stomatal limitations associated mainly with water deficit during dry spells. In the Rainy scenario, no difference was observed between treatments with and without brackish water supplementation and the values varied over time, possibly as a function of environmental changes such as air temperature and relative humidity. In the Severe Drought scenario, there were reductions on all sampling dates except for the one 27 days after planting, reflecting the response of the plants to soil moisture conditions (Figure 2).

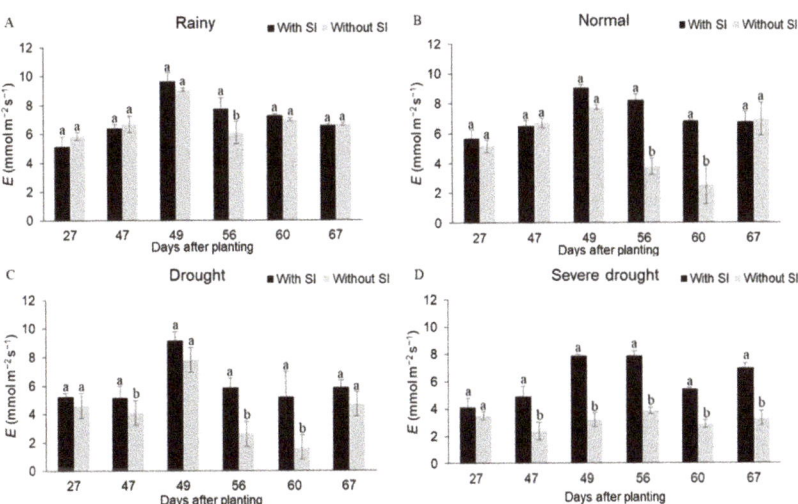

Figure 7. Transpiration rate of maize leaves under Rainy (**A**), Normal (**B**), Drought (**C**) and Severe drought (**D**) scenarios, showing sampling date and presence or absence of supplemental irrigation (SI) with brackish water of ECw = 4.5 dS m^{-1}. For each sampling date, means followed by the same letters do not differ ($p \geq 0.05$) from each other according to the Tukey test. Error bars represent the standard error of the mean (n = 4).

The values of transpiration and net photosynthetic rates also showed a rapid recovery capacity for leaf gas exchange after a dry spell, as observed in the treatment without supplementation for both Normal and Drought scenarios. This ability to recover leaf gas exchange at the end of a dry spell has also been reported in cowpea grown under rainfed farming in tropical dryland even after low-intensity rainfall [14]. However, severe water deficits associated with long dry spells can cause permanent cellular damage, making a full recovery of physiological processes and plant growth impossible [33,37].

3.3. Indicators of Salt and Water Stress

Regression analysis relating soil moisture and salinity to plant response variables also demonstrated that supplemental irrigation with brackish water reduced the deleterious effects of water stress during dry spells without causing damage associated with salt stress. The reduction of soil moisture from 7 to 2% resulted in a 58% reduction in net photosynthetic rate (Figure 8A) and an 88% reduction in biomass production (Figure 8B), considering treatments with and without supplemental irrigation. In comparison, increasing the weighted salinity of irrigation water from 1.1 dS m^{-1} (Rainy) to 2.8 dS m^{-1} (Severe Drought) reduced the photosynthetic rate and the production of total biomass by only 14 and 11%, respectively (Figure 8C,D). These reductions are expected to be even smaller or nonexistent under field conditions, because if using rainwater (ECw = 0.1 dS m^{-1}) instead of low-salinity water (ECw = 0.9 dS m^{-1}), the weighted salinity associated with supplemental brackish water would be even smaller and would promote greater leaching of salts in the soil profile, as discussed earlier.

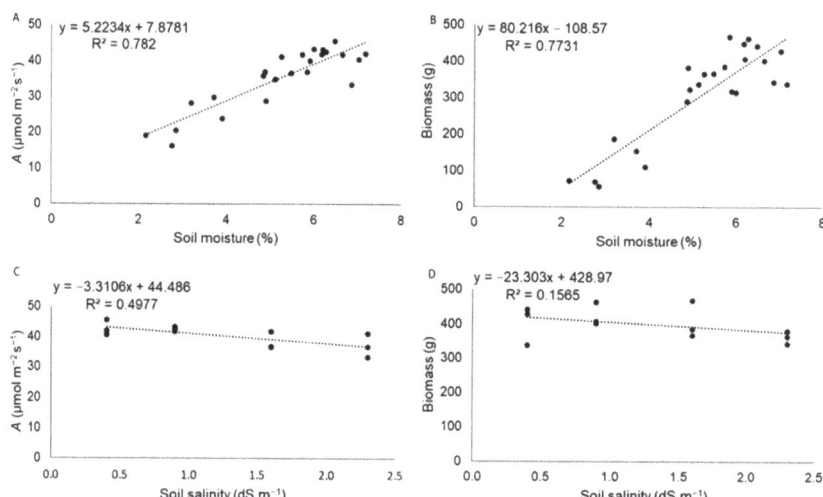

Figure 8. Photosynthetic rate A, and total biomass production per plant of maize as a function of soil moisture (**A,B**) and soil salinity (**C,D**).

The mineral analysis also suggested that the use of brackish water in supplemental irrigation did not result in an excessive accumulation of sodium in the leaves (Figure 9A). In the plots with supplemental irrigation, there was no increase in the sodium concentration in leaf tissues even when the weighted electrical conductivity of the irrigation water was increased from 1.1 dS m^{-1} (Rainy) to 2.8 dS m^{-1} (Severe Drought), and sodium contents were similar to those in the treatment without supplementation. Sodium contents were less than 2.5 g kg^{-1}, which did not result in any damage to the maize crop, as demonstrated by [40]. Irrigations with low-salinity water, simulating the occurrence of rain, reduced the accumulation of salts in the soil and consequently in the plants.

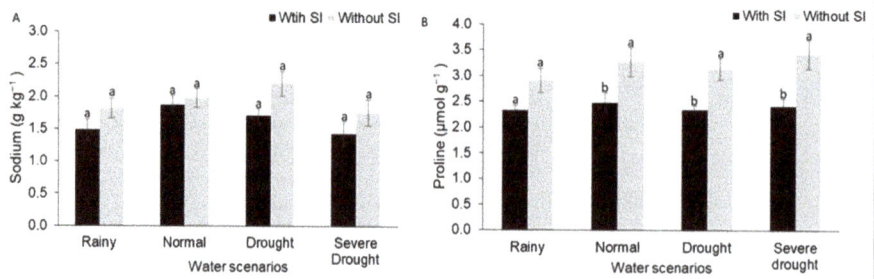

Figure 9. Sodium (**A**) and proline (**B**) concentrations in mature leaves of maize in response to presence or absence of supplemental irrigation (SI) with brackish water of ECw = 4.5 dS m^{-1}. Means followed by the same letters do not differ ($p \geq 0.05$) from each other according to the Tukey test. Sodium and proline concentrations represent the average of three sampling dates (47, 56, and 60 days after planting). Error bars represent the standard error of the mean ($n = 4$).

Proline concentration did not increase in treatments with supplemental irrigation (Figure 9B), suggesting that the salt stress caused by brackish water application (ECw = 4.5 dS m^{-1}) was not intense enough to promote proline accumulation. However, leaf concentration of proline was higher in treatments without supplementation than in treatments with supplementation for scenarios with increasing water shortage (Normal, Drought, and Severe Drought). These results indicate that proline accumulation was more affected by water stress than by salt accumulation in leaves of maize irrigated with brackish water. As for proline, although its accumulation is reported as a common response in plants under water and salt stress, there is controversy regarding the role played by this amino acid during stress. For many authors, the accumulation of proline is an important adaptive response for plants under stress, contributing to osmotic adjustment [41–44] and protecting cellular structures and functions [45–47]. For other authors, proline quantified using the method reported by [21] is an indication of osmoprotection associated with damage caused by abiotic constraints, as demonstrated for barley [48], cotton [49], and cowpea [50] under water deficits, and for rice [51] and sorghum [52,53] under salt stress. Our data presented here (Figure 9B) are in line with those who reported the highest accumulation of proline in response to damage caused by water stress [48,49], as revealed by soil moisture data (Figure 3) and net photosynthetic rates (Figure 4).

3.4. Water Use Efficiency

The similarity of the responses of net photosynthesis (Figure 4) and transpiration (Figure 7) rates resulted in minimal variations in instantaneous water use efficiency (WUEi) (Figure 10). There was also a similarity in the responses of plants under the different scenarios tested. In the present study, the lower values of WUEi in measurements performed 49, 56, and 60 days after planting may have been associated with the increase in transpiration rates in this period (Figure 7). Comparing treatments with and without supplemental irrigation with brackish water, a reduction in WUEi was observed only at 60 days after planting in the Normal and Drought scenarios, a result explained by the dramatic drop in the rate of photosynthesis during a dry spell (Figure 4).

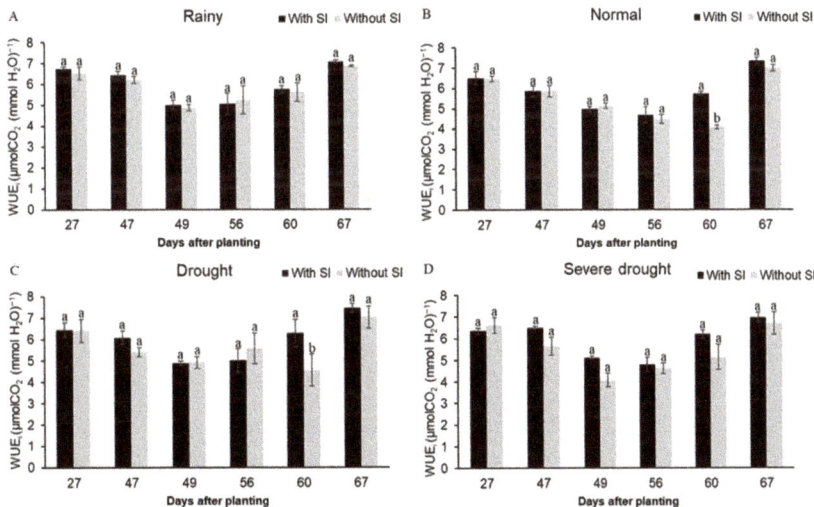

Figure 10. Instantaneous water use efficiency in maize leaves under Rainy (**A**), Normal (**B**), Drought (**C**) and Severe drought (**D**) scenarios, showing sampling dates and presence or absence of supplemental irrigation (SI) with brackish water of ECw = 4.5 dS m^{-1}. For each sampling date, means followed by the same letters do not differ ($p \geq 0.05$) from each other according to the Tukey test. Error bars represent the standard error of the mean ($n = 4$).

In the scenarios and treatments with and without supplementation, marked differences were observed in the indices of water use efficiency when considering ear or total biomass production as a function of the volume of water applied (Table 1). Our results suggest that supplemental irrigation did not significantly increase the physical water productivity (PWP) in the Rainy scenario. However, differences between treatments with and without supplementation were observed in both Drought and Severe Drought scenarios for both crop cycles. Values of PWP_{ear} and $PWP_{biomass}$ were lower without supplemental irrigation compared to supplemental irrigation for Normal, Drought, and Severe Drought scenarios. Additionally, brackish water supplementation homogenized PWP in the different scenarios tested, with no differences between them. This indicates that supplemental irrigation in years of drought and severe drought allows for an improvement in water use efficiency so that it can be compared to the results obtained in the Normal scenario.

Table 1. Physical water productivity (PWP) based on dry ear production (PWP_{ear}), total dry biomass production ($PWP_{biomass}$), and supplemental irrigation efficiency (WUE_{SI}) in maize as a function of water scenarios with and without supplemental irrigation with brackish water ($ECw = 4.5$ dS m^{-1}).

Suppl. Irrigation	Simulated Water Scenarios			
	Rainy	Normal	Drought	Severe Drought
		PWP_{ear} (kg m^{-3}) 2018		
With	1.56 ± 0.27 Aa [1]	1.77 ± 0.39 Aa	2.16 ± 0.58 Aa	1.57 ± 0.22 Aa
Without	1.49 ± 0.21 Aa	1.38 ± 0.26 Ba	1.04 ± 0.39 Bab	0.72 ± 0.21 Bb
		2019		
With	1.14 ± 0.25 Ab	1.57 ± 0.15 Aa	1.43 ± 0.24 Aa	1.53 ± 0.27 Aa
Without	0.98 ± 0.40 Aab	1.13 ± 0.55 Ba	0.73 ± 0.28 Bab	0.23 ± 0.24 Bb
		$PWP_{biomass}$ (kg m^{-3}) 2018		
With	2.78 ± 0.36 Ab	3.96 ± 0.60 Aa	3.99 ± 0.49 Aa	2.87 ± 0.61 Ab
Without	2.84 ± 0.57 Aa	2.70 ± 0.47 Ba	2.19 ± 1.17 Bb	1.10 ± 0.46 Bc
		2019		
With	2.29 ± 0.20 Ab	3.40 ± 0.16 Aa	3.31 ± 0.22 Aa	3.84 ± 0.15 Aa
Without	2.18 ± 0.32 Ab	2.77 ± 0.18 Aa	1.62 ± 0.35 Bb	1.11 ± 0.33 Bc
		WUE_{SI} (kg m^{-3}) [#]		
		Ear biomass		
	3.01 ± 0.37	3.53 ± 0.45	3.63 ± 0.11	2.53 ± 0.14
		Total Biomass		
	2.79 ± 0.30	7.99 ± 0.21	7.16 ± 0.19	5.44 ± 0.20

[1] Values represent the mean ± standard error ($n = 4$). Means followed by the same uppercase letters in columns and lowercase letters in rows do not differ ($p \geq 0.05$) from each other according to Tukey's test. [#] Values represent the mean of the two maize cycles ($n = 4$).

When considering only the water use efficiency when supplementary irrigation was applied (WUE_{SI}), the highest values were recorded for Normal and Drought scenarios for both ear and total biomass production. In general, the values of WUE_{SI} (Table 1) in the Normal, Drought, and Severe Drought scenarios were quite dramatic and higher than those obtained at the full irrigation depths (350 and 500 mm per cycle of green maize) [54,55], which were much higher than those used in supplemental irrigation in this research (102, 172 and 260 mm for Normal, Drought, and Severe Drought scenarios). However, there was a reduction of about 30% in WUE_{SI} in the Severe Drought scenario compared to the Drought scenario, which was due to the application of a greater depth of brackish water in the supplementation in the first scenario, and also to the small reduction in the biomass production observed in [7], possibly as a result of the accumulation of salts in the soil (Figure 3).

4. Conclusions

Rainfed agriculture in tropical semi-arid regions is limited by irregular rainfall during the rainy season. However, the Brazilian semi-arid region has a large number of wells with brackish water that could be used in supplemental irrigation, thus reducing the water stress in maize and other annual crops. Our results suggest that the water stress associated with dry spells is more deleterious to the carbon assimilation and water use efficiency of maize plants than the salt stress associated with the use of supplemental irrigation with brackish water. Our study showed that dry spells compromised the photosynthetic capacity of maize even under the Normal water scenario, but the effects became drastic under both Drought and Severe Drought scenarios due to stomatal and nonstomatal effects. Supplemental irrigation of maize with brackish water with an ECw = 4.5 dS m^{-1} reduced water stress and did not result in excessive salt accumulation in the sandy loam soil used in this study.

The use of brackish water did not lead to sodium accumulation in leaves and improved leaf gas exchange, with a positive impact on the CO_2 assimilation rate. Thus, supplemental irrigation with brackish water resulted in an increase in water use efficiency in different scenarios under water restriction, constituting an important strategy that can be applied in biosaline agriculture in tropical semi-arid regions to maintain crop yields, especially in years of drought and severe drought. Considering the great spatial variability of rainfall in tropical semi-arid regions and the increase in drought years associated with global climate change scenarios, future studies are required to evaluate this strategy in other important crop systems under nonsimulated conditions, as well as the long-term effects of these salts on different soil types in this region.

Author Contributions: Conceptualization, E.S.C., C.F.L. and H.R.G.; methodology, E.S.C., C.F.L., J.d.S.S. and R.O.M.; investigation, E.S.C., J.R.d.S.S. and J.d.S.S.; writing—original draft preparation, E.S.C., C.F.L., J.R.d.S.S., J.F.d.S.F. and H.R.G.; writing—review and editing, A.d.S.T., J.F.d.S.F., S.C.R.V.L. and A.S.d.M.; project administration, C.F.L.; funding acquisition, C.F.L. and S.C.R.V.L. All authors have read and agreed to the published version of the manuscript.

Funding: This research was funded by the Coordination for the Improvement of Higher Level Personnel Agency (CAPES), the State Development Agency of Ceará (ADECE) and the Foundation for the Support of Scientific and Technological Development of Ceará (FUNCAP).

Data Availability Statement: No new data were created or analyzed in this study. Data sharing does not apply to this article.

Acknowledgments: Acknowledgments are due to the State Development Agency of Ceará (ADECE), Secretariat for Economic Development and Labor of Ceará (SEDET), Institute of Technological Education Center (CENTEC), the Coordination for the Improvement of Higher Level Personnel Agency (CAPES), the Foundation for the Support of Scientific and Technological Development of Ceará (FUNCAP) and Chief Scientist Program, Brazil, for the financial support provided for this research and award of fellowship to the first author.

Conflicts of Interest: The authors declare no conflict of interest.

References

1. Frizzone, J.A.; Lima, S.C.R.V.; Lacerda, C.F.; Mateos, L. Socio-economic indexes for water use in irrigation in a representative basin of the tropical semiarid region. *Water* **2021**, *13*, 2643. [CrossRef]
2. Marengo, J.A.; Torres, R.R.; Alves, L.M. Drought in northeast Brazil: Past, present and future. *Theor. Appl. Climatol.* **2017**, *129*, 1189–1200. [CrossRef]
3. Ali, A.B.M.; Shuang-En, Y.U.; Panda, S.; Guang-Cheng, S. Water harvesting techniques and supplemental irrigation impact on sorghum production. *J. Sci. Food Agric.* **2015**, *95*, 3107–3116. [CrossRef] [PubMed]
4. Nangia, V.; Oweis, T.; Kemeze, F.H.; Schnetzer, J. Supplemental irrigation: A promising climate-smart practice for dryland agriculture. *Wagening. CGIAR/CCAFS* 2018. Available online: https://cgspace.cgiar.org/bitstream/handle/10568/92142/GACSA%20Practice%20Brief%20Supplemental%20Irrigation.pdf (accessed on 20 November 2020).
5. Hamdy, A.; Sardob, V.; Ghanem, K.A.F. Saline water in supplemental irrigation of wheat and barley under rainfed agriculture. *Agric. Water Manag.* **2005**, *78*, 122–127. [CrossRef]
6. Chauhan, C.P.S.; Singh, R.B.; Gupta, S.K. Supplemental irrigation of wheat with saline water. *Agric. Water Manag.* **2008**, *95*, 253–258. [CrossRef]
7. Cavalcante, E.S.; Lacerda, C.F.; Costa, R.N.T.; Gheyi, H.R.; Pinho, L.L.; Bezerra, F.M.S.; Oliveira, A.C.; Canjá, J.F. Supplemental irrigation using brackish water on maize in tropical semi-arid regions of Brazil: Yield and economic analysis. *Sci. Agric.* **2021**, *78*, 1–9. [CrossRef]
8. Food and Agriculture Organization of the United Nations (FAO). 1.5 Billion People, Living with Soil too Salty to be Fertile. 2021. Available online: https://news.un.org/en/story/2021/10/1103532 (accessed on 15 November 2021).

9. Masters, D.G.; Benes, S.E.; Norman, H.C. Biosaline agriculture for forage and livestock production. *Agric. Ecosyst. Environ.* **2007**, *119*, 234–248. [CrossRef]
10. Silva, J.E.S.B.; Matias, J.R.; Guirra, K.S.; Aragão, C.A.; Araujo, G.G.L.; Dantas, B.F. Development of seedlings of watermelon cv. Crimson Sweet irrigated with biosaline water. *Rev. Bras. De Eng. Agrícola E Ambient.* **2015**, *19*, 835–840. [CrossRef]
11. Hassanli, M.; Ebrahimian, H. Cyclic use of saline and non-saline water to increase water use efficiency and soil sustainability on drip irrigated maize. *Span. J. Agric. Res.* **2016**, *14*, e1204. [CrossRef]
12. Kiani, A.R.; Mosavata, A. Effect of different alternate irrigation strategies using saline and non-saline water on corn yield, salinity and moisture distribution in soil profile. *J. Water Soil* **2016**, *30*, 1595–1606.
13. Silva, F.J.A.; Araújo, A.L.; Souza, R.O. Águas subterrâneas no Ceará—Poços instalados e salinidade. *Rev. Tecnol.* **2007**, *28*, 136–159.
14. Fernandes, F.B.P.; Lacerda, C.F.; Andrade, E.M.; Neves, A.L.R.; Sousa, C.H.C. Efeito de manejos do solo no déficit hídrico, trocas gasosas e rendimento do feijão-de-corda no semiárido. *Rev. Ciência Agronômica* **2015**, *46*, 506–515.
15. Munns, R.; Tester, M. Mechanisms of salinity tolerance. *Annu. Rev. Plant Biol.* **2008**, *59*, 651–681. [CrossRef] [PubMed]
16. Xavier, T.M.B.S. *Tempo de Chuva: Estudos Climáticos e de Previsão para o Ceará e Nordeste Setentrional*; Editora ABC: Ceará, Brazil, 2001; 478p.
17. Allen, R.G.; Pereira, L.S.; Raes, D.; Smith, M. Crop Evapotranspiration Guidelines for Computing Crop Water Requirements. *Rome, FAO—Irrigation and Drainage*; FAO: Rome, Italy, 1998; Volume 300, p. 56.
18. Medeiros, J.F. Qualidade da água de irrigação e evolução da salinidade nas propriedades assistidas pelo "GAT" nos Estados do RN, PB e CE. **1992**. 173 f. Dissertação (Mestrado em Engenharia Agrícola)—Universidade Federal da Paraíba. Available online: http://dspace.sti.ufcg.edu.br:8080/jspui/bitstream/riufcg/2896/3/JOS%c3%89%20FRANCISMAR%20DE%20MEDEIROS%20-%20DISSERTA%c3%87%c3%83O%20PPGEA%201992.pdf (accessed on 15 January 2020).
19. Fernandes, F.H.F.; Aquino, A.B.; Aquino, B.F.; Holanda, F.J.M.; Freire, J.M.; Crisostomo, L.A.; Costa, R.I.; Uchoa, S.C.P.; Fernandes, V.L.B. *Recomendações De Adubação E Calagem Para O Estado Do Ceara*; Editora ABC: Fortaleza, Brazil, 1993; 247p.
20. Embrapa—Empresa Brasileira de Pesquisa Agropecuária. Manual de métodos de análise de solo. 1997, 212p. Available online: https://ainfo.cnptia.embrapa.br/digital/bitstream/item/173611/1/Pt-2-Cap-20-Sais-soluveis.pdf (accessed on 8 November 2020).
21. Cataldo, J.M.; Haroom, M.; Schrader, L.E.; Youngs, V.L. Rapid colorimetric determination of nitrate in plant tissue by nitration of salicylic acid. *Commun. Soil Sci. Plant Anal.* **1975**, *6*, 71–80. [CrossRef]
22. Bates, L.S.; Waldren, R.P.; Teare, J.D. Rapid determination of free proline for water-stress studies. *Plant Soil* **1973**, *39*, 205–207. [CrossRef]
23. Ferreira, D.F. Sisvar: Um sistema computacional de análise estatística. *Ciência E Agrotecnologia* **2011**, *35*, 1039–1042. [CrossRef]
24. Hillel, D. *Fundamentals of Soil Physics*; Academic Press: London, UK, 1980; 413p.
25. Guerra, H.C. *Física Dos Solos*; UFPB: Campina Grande, Brazil, 2000; 173p.
26. Ngolo, A.O.; Oliveira, M.F.; Assis, I.R.; Rocha, G.C.; Fernandes, R.B.A. Soil physical quality after 21 years of cultivation in a Brazilian Cerrado Latosol. *J. Agric. Sci.* **2019**, *11*, 124–136. [CrossRef]
27. Assis Junior, J.O.; Lacerda, C.F.; Silva, F.B.; Silva, F.L.B.; Bezerra, M.A.; Gheyi, H.R. Produtividade do feijão-de-corda e acúmulo de sais no solo em função da fração de lixiviação e da salinidade da água de irrigação. *Eng. Agrícola* **2007**, *27*, 702–713. [CrossRef]
28. Bezerra, A.K.P.; Lacerda, C.F.; Hernandez, F.F.F.; Silva, F.B.; Gheyi, H.R. Rotação cultural feijão caupi/milho utilizando-se águas de salinidades diferentes. *Ciência Rural* **2010**, *40*, 1075–1082. [CrossRef]
29. Lacerda, C.F.; Sousa, G.G.; Silva, F.L.B.; Guimarães, F.V.A.; Silva, G.L.; Cavalcante, L.F. Soil salinization and maize and cowpea yield in the crop rotation system using saline waters. *Eng. Agrícola* **2011**, *31*, 663–675. [CrossRef]
30. Neves, A.L.R.; Lacerda, C.F.; Sousa, C.H.C.; Silva, F.L.B.; Gheyi, H.R.; Ferreira, F.J.; Andrade Filho, F.L. Growth and yield of cowpea/sunflower crop rotation under different irrigation management strategies with saline water. *Ciência Rural* **2015**, *45*, 814–820. [CrossRef]
31. Rhoades, J.D.; Loveday, J. Salinity in irrigated agriculture. In *Irrigation of Agricultural Cropsstewart*; Stewart, D.R., Nielsen, D.R., Eds.; American Society of Agronomy: Madison, WI, USA, 1990; pp. 1089–1142.
32. Ayers, R.S.; Westcot, D.W. *A Qualidade De Água Na Agricultura. Estudos FAO: Irrigação E Drenagem, 29*, 2nd ed.; UFPB: Campina Grande, Brazil, 1999; p. 153.
33. Cavalcante, L.F.; Santos, R.V.; Hernandez, F.F.F.; Gheyi, H.R.; Dias, T.J.; Nunes, J.C.; Lima, G.S. Recuperação de solos afetados por sais. In *Manejo Da Salinidade Na Agricultura: Estudos Básicos E Aplicados*; Gheyi, H.R., Dias, N.S., Lacerda, C.F., Gomes Filho, E., Eds.; INCTSal: Fortaleza, Brazil, 2016; Volume 28, pp. 461–477.
34. Larcher, W. *Ecofisiologia Vegetal. São Carlos, Rima, Artes E Textos*; Tradução do orignal: Okoplysiologie der, Brazil, 2000; 531 p, ISBN 8586552038.
35. Osborne, C.P.; Sack, L. Evolution of C4 plants: A new hypothesis for an interaction of CO_2 and water relations mediated by plant hydraulics. *Philos. Trans. R. Soc.* **2012**, *367*, 583–600. [CrossRef] [PubMed]
36. Souza, M.L.C.; Silva, A.Z. Starling, C.; Machuca, L.M.R.; Zuñiga, E.A.; Galvão, Í.; Jesus, G.J.; Broetto, F. Biochemical parameters and physiological changes in maize plants submitted to water deficiency. *SN Appl. Sci.* **2020**, *2*, 1–9. [CrossRef]
37. Rahnama, A.; James, R.A.; Poustini, K.; Munns, R. Stomatal conductance as a screen for osmotic stress tolerance in durum wheat growing in saline soil. *Funct. Plant Biol.* **2010**, *7*, 255–269. [CrossRef]
38. Taiz, L.; Zeiger, E.; Moller, I.M.; Murphy, A. *Fisiologia E Desenvolvimento Vegetal*; Editora Artmedl: Porto Alegre, Brazil, 2017; 888p.

39. Lacerda, C.F.; Oliveira, E. Victor.; Neves, A.L.R.; Gheyi, H.R.; Bezerra, M.A.; Costa, C.A.G. Morphophysiological responses and mechanisms of salt tolerance in four ornamental perennial species under tropical climate. *Rev. Bras. De Eng. Agrícola E Ambient.* **2020**, *24*, 656–663. [CrossRef]
40. Barbosa, F.S.A.; Lacerda, C.F.; Gheyi, H.R.; Farias, G.C.; Silva Junior, R.J.C.; Lage, Y.A.; Hernandez, F.F.F. Yield and ion content in maize irrigated with saline water in a continuous or alternating system. *Ciência Rural* **2012**, *42*, 1731–1737. [CrossRef]
41. Verslues, P.E.; Bray, E.A. Role of abscisic acid (ABA) and *Arabidopsis thaliana* ABA—Insensitive loci in low water potential-induced ABA and proline accumulation. *J. Exp. Bot.* **2006**, *57*, 201–212. [CrossRef]
42. Chen, Z.; Cuin, T.A.; Zhou, M.; Twomey, A.; Naidu, B.P.; Shabala, S. Compatible solute accumulation and stress-mitigating effects in barley genotypes contrasting in their salt tolerance. *J. Exp. Bot.* **2007**, *58*, 4245–4255. [CrossRef]
43. Chun, S.C.; Paramasivan, M.; Chandrasekaran, M. Proline accumulation influenced by osmotic stress in arbuscular mycorrhizal symbiotic plants. *Front. Microbiol.* **2018**, *9*, 2525. [CrossRef]
44. Li, J.; Ma, J.; Guo, H.; Zong, J.; Chen, J.; Wang, Y.; Li, D.; Li, L.; Wang, J.; Liu, J. Growth and physiological responses of two phenotypically distinct accessions of centipede grass (*Eremochloa ophiuroides* (Munro) Hack.) to salt stress. *Plant Physiol. Biochem.* **2018**, *126*, 1–10. [CrossRef] [PubMed]
45. Kishor, P.B.K.; Hima Kumari, P.; Sunita, M.S.L.; Sreenivasulu, N. Role of proline in cell wall synthesis and plant development and its implications in plant ontogeny. *Front. Plant Sci.* **2015**, *6*, 1–17. [CrossRef] [PubMed]
46. Wang, H.; Tang, X.; Wang, H.; Shao, H.B. Proline accumulation and metabolism-related genes expression profiles in *Kosteletzkya virginica* seedlings under salt stress. *Front. Plant Sci.* **2015**, *6*, 792. [CrossRef] [PubMed]
47. Zhang, L.; Becker, D.F. Connecting proline metabolism and signaling pathways in plant senescence. *Front. Plant Sci.* **2015**, *6*, 552. [CrossRef] [PubMed]
48. Hanson, A.; Nelsen, C.E.; Everson, E.H. Evaluation of free proline accumulation as an index of drought resistance using two contrasting barley cultivars. *Crop Sci.* **1977**, *17*, 720–726. [CrossRef]
49. Ferreira, L.G.R.; Souza, J.G.; Prisco, J.T. Effects of water deficit on proline accumulation and growth of two cotton genotypes of differing drought resistance. *Z. Pflanzenphysiologie.* **1979**, *93*, 189–199. [CrossRef]
50. Melo, A.S.; Melo, Y.L.; Lacerda, C.F.; Viégas, P.R.A.; Ferraz, R.L.S.; Gheyi, H.R. Water restriction in cowpea plants [*Vigna unguiculata* (L.) Walp.]: Metabolic changes and tolerance induction. *Rev. Bras. De Eng. Agrícola E Ambient.* **2022**, *26*, 190–197. [CrossRef]
51. Lutts, S.; Kinet, J.M.; Bouharmont, J. Effects of salt stress on growth, mineral nutrition and proline accumulation in relation to osmotic adjustment in rice (*Oryza Sativa L.*) cultivars differing in salinity resistance. *Plant Growth Regul.* **1996**, *19*, 207–218. [CrossRef]
52. Lacerda, C.F.; Cambraia, J.; Cano, M.A.O.; Ruiz, H.A.; Prisco, J.T. Solute accumulation and distribution during shoot and leaf development in two sorghum genotypes under salt stress. *Environ. Exp. Bot.* **2003**, *49*, 107–120. [CrossRef]
53. Lacerda, C.F.; Cambraia, J.; Cano, M.A.O.; Prisco, J.T. Proline accumulation in sorghum leaves is enhanced by salt-induced tissue dehydration. *Rev. Ciência Agronômica* **2006**, *37*, 110–112.
54. Embrapa—Empresa Brasileira de Pesquisa Agropecuária. A Cultura Do Milho-Verde. 2008. Available online: https://ainfo.cnptia.embrapa.br/digital/bitstream/item/11921/2/00082390.pdf (accessed on 10 November 2020).
55. Pereira Filho, I.A.; Silva, A.R.; Costa, R.V.; Cruz, I. Milho Verde. 2010. Available online: https://www.agencia.cnptia.embrapa.br/gestor/milho/arvore/CONT000fy779fnk02wx5ok0pvo4k3c1v9rbg.html (accessed on 10 November 2020).

MDPI
St. Alban-Anlage 66
4052 Basel
Switzerland
www.mdpi.com

Agriculture Editorial Office
E-mail: agriculture@mdpi.com
www.mdpi.com/journal/agriculture

Disclaimer/Publisher's Note: The statements, opinions and data contained in all publications are solely those of the individual author(s) and contributor(s) and not of MDPI and/or the editor(s). MDPI and/or the editor(s) disclaim responsibility for any injury to people or property resulting from any ideas, methods, instructions or products referred to in the content.